Global Environmental Changes in South Asia

A Regional Perspective

Global Environmental Changes in South Asia

A Regional Perspective

Edited by

A.P. Mitra (deceased)

and

C. Sharma

National Physical Laboratory
New Delhi, India

Springer

CP

A C.I.P. Catalogue record for this book is available from the Library of Congress.

ISBN 978-1-4020-9912-0 (HB)
ISBN 978-1-4020-9913-7 (e-book)

Copublished by Springer,
P.O. Box 17, 3300 AA Dordrecht, The Netherlands
with Capital Publishing Company, New Delhi, India.

Sold and distributed in North, Central and South America by Springer,
233 Spring Street, New York 10013, USA.

In all other countries, except India, sold and distributed by Springer,
Haberstrasse 7, D-69126 Heidelberg, Germany.

In India, sold and distributed by Capital Publishing Company,
7/28, Mahaveer Street, Ansari Road, Daryaganj, New Delhi, 110 002, India.

www.springer.com

Cover photo credit: Photograph taken by R. Nagarajan, at Kolar District,
Karnataka, India.

Printed on acid-free paper

Printed in India.

Foreword

The Fourth Assessment Report of IPCC having clinched in 2007 the evidence of global warming on account of anthropogenic activities, backed with scientific data gathered and analyzed globally, has made it mandatory world over to focus efforts on delineation of the anticipated adverse impacts of global warming on regional temperature and moisture regimes and the linked hydrologic, climatic and biospheric processes. First and foremost is the requirement to understand vulnerability to food and livelihood security in various ecosystems—on mainland, mid-range and high mountains as well as coastal areas including CEZs. The projected global temperature rise of the order of about two degrees or more and further rise at a decadal rate of around 0.2°C is sufficient to make grievous changes in sea surface level and submerge many low lying coastal areas around the world thereby possibly causing unprecedented losses to human habitat and livelihood in the coming years. A rise in climate variability is also becoming increasingly evident with potential direct impact on agricultural performance, on water accessibility and on weather extremes.

Developing countries due to their poor infrastructure, limited resources and large impoverished population are likely to face more intense and widespread adverse impact of climate change than the developed world and also have limited adaptation capacity. Keeping this in view the Global Change System for Analysis Research and Training (START), of which the South Asian START Regional Centre (SAS-RC) operated by National Physical Laboratory at New Delhi is an organ, had initiated in 2003, with cooperation of its member countries, an exercise under their MAIRS project to enable Rapid Assessment of the present state of the environment in the developing regions of Asia and highlight vulnerability of the prevalent human systems, of the deliverables from various sectors and the livelihood security. The objective of the exercise was to assess and later facilitate to build or enhance the existing capacities in these countries needed to adequately understand and tackle the extent of likely impact of climate change on ecosystems.

Possible pathways to adapt to upcoming climate variability in order to minimize damages to human support systems were also to be explored in different sectors. The Regional Centres of the three Asian networks within START, namely, the South Asian Regional Network, the Temperate East Asian Regional Network and the South-East Asian Regional Network undertook this exercise beginning 2003. The Asia Pacific Network for Global Change Research (APN) facilitated this by sponsoring scoping meetings of Rapid Assessment activity.

The NATCOM exercise implemented by the Indian Ministry of Environment and Forest for the purpose of submitting in 2004 the First National Communication to UNFCCC from India also succeeded in establishing an extensive country-wide network of experts and knowledge bases for vulnerability assessment. This network has continued to strengthen in the following years with NATCOM-II programme currently underway. This and similar experience of the other countries of the South Asian Region has made the Rapid Assessment exercise initiated feasible. The result is the present book "Global Environmental Changes in South Asia". The "Introduction" and the first chapter of the book titled "Human Dimensions of Changing Environment", record well the vision of the leader of this cooperative effort, Dr. Ashesh Prasad Mitra, who is no more with us. He was the founder Director of SAS-RC and the first Chairman of its Planning Committee, SASCOM and has edited several chapters of this book. Dr. Chhemendra Sharma, his long term associate and co-editor of this publication, has gathered vast experience in various assessment exercises linked with climate change including budgeting of greenhouse gases as well as vulnerability to various sectors. The authors of other chapters of the book are all well established in their respective areas. Many of them have led internationally significant initiatives. They have addressed all the significant issues in the book, such as reliability of the monsoon system, altering C&N pools in quickly transforming lands, floods and droughts frequency, deteriorating air quality in the urban regions, changing oxidizing capacity of the atmosphere, impact on mangrove biotic structure, limited denitrification capacity of the Bay of Bengal, etc. The need for quickly implementing new sustainable development pathways are also elaborated upon.

This book is a true amalgamation of the region's capabilities to capture environmental changes of a global nature in the region and its vulnerability to climate change. It is indeed possible based on this publication to identify important capacity gaps in adapting to climate variability, bridging of which over a period of time is necessary in the region.

New Delhi
November 2009

M.K. Tiwari
Director
South Asian START Regional Centre (SAS-RC)
National Physical Laboratory, New Delhi, India

Introduction

THE SOCIO-ECONOMIC SETTINGS

South Asia is a unique region of the world. The land area of this region is only 3% of the world but the 1990 population was 21.3% and 2025 population is expected to rise to nearly 24% of the world. It is one of the most densely populated regions of the world ranging from 250 persons/km^2 in Nepal to over 800 persons/km^2 in Bangladesh. The scenario for 2025 is further alarming: as much as over 1300 persons/km^2 in Bangladesh and between 300-400 persons/km^2 in other countries in the subcontinent. A positive aspect of this increasing population is, however, the increasing dominance of the youth. People of the age of 60+ range from 5% in Bangladesh to 9.5% in Sri Lanka in 2001. In comparison the developed countries in eastern half of the globe has larger share of 60+ age group population (e.g. 24% in Japan, 16.5% in Australia and 16% in New Zealand).

Urbanization in South Asia – a major source for emissions of trace gas and aerosol emissions – is rapidly increasing. From 1995 to 2005, the rise has been from 27 to 45% in India, 18 to 40% in Bangladesh, 22 to 43% in Sri Lanka and 14 to 34% in Nepal. The middle class in this region is mainly professional and is a major consumer shaping the developmental pathways of the region. In India, the 200-250 million professional middle class is at the base of industrial transformation and, also as a result, of conspicuous consumptions. The major socio-economic features of South Asia are country-wise summarized in Table 1.

THE GEOPHYSICAL SETTINGS

The key geophysical features of the South Asia (Table 2) are (a) the presence of mountains in the north, (b) the Indo-Gangetic Plains which is

Table 1: Socio-economic features of South Asia

	Bangladesh	*Bhutan*	*India*	*Nepal*	*Pakistan*	*Sri Lanka*
GDP (current US$) (billions)	56.6	0.7	691.2	6.7	96.1	20.1
Population, total (millions)	139.2	0.9	1079.7	26.6	152.1	19.4
Per capita GDP (US$)	406.6	777.8	640.2	251.9	631.8	1036.1
GNI per capita, Atlas method (current US$)	440	760	620	250	600	1010
Population under poverty line (% of total population)	49.8	-	28.6	30.9	32.0	22.0
Life expectancy at birth, total (years)	63.5	63.5	63.5	62.2	64.9	74.4
Population growth (annual %)	1.9	2.5	1.4	2.0	2.4	0.9
Surface area (thousands sq. km)	144.0	47.0	3287	147.2	796.1	65.6

Table 2: Major physical parameters

	India	*Pakistan*	*Bangladesh*	*Sri Lanka*	*Nepal*
Population (Millions)					
1990	853	123	112	17	19
1995	929	130	120	18	22
2001	1025	145	140	19	24
2025 (Projected)	1442	267	235	25	35
Annual Growth Rate (1901-2001)	1.8	2.6	22	1.0	2.4
Population density (per sq km)					
1990	287	159	888	266	140
1995	285		836	380	256
2025	423	345	1362	382	289
P.C. of population aged 60^+ (2001)	7.7	5.8	5.0	9.5	5.9
EEZ ('000 km^2)	2015	319	77	517	-
Length of coast (km)	12700	1046	580	1340	-
EEZ/land (%)	68%	41%	59%	795%	-

bread-basket of the region, (c) the perennial river system which is under threat now due to increasing stress from population and climate, (d) large expanses of wetlands and coral reef providing eco-system services to the region, (e) the large area under Exclusive Economic Zone (EEZ) and (f) the seasonal alteration of the atmospheric flow patterns associated with the monsoons – the key parameter influencing the food security issues of the region. The EEZ for the region totals around 29,25,000 km^2 compared to land area of 44,30,000 km^2 which is as much as 66%. However there is a wide variation for different countries in South Asia e.g. while Sri Lanka's EEZ is as high as 795% and Nepal is a land-locked country. EEZ is, therefore, a major natural resource for this region.

The mountains in the North are dominated by the Himalayas. The Himalayas extend from 35°N, 74°E to 30°N, 95°E stretching over 2400 km and over an area of 460,000 km^2. This region is rich in biodiversity, a storehouse of unique gene pool, is provider of a large fraction of water to the rivers Indus and Ganges, isolates the subcontinent geographically and meteorologically. Its rich forests have in recent years been greatly depleted catalyzing large sediment flow, floods and climate events. The physical and socio-cultural parameters of the Himalayas are given (K.L. Shrestha, Personal Communication) in Table 3.

The Indo-Gangetic Plain (IGP) region is the bread-basket of this region (Fig. 1) but it is also a major emitter region of nearly all the climate forcing gases and particles: CO_2, CH_4, N_2O, CO, NO_x, BC and OC. The extended urban sprawl in this region merges with that of the Mekong delta and eastward combining with the Yangtse river delta; it provides a massive sprawl over the entire Asian region dominating industrialization, urbanization, city growths, and is the home of some of the most populated megacities of the world.

Table 3: Physical and socio-cultural parameters of Himalaya region

Description	*Western Himalayas*	*Central Himalayas*	*Eastern Himlayas*
Latitude	37°N	28°N	28°N
Longitude	72°E	81°E	97°E
Snowline	4,800 m	5,000 m	4,900 m (3,500 m)
Treeline	4,000 m	3,900 m	4,200 m
Predominant species (lower elevation)	Pinus variety	Shorea to Pinus	Shorea Robusta
Vegetation regime	Xerophytic	Transitional	Hydrophytic
Annual average precipitation	700-2000 mm	3500 mm	
Main river system	Indus	Ganges	Brahmaputra
Drainage area	1,263,000 sq.km	1,075,000 sq.km	940,000 sq.km
Average annual runoff	3,850 cu. M/s	15,000 cu.m/s	20,000 cu.m/s
Specific Runoff	3.05	13.95	21.28
Number of Languages	Indo Aryan: 11 Tibeto Burman: 9	Indo Aryan: 5 Tibeto Burman: 11	Tibeto-Burman:11
Culture	Caucasoids	Mixed	Mongoloids

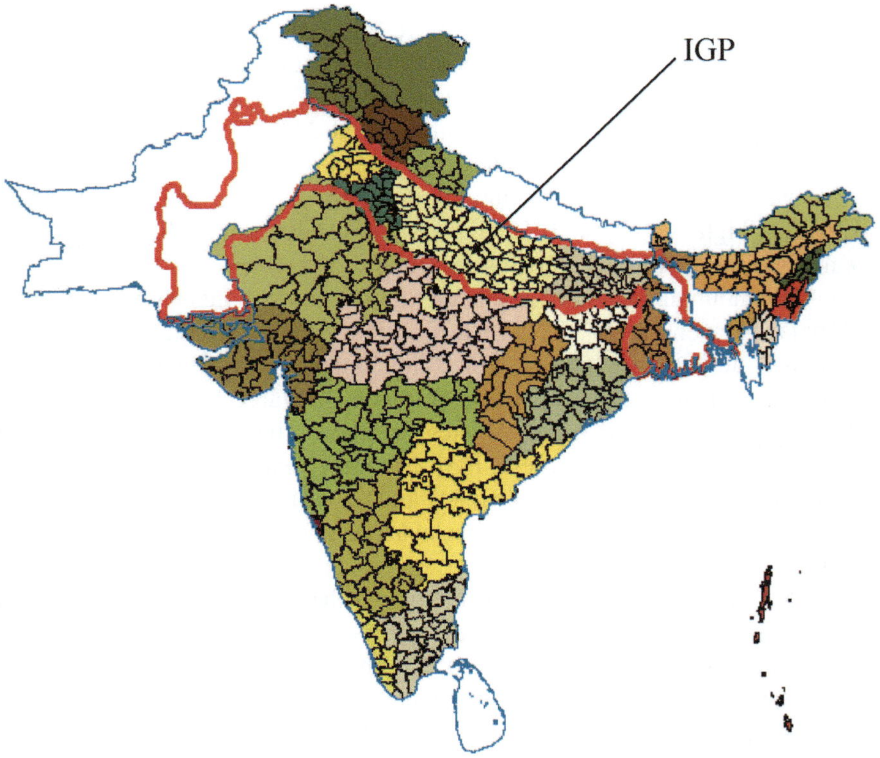

Fig. 1: Indo-Gangetic Plain (IGP) region of South Asia.

Monsoon is the major dominating factor for agriculture sustainability in this region. The seasonal alteration of the surface wind flow is associated with monsoon. The inter-tropical convergence zone (ITCZ – the so-called meteorological equator) moves from its northern location in summer to the south location to around 5°S in winter. The movements and locations of the ITCZ vary from year to year and not only respond to climate change but also contribute to it. The position of ITCZ determines the nature and direction of transport to and from the India land mass. Another special feature is that rainfall is principally confined to four months in a year.

One of the special features is the presence of large wetlands including coral reefs in the region. There is a rich variety of as many as 3959 coastal wetland sites in India alone. Of these, the most important is the Sunderbans, stretching over an area of 10,000 km^2 which is the largest mangrove in the world and an UNESCO heritage area but now heavily degraded. About 38% of Sunderbans area is in West Bengal (India) and the rest is in Bangladesh. Coral reefs, in six major regions in the Indian coast, are also undergoing changes from anthropogenic pressures from fishing, mining, sedimentation, invasion by alien species, exploitation and deforestation of islands.

Other features include presence of appreciable arid and semi-arid areas, of high sedimentation rate, of a warm pool in the Indian Ocean, of differing rainfall patterns in different areas and of frequent occurrences of floods and droughts.

THE DRIVERS

The key drivers sourcing from the developmental paths chosen by the institutions of governance are influenced by the historical path, culture, dominant ideologies and political will in almost all the setup. This is also true for the South Asian region. The key direct and indirect drivers responsible for the developmental pathways steering the consumption and production patterns, influence the key sectors like energy, industries, landuse change and forestry and waste sectors. These sectors are responsible for emissions of greenhouse gases and other pollutants which are responsible for climate change. These emissions, in turn, also impact the aforementioned sectors.

The major drivers operating in the South Asian region are population, consumption pattern and changing atmosphere. Historical analysis shows that the developmental processes in the South Asia over past century has been guided by regimes and institutions that led to development trajectory which are no different from predecessors and thus converging (unless intervened) towards high energy intensive emission loaded pathways. However in recent years, the governance is undergoing rapid changes as governance has lost its territories and Millennium Development Goals are dominating the developmental agenda. The global common framework for policy thoughts and action is allowing the nations to adjust national policies to global regimes. Despite this upward movement in institutional convergence, attempts to streamline local micro institutions towards management of local resources are also a predominant trend in South Asia.

The changing atmospheric composition comes from a variety of sectors: energy, industrial processes, agriculture, forests and landuse changes and waste. These contribute to the climate forcing gases and particles which can be classified in three categories: Long-lived gases (CO_2, CH_4, N_2O principally), short-lived gases (O_3, CO, NO_x, VOC, SO_2) and aerosols (two types; scattering aerosols and absorbing aerosols). The short-lived gases (SLG) and aerosols have life times of a week or so, and have in the past been considered mainly as health hazard pollutants. Their role in climate change has been recognized only recently. Ozone forms a category of its own. It is a greenhouse gas. CO, $NO_{x,}$ HC and VOCs are precursor gases that contribute to photochemical production of ozone. Although because of their short life-times their immediate impact is sub-regional or regional, the life-times are long enough for transboundary transport across countries and sometimes across continents.

In past years, efforts for estimation of emissions from the various sectors for South Asian countries have been made which included the following efforts:

India	First Inventory in 1992; ALGAS; MAC 98 (for methane) INDOEX, National Communication to UNFCCC
Bangladesh	National Communication to UNFCCC; ALGAS
Pakistan	National Communication to UNFCCC; ALGAS
Sri Lanka	National Communication to UNFCCC; ALGAS

Except for India, in most other South Asian countries, the emission inventories were limited to CHGs only. In estimating South Asian emissions for other pollutants, provisional estimates have been earlier made by National Physical Laboratory, India using emission factors derived for India.

The key points which emerged from South Asian region are:

(i) Although this region is inhabitated by more than 20% of population of the world, the greenhouse gas emissions are small. In fact these are considerably lower: around 2.7% for CO_2 (all sources) and 7% for CH_4 (all sources). For fossil fuel alone, the emission from this region is 3% (of the order of 175 Tg/Yr) and is expected to rise to 11-12% (690-800 Tg/Yr) in 2025 under baseline scenario. A point to note is that the per capita emission is much lower than the global average of 1.2 T/capita for all cases. For several South Asian countries it is unusually low (0.05 T/capita for Bangladesh, 0.06 for Sri Lanka, and 0.016 for Nepal).

(ii) For some countries, the CH_4 role is dominant. The CO_2 equivalent value of CH_4 emission is 21 times that of CO_2 for Nepal, 1.4 times in Sri Lanka and 0.63 times for India. Thus for mitigation measures CH_4 emission needs special attention especially emissions from rice and animals which are major sources of CH_4 emission.

(iii) The agricultural sector is a source of major emissions. In terms of CO_2 equivalent values, agricultural sector emissions are 50% of the energy and industry sector for India, as large as 126% for Bangladesh and 325% for Sri Lanka. Thus policy approach should be appropriately re-oriented.

(iv) Atmospheric SO_2 loading from anthropogenic sources is no longer negligible for this region. Also this loading has been growing rapidly.

(v) It is important to note that aerosols as well as ozone-precursor gases like CO, NO_x, NMVOC can travel long distances, several thousand kilometres specially in northern winter.

(vi) From INDOEX Programme carried out in 1998 and 1999, an aerosol model for the tropical Indian Ocean during Northeast Monsoon, prepared by Satheesh et al., shows largest contribution coming from non-sea salt sulfates and ammonia (29%), sea salt and nitrates (17%), missing organics (20%), dust (15%) and soot (11%). A surprising element is the presence of a large haze cloud in winter over North India Ocean and in East Asia, and its transport. The MODIS data show large AOD values in South Asian region (Fig. 2).

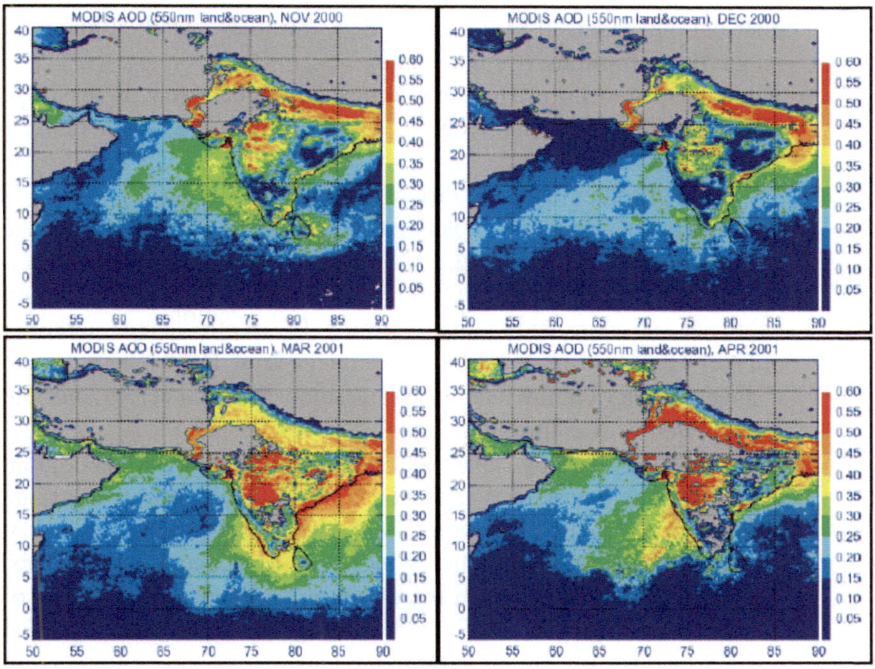

Fig. 2: The MODIS data showing large AOD values in South Asian region.

(vii) One of the special observation which emerged from the INDOEX study is the presence of large amount of carbonaceous aerosols which are absorbing in nature and thus contribute to the changing radiations budget in the region.

(viii) Various regions of South Asia experiences high climate variability, both spatially and temporally. The hydrological regime of major parts of the region is predominantly influenced by monsoon, which brings 70-80% of total annual rainfall during early June to September. The post-monsoon months become dry and there is hardly any appreciable rainfall during winter months (December to February). Moreover, the western parts of Bangladesh and India generally receive significantly lower amounts of rainfall compared to the eastern parts of both the countries, which is a manifestation of high spatial distribution of rainfall. Rainfall in Nepal is also higher in the eastern part compared to the western region. Topographically, in the Terai region with flatter topography along the Indian border, rainfall is very high. In a warmer climate in future, the overall pattern of rainfall/precipitation is expected to change spatially as well as temporally. Water resources of the South Asia region, however, are highly sensitive to climate variability and change. Therefore, an anticipated change in climate system – as a consequence of global warming and subsequent sea level rise –

could considerably affect both the hydrological cycle as well as distribution, which in turn would affect the lives and livelihoods of hundreds of millions of inhabitants.

(ix) A matter of special interest, for atmospheric chemistry, is the large OH concentration in this region. This means a large rate of CH_4 destruction.

(x) The ozone problem is quite different for this region. The problem is not of stratospheric ozone depletion (there has been virtually no change over the last decade, see Fig. 3), but of tropospheric ozone increase principally from CO and NO_x injection into the atmosphere due to vehicular emissions and biomass burning. UV-B effects are, therefore, only of academic interest, but consequences of tropospheric ozone changes on agriculture and health assume importance. In this the region itself is a contributor.

(xi) The changing oxidising capacity of the tropical atmosphere is of great concern. The increased emissions of NO_x and non-methane hydrocarbons from combustion, along with higher penetration of UV radiation tend to increase the ozone and OH concentrations in the tropical-free troposphere. On the contrary, the increased emission of CO from biomass and fossil fuel burning tends to reduce the OH concentrations. Also an increase in methane emission due to increase in natural gas usage, cattle and paddy field also contribute to a reduction in the global OH concentration. Model estimates, on the decrease in OH concentration since the pre-industrial era, differ anywhere between 5 and 30%. Future changes in the oxidising or the cleansing capacity of our atmosphere is going to depend critically on the future anthropogenic emissions.

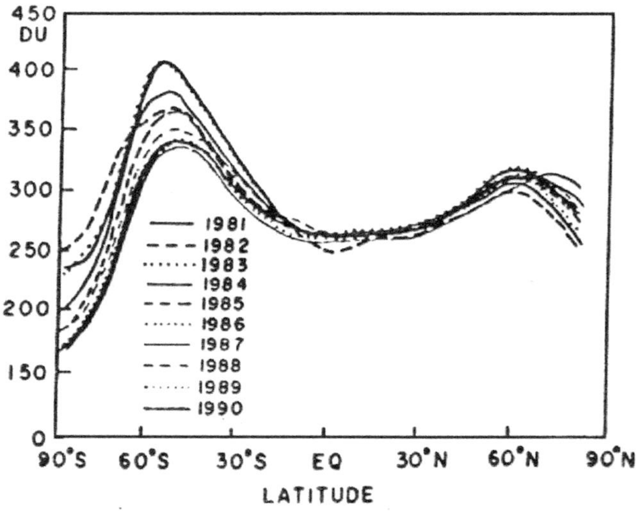

Fig. 3: Average October ozone amounts (1981-1990).

(xii) The two northern basins of the North Indian Ocean – the Arabian Sea and the Bay of Bengal – though situated at similar latitudes and subject to similar physical processes, exhibit widely differing biogeochemical features. Processes like upwelling during summer monsoon and convective mixing during winter make the Arabian Sea one of the most productive regions in the world. High biological production at surface causes heavy fluxes of organic matter and, consequently heavy oxygen demand in intermediate waters, leading to formation of the intense oxygen minimum layer. Renewal of oxygen is slow due to the high rates of respiration, aided by moderately slow water circulation and lower oxygen content of the source waters. Presence of a 5 m thick low salinity lens at the surface in the coastal waters of eastern Arabian Sea, caused by the heavy freshwater flux during monsoon, prevents vertical mixing in the water column. Aided by the high biological production, this leads to anoxia in subsurface waters and emission of H_2S, N_2O, CH_4 and DMS gases to the atmosphere at rates, which are highest ever recorded in the oceans. Similarly, the upwelling areas of coastal and central Arabian Sea give rise to high fluxes of these gases, especially CO_2, round the year, thus making the Arabian Sea, on the whole, a perennial source of atmospheric CO_2.

(xiii) The Bengal delta is formed by the continued deposition of sediment by the Ganga-Brahmaputra-Meghna (GBM) river system. Its hydrology is characterized by "too much water" during monsoon followed by "too little water" in dry season giving rise to extreme temporal distribution of flows posing a complex hydrological challenge of managing both flood and drought in one hydrological cycle. The delta is prone to high intensity floods as well as droughts and low-flow induced salinity ingress. In addition to surface water, its ground water is a major source of water for agriculture, industry and municipal uses. The delta is morphologically very active, e.g., the Sylhet basin is subsiding at 1-3 $mm.y^{-1}$ while the coastal plains exhibit both accretion and erosion. Until recently, there was only one predominant land use – paddy cultivation – which formed the main economic activity. Urban centres started flourishing since 1970s leading to gradual change in pattern of land use. Large cities, including the two megacities of Kolkata and Dhaka, having high population densities, are built either on river banks or along main roads.

(xiv) Sunderbans, with its land area shared between India (38%) and Bangladesh (62% or about 577,000 ha) is the largest single patch of mangroves in the world. Located at the fag end of the Ganges basin, it receives fresh water and sediment from a number of distributaries of Ganges. About 1/3 of the forest consists of water bodies like rivers, channels and tidal creeks. The present geomorphological features were formed over millennia due to continuous deposition of weathered materials by the river system. Mudflats along the rivers are subject to direct wave action, flow and turbulence of water currents. During high tides the lower parts of mudbanks remain submerged with brackish water. Gradual sediment deposition along the river banks gives rise to formation of ridges and levees. It is influenced by semi-diurnal tides with tidal variation ranging from 3.5 to 5.0 m. The soil is saline due to tidal interactions. The soil salinity influences the floral distribution. It hosts one of the richest natural genepools for forest flora and fauna species in the world. It is also endowed with a number of commercially important mangrove species. In addition to the Bengal Tiger, it supports a variety of wild animals, birds, reptiles, sweet and brackish water fish, shrimps, crabs, mollusks, shellfish, turtles and snakes including the King Cobra and Vipers. The creeks serve as nurturing grounds for shrimp larvae and fries. Giant estuarine crocodiles, wild boars and the Indian Otters are also found. The forest provides about 0.3 million tons of fuel wood annually.

(xv) Major mangrove formations in South Asia occur in the deltaic regions of Indus, Mahanadi, and Sunderbans, in the Gulf of Kutchchh and Andaman & Nicobar group of islands. In Sri Lanka they are found in Jafna peninsula and along the west coast. These ecosystems are at serious risk due to rise in sea levels as a result of global warming. Additionally, individual components of the habitat may change their response pattern to the increased temperature and CO_2 levels, and alterations of the hydrological regime in the ambience. Among the various types of biota associated with the ecosystem, the soft bodied animals and bivalves will be the most adversely affected by rise in temperature and salinity. While the biota with specific tolerances within the tidal spectrum will migrate farther landward, in regions with limited land margins there will be no scope for further expanse in response to the sea level rise and change in the hydrological regime of the ambience.

(xvi) For most of the climate parameters, the South Asian region is highly vulnerable. The low capacity of this region to really have a sufficient adaptability to such climate impacts makes this region further vulnerable.

THE KEY REGIONS

The key region in South Asia is the Indo-Gangetic Plains, cutting across Pakistan, India, Nepal and Bangladesh, the bread-basket of the region, and the one which has gone through much changes, especially in landuse, in the last century principally from human activities. The cumulative CO_2 today from landuse change from this region from 1850 to 1985 is as much as half of that from fossil fuel. Although, endowed with mighty rivers (Ganga,

Brahmaputra, Indus etc.), abundant natural resources, and biodiversity, with intense human activities and excessive intensification of agriculture, IGP ecosystem has degraded drastically and poses a major environmental threat not only for itself but for the entire region.

The IGP region stretches from the Arabian Sea to the Bay of Bengal and from the Himalayan foothills to the Indian peninsula (Fig. 1). For India, the region covers 21% of the geographical area; for Nepal 14% of its area; for Pakistan it covers 24% and for Bangladesh 100%. Population residing in this region is high: 40% of Nepal's population, 40% of India's, and 86% in Pakistan. The dominant landuse is agriculture: 70% in India, and 60% in Nepal.

The IGP region affects the climate drivers in a number of ways:

(i) From forestry changes and landuse: Large scale deforestation contributing to a reduction in its capacity to serve as a "sink".

(ii) Agricultural practices contribute to atmospheric loading of CH_4 and N_2O: the key sector are rice production and inadequately fed ruminant animals. This is the single most important contributor to CH_4.

(iii) The large urban sprawl that constitutes the IGP is replete with many large cities. This sprawl extending to the Mekong urban sprawl and that of the Yangtse river is one of the largest sprawls in the world (Fig. 4). Urbanization is the key driving factor. We have in this sprawl the mega cities of Karachi in Pakistan, Mumbai, Delhi and Kolkata in India, Kathmandu in Nepal and Dhaka in Bangladesh. The estimated CH_4 emission from this region constitutes over 50% of the GHG emissions for the entire South Asia. Cities in South Asia included several amongst world's largest cities in 1995. These are also some of the most polluted ones in the world.

THE CHANGING SCENARIO

In South Asia, the origin of agriculture and domestication of animals have been linked to changes in the monsoon precipitation, which in turn would have driven the beginning of agriculture and human civilizations in the region. As the arid phase (weak summer monsoon) began ~5,000 cal yrs BP, the societies in India migrated to more productive areas to the east and south and some may have developed mechanisms of adaptation to climate change. This can be viewed from the presence of ponds, tanks, and artificial reservoirs constructed during the late Holocene across India when the monsoon reached its Holocene minimum, and we find correlation between heightened historical human efforts for adaptation and the most recent minima in the monsoon that occurred during the Maunder Minimum (1600 AD). The monsoon record supports an emerging paradigm that at least in the tropics, the largest climate changes and societal responses were driven by changes in precipitation rather than surface temperature. A key factor for the agriculture is the prediction of

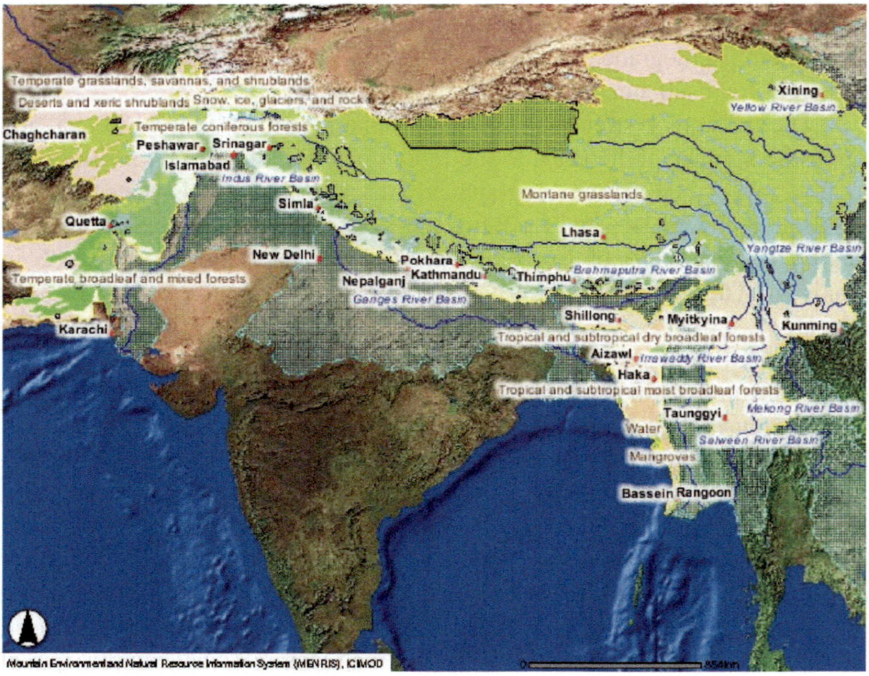

Fig. 4: The urban sprawl extending from Ganges to Yangtse river basins

the onset date of monsoon and distribution of rainfall. These factors determine the agricultural yield.

The land use change pattern is undergoing a rapid phase of transition. The increasing urbanization in South Asia is putting immense pressure primarily on agriculture land and, thus catalyzing a change in socio-economic character of the region which has been historically agrarian. But now the agriculture's share in national GDP is gradually declining with other sectors like service/industries picking up. The urbanization has altered the carbon-flow in the society.

The atmospheric properties in the South Asia are also changing very significantly. While the ambient air quality – especially in many urban and industrial areas – have assumed chronic proportion, the oxidizing capacity of the atmosphere is also being altered due to significant increase in emissions of various gases and particulates in recent years. The anthropogenic activities in this region is also responsible for the significant winter-time clear sky aerosol radiative forcing at the surface in the range of 23-27 W/m^2 over the polluted ocean region and a sufficiently large atmospheric absorption, in the range of 16-20 W/m^2.

Water resources availability in South Asia is highly sensitive to climate. Monsoon contributes maximum share to the annual runoff. However, high runoff availability is more or less useless as the water cannot be stored for dry season. There is simply not enough water in the dry season to meet present demands of various economic sectors. The paucity of water has led to inter-nation and intra-nation water conflicts. Many of the South Asian nations have drafted their national water plans or water policies. Unfortunately, however, the very issue of climate change has not been featured in the plans/ policies in an appreciable manner. It is important to take notice of the changes in climate system and find correlation between climate parameters and timely availability of water resources of a country in order to avoid future complications.

The changing climate has significant bearing on the coastal and marine ecosystems in the South Asian region as well, and the rising atmospheric CO_2 levels influences the sea surface temperature (SST) and sea level.

The South Asia is one of the most vulnerable areas of the world. The vulnerability of climate change and more specifically from extreme climatic events is further exacerbated due to low coping capacity and adaptability of the region. Although policies for coping with the extreme climatic changes in South Asian region exist but the same remain by and large ineffective due to reasons like apathy, uncoordinated institutional approach, improper dissemination of relevant information to the stakeholders and non-availability of impact assessment information. Therefore, the region requires a shift from policy to action by development of adequate realistic adaptation plans that are integrated into existing development initiatives.

The present efforts by regional scientists of South Asia to synthesize the current understanding of various issues of global change relevant for the region has been undertaken under the auspices of STARTS MAIRS programme in which regional scientists of different disciplines pooled up their own resources and intellect which is reflected in the chapters in this book.

The preparation of this book is facilitated by the discussion opportunities provided through support by the global change **S**ys**T**em for **A**nalysis, **R**esearch and **T**raining (START) and **A**sia-**P**acific **N**etwork for Global Change Research (APN) which is thankfully acknowledged. The help received from Dr. Amir Muhammed, Dr Janaka Ratnasiri, Dr. K.L. Shrestha and Prof. Sibaji Raha in organization of discussions during the course of the preparation is also gratefully acknowledged. Editors are also grateful to Dr. P. Banerjee, Acting Director, National Physical Laboratory (NPL), New Delhi for supporting this activity.

About the Editors

Late Dr. A.P. Mitra, a Fellow of Royal Society of London, was born on 21 February, 1927 in Kolkata, India. He obtained his D.Phil. in 1955 from University of Calcutta. Dr Mitra was an outstanding scientist and began his research career in the area of ionospheric and related atmospheric research and became the doyen of upper atmospheric research in India. Over the last one and a half decade, he had concentrated on the scientific aspects of global environmental hazards of human activities. His scientific contributions to ozone problem, to the atmospheric chemistry and measurement of greenhouse gases in India, and to global environmental chemistry have had international impacts. He led the global change science programme in India and South Asia and was instrumental in the initiation and execution of several regional collaborative global change research programmes in Asia-Pacific region. Dr Mitra was the Fellow of all the three Academies of India and Fellow of Third World Academy of Sciences (TWAS) and of the International Academy of Astronautics. He was also the Past President of the National Academy of Sciences. He was also the honorary president of International Union of Radio Science (URSI). He had over 200 publications to his credit and also written and edited several books and monographs. Dr Mitra had been bestowed with several awards and medals including the Padma Bhushan by Government of India.

Dr. C. Sharma, is presently working as a scientist at the National Physical Laboratory, New Delhi. Dr Sharma's research interests include climate change, vulnerability and adaptation studies, measurements of greenhouse and other trace atmospheric gas emissions, emission inventory preparations and air quality measurements and impacts. Dr Sharma has extensively worked for the generation of country specific emission factors of various greenhouse and other trace gases from different sources like agriculture, biomass burning, landfill etc. He has participated in a number of international and national global-change related projects and has published more than 80 research papers in various national and international journals. Dr Sharma has contributed to the Intergovernmental Panel on Climate Change (IPCC) process as well and has been provided the recognition certificates by the IPCC as well as by the Prime Minister of India for his contribution to the work of IPCC, which co-shared the Nobel Peace Prize 2007.

Contributors

A. Jayaraman
National Atmospheric Research Laboratory, Department of Space
Govt of India
Gadanki 517 112, India
jayaraman@narl.gov.in

A.K. Singhvi
Physical Research Laboratory
Ahmedabad 380 009, India
singhvi@prl.res.in

A.P. Mitra (Late)
National Physical Laboratory
Dr. K.S. Krishnan Marg
New Delhi 110 012, India

A. Velmurugan
Indian Institute of Remote Sensing (NRSA)
4, Kalidas Road
Dehradun 248 001, India
vels@iirs.gov.in

A. Hassan
Center for Environmental Geographic Information Services (CEGIS)
House 6, Rd. 23/C, Gulshan-1
Dhaka 1212, Bangladesh

Ahsan Uddin Ahmed
Centre for Global Change (CGC)
12-Ka/A/1 Shaymoli Second Lane
Dhaka 1207, Bangladesh
ahsan.ua@gmail.com

A. Bhattacharyya
Birbal Sahni Institute of Palaeobotany
53, University Road
Lucknow 226 007, India

A.K. Gupta
Department of Geology & Geophysics
Indian Institute of Technology
Kharagpur 721 302, W.B., India
anil_k_gupta1960@yahoo.co.in

Anoja Wickramasinghe
Dept. of Geography
University of Peradeniya
Peradeniya, Sri Lanka
niluwick@slt.lk>niluwick@slt.lk

Arshad M. Khan
Global Change Impact Studies Centre (GCISC)
National Centre for Physics (NCP) Complex
Quaid-i-Azam University Campus
Islamabad, Pakistan
arshad.khan@gcisc.org.pk

A.B. Shrestha
International Centre for Integrated Mountain Development (ICIMOD)
Khumaltar, Lalitpur, Nepal
abshrestha@icimod.org

Brenna Enright
Department of Civil Engineering
University of Toronto
35, St. George St.
Toronto, ON, M5S 1A4
brenna.enright@gmail.com

C. Venkataraman
Department of Chemical Engineering
Indian Institute of Technology Bombay, Powai
Mumbai 400 076, India
chandra@iitb.ac.in

Chetana Chaudhuri
Jawaharlal Nehru University
New Delhi

C. Sharma
Radio and Atmospheric Sciences Division
National Physical Laboratory
Dr. K.S. Krishnan Road
New Delhi 110 012, India
csharma@mail.nplindia.ernet.in

Chhonda Bose
Department of Economics
La Martiniere for Boys
11 Loudon Street
Kolkata 700 017, India
chhonda_bose@yahoo.com

D.A. Quadir
Apt B-1, House 62, Road 4-A, Dhanmondi R/A
Dhaka 1209, Bangladesh
dquadir@yahoo.com

G. Beig
Indian Institute of Tropical Meteorology
Dr. Homi Bhabha Road, Pashan
Pune 411 008, India
beig@tropmet.res.in

J. Ashraf
Global Change Impact Studies Centre (GCISC)
National Centre for Physics (NCP) Complex
Quaid-i-Azam University Campus
Islamabad, Pakistan
javeria.ashraf@gcisc.org.pk

Joyashree Roy
Department of Economics
Jadavpur University
188, Raja S.C. Mallik Road
Kolkata 700 032, India
joyashreeju@gmail.com

K. Rupakumar
World Climate Applications and Services Division
Climate Prediction and Adaptation Branch
Climate and Water Department
World Meteorological Organization
7bis, Avenue de la Paix
Case Postale No. 2300
1211 Geneva 2, Switzerland
RKolli@wmo.int

L.P. Devkota
Central Department of Hydrology & Meteorology
Tribhuvan University
Kathmandu, Nepal
devkotalp@hotmail.com

M.D. Kumar
National Institute of Ocenography
Dona Paula, Goa 403 004, India
dileep@nio.org

M.S. Chauhan
Birbal Sahni Institute of Palaeobotany
53, University Road
Lucknow 226 007, India
mschauhan_2000@yahoo.com

M.R. Ray
Neuroendocrinology Department
Chittaranjan National Cancer Research Institute, Kolkata, India

Manaswita Konar
HM Treasury (Labour market policy team)
I Horse Guards Road
London SW1A 2HQ, UK
manaswita.konar@gmail.com

Md Rashed Chowdhury
Pacific ENSO Applications Climate Center
Affiliate Faculty - Urban and Regional Planning
University of Hawaii at Manoa
2525 Correa Road (HIG 350)
Honolulu, HI 96822, USA
rashed@hawaii.edu

M. Thamban
National Centre for Antarctic & Ocean Research
Headland Sada, Vasco-da-Gama, Goa 403 804 India
meloth@ncaor.org

Mitali Dasgupta
Resources and Global Security Division
The Energy and Resources Institute (TERI)
741, Jessore Road
Kolkata 700 028
mitalipal@gmail.com

M. Monirul Qader Mirza
Adaptation and Impacts Research Division (AIRD)
Environment Canada
c/o Department of Physical and Environmental Sciences
University of Toronto at Scarborough, 1265 Military Trail
Toronto, ON M1C 1A4, Canada
monirul.mirza@utoronto.ca

Moumita Roy
Economics Department
Krannert School of Management
Purdue University, Krannert Center
425 West State Street
West Lafayette, Indiana 47907

M. Adnan
Global Change Impact Studies Centre (GCISC)
National Centre for Physics (NCP) Complex
Quaid-i-Azam University Campus
Islamabad, Pakistan
m.adnan@gcisc.org.pk

M.M. Sheikh
Global Change Impact Studies Centre (GCISC)
National Centre for Physics (NCP) Complex
Quaid-i-Azam University Campus
Islamabad, Pakistan
munir.sheikh@gcisc.org.pk

N.R. Phadtare
Palynology Laboratory
Wadia Institute of Himalayan Geology
33 GMS Road
Dehradun 248 001, India
phadtarenr@wihg.res.in

N. Manzoor
Global Change Impact Studies Centre (GCISC)
National Centre for Physics (NCP) Complex
Quaid-i-Azam University Campus
Islamabad, Pakistan
naeem.manzoor@gcisc.org.pk

O.N. Dhar
Indian Institute of Tropical Meteorology
Pashan, Pune 411 008, India

O.S. Chauhan
National Institute of Oceanography
Dona Paula, Goa 403 004, India
onkar@nio.org

P.D. Roy
Departamento de Geoquimica
Instituto de Geología
Universidad Nacional Autonoma de Mexico, Ciudad Universitaria
C.P. 04510
Coyoacan, Mexico D.F., Mexico
p_debajyotiroy@yahoo.com

R.R. Yadav
Birbal Sahni Institute of Palaeobotany
53, University Road
Lucknow 226 007, India
rryadav2000@gmail.com

Ranjan Bose
Energy and Transport Specialist
Energy Sector Management Assistance Program (ESMAP)
World Bank, Washington DC
1818 H Street, NW

Rucha Ghate
SHODH: The Institute for Research and Development
106, K.T. Nagar, Katol Road
Nagpur 440 013, India
www.shodh-research.org

S.K. Satheesh
Centre for Atmospheric & Oceanic Sciences
Indian Institute of Science
Bangalore 560 012, India
satheesh@caos.iisc.ernet.in

Saikat Sinha Roy
Jadavpur University
188, Raja S.C. Mallick Road
Kolkata 700 032, India
saikat@ris.org.in

Sarmistha Das
Jadavpur University, Kolkata
188, Raja S.C. Mallick Road
Kolkata 700 032, India

S. Nandargi
Indian Institute of Tropical Meteorology
Dr. Homi Bhabha Road, Pashan
Pune 411 008, India
nshobha@tropmet.res.in

Shobhakar Dhakal
Global Carbon Project
National Institute for Environmental Studies
16-2, Onogawa
Tsukuba, Japan 3058506
shobhakar.dhakal@nies.go.jp

S.N. de Sousa
Goa State Pollution Control Board
Dempo Tower (1st floor)
EDC Plaza, Patto, Panjim, Goa
simao@rediffmail.com

S. Sardessai
National Institute of Oceanography
Dona Paula, Goa 403 004, India
sugandha@nio.org

Sumana Bhattacharya
NATCOM Project Management Cell
Ministry of Environment and Forests, Government of India
Winrock International India
S-212, Panchsheel Park
New Delhi 110 017
sumana@winrockindia.org

S. Chakravorty
Climatology and Hydrometeorology Division
Indian Institute of Tropical Meteorology
Dr Homi Bhabha Road, Pashan
Pune 411 008, India
supriyo@tropmet.res.in

T.G. Jagtap
National Institute of Oceanography
Dona Paula, Goa 403 004, India
tanaji@nio.org

T. Lahiri
Neuroendocrinology Department
Chittaranjan National Cancer Research Institute, Kolkata, India

U.C. Kulshrestha
School of Environmental Sciences
Jawaharlal Nehru University
New Delhi 110 067, India
umesh@mail.jnu.ac.in

V.K. Dadhwal
Indian Institute of Remote Sensing (NRSA)
4, Kalidas Road
Dehradun 248 001, India
vkdadhwal@iirs.gov.in

V.S. Kale
Department of Geography
University of Pune, Pune 411 007
India
vskale@unipune.ernet.in

Contents

1 Human Dimensions of Changing Environment

A.P. Mitra

1. INTRODUCTION

There is an increasing global concern resulting from human activities affecting not only the human development but also endangering the planet earth. One of the important component of this hazard is that of its environment. It is interesting to investigate how it has been changing in the past before human activities began to alter it and how it is changing now and the extent to which recent changes have been contributed by human activities. These changes are caused by and in turn influence other components of the earth system – land, atmosphere, oceans – and the processes that relate them. There are concerns about human well-being and the earth system stability. Global change programme has now been redesigned to cover this entire earth system. This connectivity is shown in Fig. 1.

Global change and human society are interlinked. The linkages have several aspects:

- Human societies are part of ecosystem but also more than ecosystems,
- Human societies do not change via Darwinian mechanism alone, but depends also on human choice.
- Carrying capacity of the earth is determined not only by the number of people on Earth that drives global change but also per capita levels of resource consumption.
- Collapse of societies: Societies choose to fail or survive.
- There are reverberations of such changes in the responses of the Earth System. Figure 2 shows the reverberations of fossil fuel combustion through the earth system. This is only one of the examples.

The responses include:

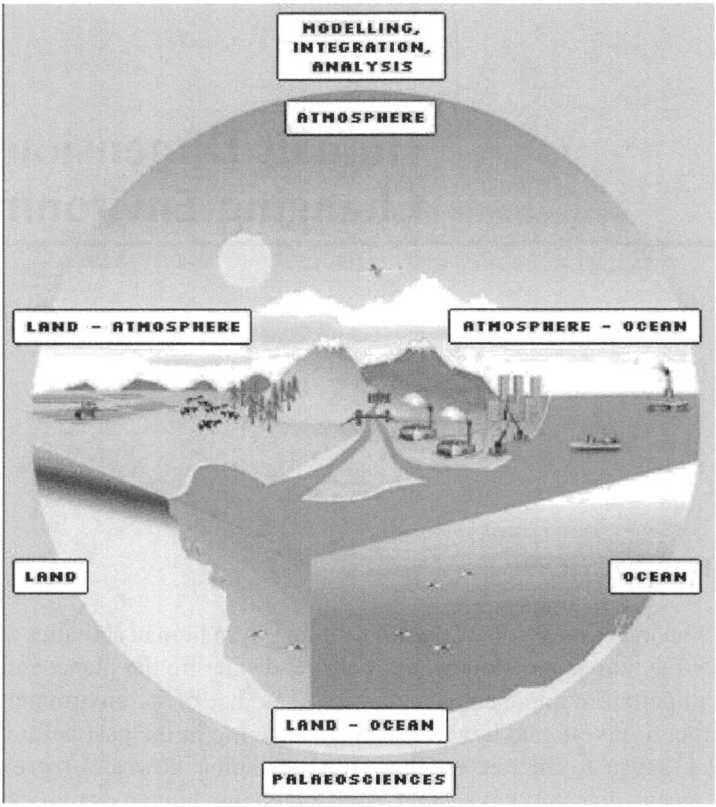

Fig. 1: Global change programme covering earth system.

- Responses of the earth system to fossil fuel combustion
- Forest responses to a future CO_2-enriched atmosphere
- Atmospheric reactive compounds and the climate system
- Response of the earth system to land-use and land-cover change
- Multiple and interacting systems

The Special Role of the Atmosphere

In this matter of changing Earth, the role of the changing environment is crucial. Recently International Global Atmospheric Chemistry group (IGAC) has made the following statement:

"The story of the importance of atmospheric chemistry begins with the origin and evolution of life on Earth. The accumulation of greenhouse gases in the atmosphere of the prebiotic Earth allowed surface temperatures to be maintained above the freezing point of water. Reactions involving carbon, hydrogen and nitrogen compounds in the primeval soup led to the formation of self-replicating molecules, and 400 million years ago, the rise of atmospheric oxygen led to the formation of the stratospheric ozone layer

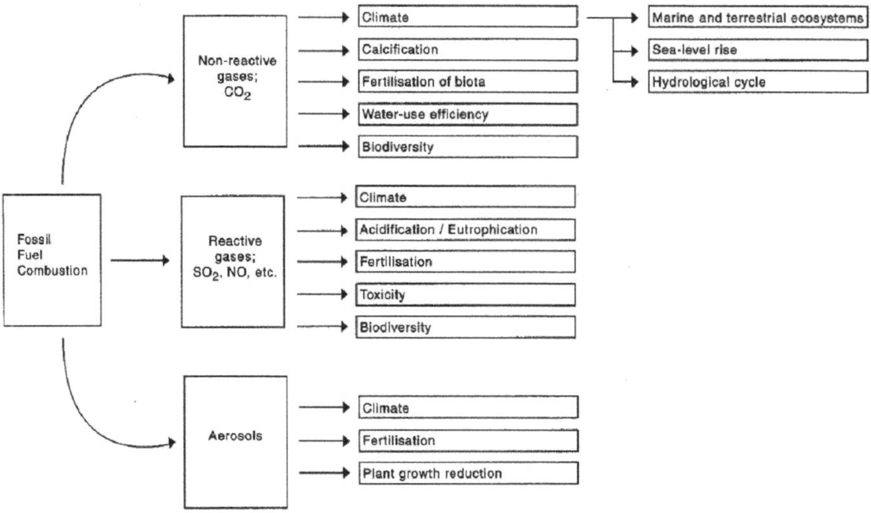

Fig. 2: Reverberations of the effects of fossil fuel combustion through the Earth System.

which protects life on Earth from extremely harmful levels of solar ultraviolet radiation."

At this stage, one should note the interesting GAIA concept – that of self-regulation. The hypothesis introduced by James Lovelock in 1960 visualizes the earth's atmosphere to be regulated by life (atmospheric homeostatics by and for the biosphere). This means that any changes in the global environment and that of the Earth System will be regulated internally by the biosphere. This "healing hypothesis" has, however, been questioned. Paul Crutzen, the Nobel Prize winning chemist, thinks that the evolving role of the biosphere is difficult to conceive in the context of major disturbances and instability regimes in the Earth.

How Do We Define the Environment?

Over the decades the definition has seen continuous changes. The atmospheric environment, as we consider now, extends from the surface of the earth, through troposphere, stratosphere and thermosphere, including the ionosphere, to distances of several earth radii. Changes occur at all levels, but impact of human activities are perceived at heights lower than 1000 km.

The early focus was on the upper atmosphere and the ionosphere (resulting from the successful transatlantic communication by Marconi). It shifted first to higher levels of space environment with the emergence of satellites, and then in recent years moved nearer the ground as we began to consider the possibility of climate change. We are now coming to a new concept of integrating the entire atmospheric system from surface to outer space.

In this path of many milestones and changing focus, there is now another new element. This is the emergence of the view that the entire earth system – the solid earth, the oceans, biosphere and atmosphere – is now under assault. This assault on the natural system began from the introduction of electricity by James Watt, continuing with damages of the entire components of the earth system including the upper atmosphere and near space. This new period has been called the 'Anthropocene era" by the Nobel Prize winner scientist Paul Crutzen.

Human Influence to Higher Atmospheric Levels

While concern from climate change has rivetted our attention to levels near the earth, human influence can, and has been observed, to alter levels upto 1000 km and beyond. For one thing, GHGs cause cooling at heights beyond about 15 km, and this cooling extends to upper atmosphere causing substantial reduction in atmospheric density. One effect of this is the increase in satellite lifetime because of reduced drag. And then, in addition, one encounters changes caused by rocket exhausts, from heating through high power radio waves, from deliberate injection of chemically active materials into the upper atmosphere. The Nobel Prize winning work of ozone (O_3) depletion was initially prompted by concern over the destruction of ozone – catalytically – by NO_x and ClO: the first arising from exhausts from supersonic aircrafts and the second from rocket exhausts and CFCs used for refrigeration.

It is not that the environment is only changing now. It has always been changing, sometimes even drastically, excepting that we have now reached a no-analogue state, beyond the ranges of variabilities in the past. Today's global system is being changed in mere 300 years or so. Man has now begun to replace nature as the engine of climate change.

Environment Prior to Human Influence

Let us now look at how things have changed in the past. Oxygen evolved some two billion years ago. Ozone formed much later when molecular oxygen became abundantly available since ozone is a photochemical product resulting from a combination of O_2 and O. Ozone is a minor constituent of the atmosphere, currently around 40-100 ppbv at the surface. The main atmosphere O_2 and N_2 have not changed, but the changes we see and are concerned about all refer to constituents which are only millionths or billionths, and sometimes even trillionths of the main constituents. It is strange that constituents which are present in such small qualities have such critical roles in modifying the earth system.

There is direct information on the nature of changes of the earth's atmosphere from air bubbles trapped in ice cores. These go back to nearly 400,000 years. Over this period there were periods of warming and cooling, as well as changes in CO_2 (Fig. 3; IPCC AR4, 2007). These go up and down together, but have been within a range of values during this period.

In the recent times of earth's history, there were three major periods of interest (Fig. 4; IGBP, 2001). First, the glacial period of 20,000 years ago when the main surface temperature was 4-5°K lower and CO_2 concentration

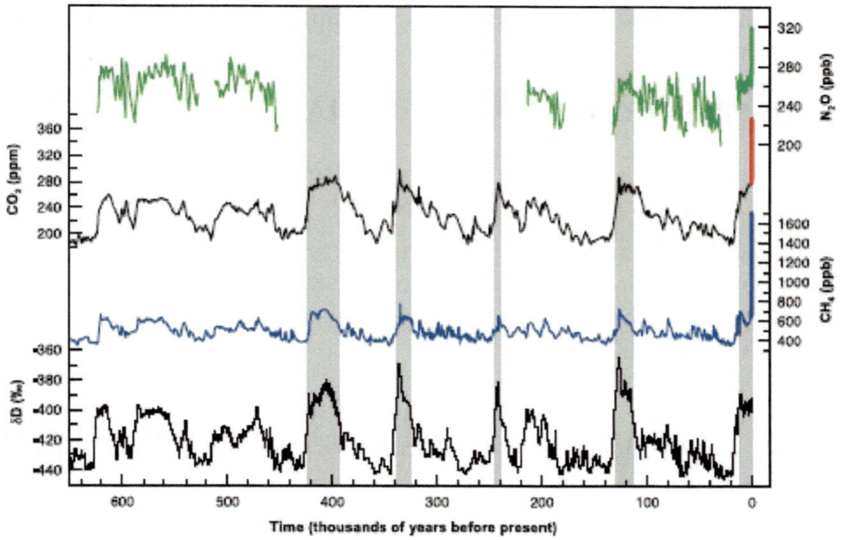

Fig. 3: Variations of deuterium (^{TM}D) in antarctic ice, which is a proxy for local temperature, and the atmospheric concentrations of the greenhouse gases carbon dioxide (CO_2), methane (CH_4), and nitrous oxide (N_2O) in air trapped within the ice cores and from recent atmospheric measurements. Data cover 650,000 years and the shaded bands indicate current and previous interglacial warm periods.

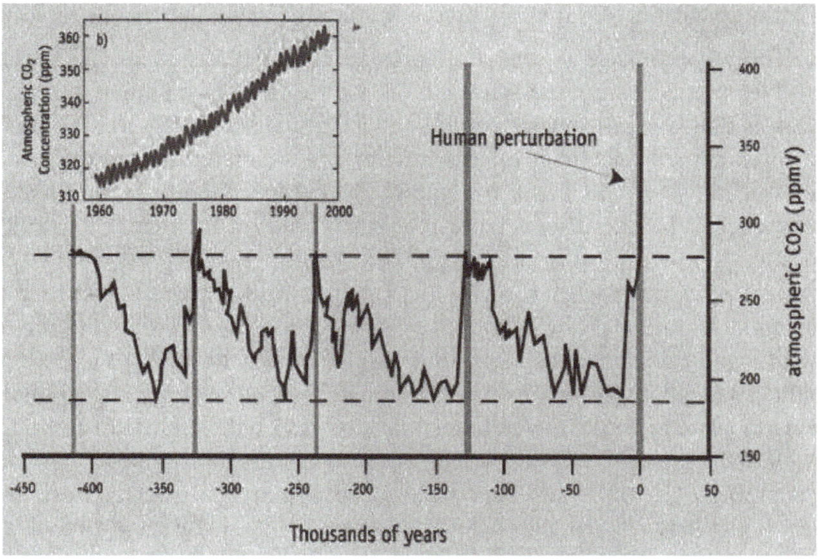

Fig. 4: The recent human influence on the carbon cycle.

was 200 ppm. Second, the medieval warm period from 1000-1200 AD, although that warm period was not as warm as during the last decade. Third, the Maunder minimum from 1640-1710 AD, when the earth experienced a mini-glacial stage. An interesting feature was that solar activity was totally absent during this 70 years period indicating a possible sun-climate relationship.

2. THE PRESENT: THE ENDANGERED PLANET

The present gives a picture of frightening degradation of the atmosphere, land and oceans. Some key changes include:

- Population – Increased ten-fold during three centuries
- Urbanization – Increased ten-fold since 1980
- Fossil fuel – Exhausting in a few generations which have been generated for several hundred million years
- Land surface – 50% of land surface transformed
- Nitrogen fixation – More fixed synthetically than naturally
- Fresh water – More than half of accessible water used by man and depletion of underground water
- Greenhouse Gases (GHGs) – Rapid increase in GHG emission and of their concentrations. Discernible human influence on global climate (in surface temperature, in the occurrence of floods, in loss of tropical rain forests etc.)
- Coastal and marine boundaries - 50% mangroves removed, wetlands shrunk by half
- Marine fisheries – 22% of recognized fisheries over-exploited or already depleted
- Marine and terrestrial ecosystem – Extinction rates increasing rapidly.

The key message is that the earth is now threatened by the human activities that needs immediate attention. An example of a human corrective action is that of stratospheric ozone – ozone at heights of 20-30 km. This is one of lucky escape from an unforeseen and unintended consequence of widespread use of CFCs as refrigerants. The escape came from a number of lucky events: (a) the discovery of the ozone hole by Farman et al. (1985), (b) the 70's work of Crutzen (1970), Molina and Rowland (1974), nobel-prize winning work showing the catalytic role of CO and NO_x and (c) the implementation of the Montreal Protocol gradually withdrawing the use of CFCs and halons. Stratosphere ozone depletion is an example of a powerful, nonlinear feedback system as well as human induced chemical instability of the earth system. Paul Crutzen notes "This brings up the nightmarish thought that if the chemical industries had developed organobromine compounds instead of CFCs, we would have been faced with a catastrophic ozone hole everywhere and at all seasons". The adoption of Montreal Protocol and phasing out of ozone-depleting substances has stopped the precipitous collapse of stratospheric ozone. However, surface ozone continues to increase from increasing air pollution – a matter ignored in the Indian air pollution standards or in discussions on impact on health and agriculture.

A second aspect is the "carrying capacity of the Earth". It is influenced by lifestyle, social organization, distribution of poverty, cultural heritages and several other socio-economic parameters. One has to identify the multidimensional range of threshold. When these thresholds are exceeded, one may have a triggering shock disturbing Earth's life support system. The global climate system is an interaction of the land, ocean and the atmosphere. Changes in these components also change the climate system, and the changes are interactive.

The Drivers

The drivers of the changing environment are increasingly anthropogenic in addition to natural sources like solar activity and volcanic eruptions. The key drivers are atmospheric constituents, present only in small quantity, but increasing rapidly.

Four special aspects are only beginning to be recognized: one is that the climate forcing constituents of the air are of three distinct categories: the long lived gases (CO_2, CH_4, N_2O), the short lived gases (CO, NO_x, SO_2) and the particulate materials of different sizes and composition including climatically important soot carbon (Table 1). All these have been increasing very rapidly in South Asian region with growth rates of 4-6% in most cases, 3-5 times larger than in most industrialized countries. Of the particulates, a new entrant is the soot carbon, the absorbing aerosols. Soot carbon has now entered the center stage in climate forcing scenario, and that needs to be investigated thoroughly.

The second aspect is the recognition of the need to predict regional scenarios of pollution (of all these categories) and of their impacts on different time scales from seasons to years to decades. Climate change models usually have coarse grid resolution. Regional models with higher resolution are needed for regional and sub-regional studies. Such models are now becoming available which are useful for many calculations of climate change impacts.

Table 1: Three distinct categories of climate forcing constituents

Protocol GHGs			*Short Lived Gases*	*Particulates*
1.56 Wm^{-2}	CO_2	1	O_3	• Cooling Aerosols
0.47 Wm^{-2}	CH_4	21	CO	• Heating Aerosols
0.14 Wm^{-2}	N_2O	310	NO_x	(*transport over*
	HFC	140	SO_2	*long distances*)
Little now	SF_6	23,900		
	PFC	Large		
Scale:	Global		Scale: Local to Regional	
Time Scale:	Decades and Centuries		Time Scale: Weeks and Days	Time Scale: Days

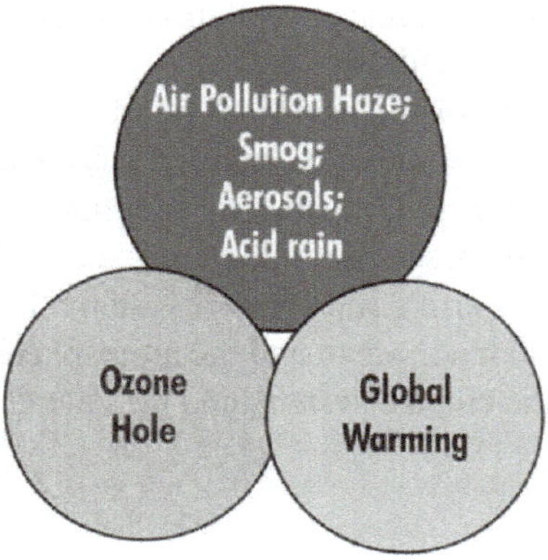

Fig. 5: Major environment problems: Intersecting issues.

The third aspect is that air pollution, global warming and ozone depletion need one common framework to assess impacts. Often, these are dealt with by different groups, totally isolated from each other, and driven by external pressure points (e.g. protocols). Examples include: The Montreal Protocol controlling elimination of ozone depletion substances, does not consider ozone as a climate forcing gas; the UNFCCC concentrating on CO_2 , CH_4, N_2O (and not on the other climate forcing parameters); and air pollution looked at only from health considerations and covered for South Asian region by Male Declaration. A new paradigm must consider all three together in an interactive way, figuratively shown in Fig. 5.

While warming comes principally from the first category of constituents in Table 1 (long lived gases), the others also force the climate in different ways. Some warm the atmosphere; some cool it (Fig. 6; IPCC AR4, 2007) but these have different lifetimes in the atmosphere. The fourth aspect is that

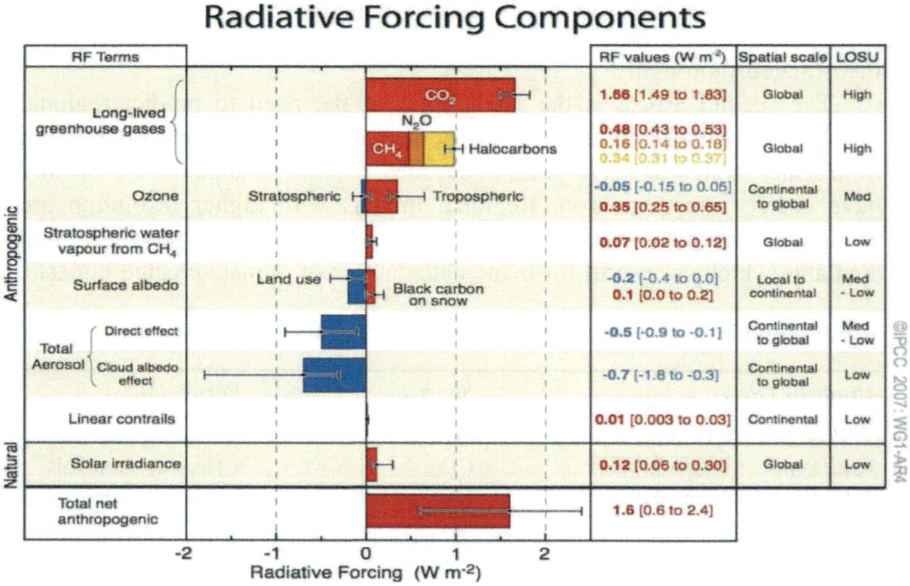

Fig. 6: Schematic comparing several factors that influence Earth's climate on the basis of their contribution to radiative forcing (IPCC AR4).

GEC – global environmental change – covers much more than climatic change, and interacts with forests, oceans, wetlands, and agricultural systems – different components of the Earth System. The interaction is both ways. For example, while environmental change affects agriculture, agricultural system (through methane emission from paddy fields and ruminating cattle) also change the environment.

Jolting the Earth System: Critical Threshold

The addition of infrared energy from greenhouse gases is a very small fraction of the total solar input. Currently this amounts to 2.5 W/m^2, which is only one hundredth of the energy from the sun. How does the earth deal with this extra energy (added in a relatively short time)? The new equilibrium results in a warmer atmosphere. This is not in dispute: but how much? Is this limited change so critical and does it have implications for the entire earth system? To what extent do we believe that there may be irreversible changes?

We have examples in the past where multiple stresses interact to jolt the earth to cross a crucial threshold and a step-change occurs in its functioning. A major example was the catastrophic event in the region of the Akkadian empire some 4000 years ago connecting the incidence of drought with the collapse of its civilization (IGBP, 2001). Another example was the transformation of the North Africa (Sahelian region) from a moist savanna to dramatic desertification about 5500 years ago from a small change in the earth orbit jolting the earth to cross threshold and triggering a series of biophysical feedbacks leading to a drying climate. Survival or collapse depends on how we cope with such sudden jolts.

The population is increasing at an alarming rate, especially in developing countries and so is the consumption rate which is already very high in developed world (and consequently also emissions). These add substantially to atmospheric pollutants of all categories, alter forests and landuse, oceans and coastal regions, and add to local, regional and global environmental stresses. The infrared energy added to the planet may rise to 4 Wm^{-2} or more, with warming between 1.5 and 4.5°C also perhaps switching on some of the key instability parameters. There are a number of potentially critical points in the Earth System with major thresholds, bottlenecks and switch points existing. Perturbation to the system may jolt it to produce abrupt changes either repeating conditions that existed in the past or even causing a shift to a different state.

3. INDIA AND THE CHANGING ENVIRONMENT

Indian society is concerned in two major ways. First, its own role in degrading the earth. Second, the impacts of the changing earth on climate, agriculture, ecosystem and health – the question of living in the changing earth system.

The first has multiple dimensions. At the level of GHG emissions, Indian emission is low although increasing at a high rate of 5-6% and also some of the emissions (CH_4) are related to our food security. However, at the level of short-lived gases and particulate emissions, the distribution is different. Surface ozone is still low at 40-60 ppbv but is increasing as a result of increasing air pollution. Stratospheric ozone has already decreased. Carbon monoxide, coming primarily from biomass burning and the transportation sectors, is already high. The most serious is the matter of black carbon, which in cities like Delhi is abnormally high.

In regard to methane there were in the past statements that the developing world is the major contributor (India and China together being responsible for 90% of the emission from rice paddy fields). In the 1990 IPCC Report, out of a total of 500 Tg/yr for methane emissions as much as 100 was attributed to global emission from the rice paddy fields. In late eighties, an US-EPA study attributed 37 Tg/yr to India alone (Ahuja, 1990). A very detailed campaign undertaken in 1991 in India showed drastically different results; emission ranging from 4 to 6 Tg/yr that is almost $1/10^{th}$ of this value (Parashar et al., 1996). Another important case concerns emission from the increasing destruction of Amazonian forest. In this case there were major differences between earlier FAO and WRI estimates and those obtained from high resolution satellite data obtained by Brazilian space scientists. We learn a major lesson here: that inventories must be made by countries themselves.

The Indo-Gangetic plains have been severely damaged and is a major emitter. The Indian Ocean is a net emitter of CO_2 and N_2O (instead of a sink), soil erosion is of concern, forest cover depletion in the Himalayan region is inhibiting the "sink" character of CO_2. Coastal zones are experiencing and will increasingly experience large-scale biogeochemical modification which may extend to open sea.

Much of Indian pollution comes from low level of technology. Incomplete combustion of coal used in small scale industries produce particulates including black carbon; so is the case with diesel driven trucks and buses and two-stroke systems such as motor cycles and scooters. The positive aspect is the gradual disappearance of coal as cooking material in the cities. Increasingly coal in India is being used at the high technology end (power and steel) and transportation sector is going through major overhaul. These will contribute to an improvement in the emission scenarios of CO and particulates. However, it may be noted that the pollutants loading over any region may include an appreciable portion transmitted from elsewhere – sometimes across continents.

There is another frightening picture: indoor air pollution from the use of fuel wood and cow-dung in villages' households. Key pollutants are: CO_2, CO and particulates. The risk factor is high. There is also, additionally, the question of impact of such pollutants on the climate system. We tend to shrug off these emissions as "survival" emissions. But simple modifications of the chullahs, not only for good combustion but also for emission efficiency, are not difficult to introduce. There are 2-3 million households using improved chullahs. Such health-damaging pollution can be addressed reasonably quickly and adequately through increasing use of solar power, biogas system and modification of chullahs.

The second aspect is the question of vulnerability and adaptation. Like other developing countries, India is at the receiving end of the global change situation caused primarily by industrialized countries. Monsoons, water resources and agriculture form one such aspect which are interrelated. Future survival will depend on the proper understanding of their relationships and of how each is changing with the changing earth system and whether and to what extent, we can adapt to these changes.

The major concern is the monsoon variability and the occurrences of extreme climatic events – droughts, floods and cyclones. The recent unexpected event of Tsunami shows the value of creating a mechanism for early warning and for disaster management. We are aware of the consequences of drought in recent years. Fortunately, the Indian monsoon system has been relatively stable over the last century, varying within a range of 10%. The question is whether this variability itself is changing. There is some evidence that ENSO-monsoon relationship is weakening which itself is a pointer.

For cyclones, there appears to be a slightly decreasing trend with time. Over the Indian Ocean, Bay of Bengal and Arabian Sea significant and consistent warming of the sea surface has occurred during the twentieth centuries. This should have a major effect. There are large tracts in northwestern India and interior peninsula and in Pakistan with arid conditions. However, there is a decreasing trend of the total arid area during the period 1871-1984 over India as a whole. This is possibly a result of a westward shift in the monsoon rainfall activities and the changes in the landuse pattern over Northwest India.

Impacts of floods and sea-level rise (for low lying regions and for small island states) is another serious concern. Sea-level rise of 45 cm will submerge about 11% of land area of Bangladesh (16,000 km) including nearly 75% of Sundarbans mangrove forests, and for 100 cm rise about 21% land area will be inundated. For Sri Lanka, for 100 cm rise one expects 100 m shoreline recession drowning most of the coastal wetlands. The Maldives are mostly composed of coral reef no higher than 1 to 2 metre – the danger from storm surges and from high sea level rise are very real.

Forestry continues to be a major concern for many of the developing countries. As mentioned earlier, carbon dioxide emission due to deforestation comes almost entirely from the developing countries. Hence the thrust for mitigation should primarily concentrate in this area.

For countries like India, where food security is at the core of the development process, the effects arising from haze-induced reduction of sunlight is a new parameter that should now be added to previously examined efforts from changes in temperature and precipitation and CO_2 enrichment. The effect of such reduction is a decrease in agriculture yield. However two points need special emphasis. First, GHG related effects are slow processes acting over decades, whereas changes through aerosols and through CO, NO_x, O_3 etc. are more rapid. Second, most of the effects are negative. The positive effects are from precipitation (excess or deficit of rainfall of 100 mm produces corresponding excess on deficit of 0.7 MT of crop production) and CO_2 increase. The strategy, therefore, should be to optimize the combined effects of these perturbations through selective mitigation efforts. A reduction in aerosol loading appears to be one such effort.

We need to know how CO_2 fertilization will work. For this, Indian Agricultural Research Institute (IARI) and National Physical Laboratory (NPL) jointly established a facility called FACE (Free Air Carbon dioxide Enrichment) at the IARI grounds in Pusa which has been used to conduct a number of experiments principally on rice. Additionally, this effort has led to the installation of a number of OTCs (Open Top Chambers) in India, Pakistan, Nepal, Bangladesh and Sri Lanka. These efforts are currently continuing and also being further strengthened.

Water availability is a direct consequence of rainfall. There are three major watersheds in the South Asian river system: the Himalayas, the Central highlands of India and the Western Ghats of peninsular India. The Himalayan river system is the main system providing perennial water for India, Pakistan, Nepal, Bhutan and Bangladesh. South Asia is a region with increasing water stress. Per capita availability of water has decreased dramatically in India and Pakistan: for India from 5200 m^3/yr/cap in 1950 to 1860 in 2000, just above the water stress level and is expected to come down further to 1600 m^3/cap/yr in 2025 (water stress regime) and around 950-1500 in 2050 (water scarcity regimes).

For India and for many countries in the third world, agriculture takes up a major part of the water resource. Consumption is the highest in Asia, Africa and Central America i.e. areas of low income nations, predominantly dependent on agricultural income. For Asia it is as high as 85%. A serious question is about whether the large water use in agriculture can be reduced and the wastage avoided. The concept of "virtual water" needs careful attention.

Global warming has affected Himalayan glaciers more than the Alps or South America. The glaciers are melting/receding faster than those elsewhere being located in tropical zone. For example, the Dokriyani glacier, from Uttarakashi in Garhwal, at an altitude of 13,000 feet, is receding (as other glaciers in the Himalayas) at a rate of about 30 m/yr. Similarly other Himalayan glaciers like Gangotri Glacier is also shrinking over time (Hasnain, 2002). Scientists say that 50 years from now some of these will vanish.

4. CONCLUSIONS

Thus, the changes in the environment – natural in the past and increasingly anthropogenic now – have affected human societies, producing collapse of entire societies in some cases. There is a need for thorough scientific investigation on how the present situation is affecting our well-being and the sustainability of the planet. The mankind's current social and political vision is still limited to short term perspectives forgetting that the survival or collapse depend on how we manage our resources and the earth system. When we consider these changes, we should not forget that there is a whole lot of environment space outside the limited region near the surface that is also going through changes with effects that have not yet been adequately examined. The scientific capacity, especially in the developing world like the South Asian region, is still very low to investigate such issues and therefore, there is an urgent need to address this by sharing the international expertise with regional researchers.

REFERENCES

Ahuja, D.R., 1990. Estimating regional anthropogenic emissions of greenhouse gases, US Environmental Agency Technical Series, Washington D.C. U.S.A.

Crutzen, P.J., 1970. The influence of nitrogen oxides on the atmospheric ozone content. *Q. J. R. Meteorol. Soc.*, **96,** 320-325.

Farman, J.C., Gardiner, B.G. and Shanklin, J.D., 1985. Large losses of total ozone in Antarctica reveal seasonal ClO_x/NO_x interaction, *Nature,* **315,** 207-210.

Hasnain, S.I., 2002. Himalayan glaciers meltdown: impacts on South Asian Rivers. *FRIEND 2002-Regional Hydrology: Bridging the Gap between Research and Practice*, H.A.J. van Lanen and S. Demuth, Eds., IAHS Publications,Wallingford, No. **274,** 417-423.

IGBP Science No. 4, Global Change and the Earth System: A Planet under pressure. 2001, Editors: Will Steffen and Peter Tyson, IGBP Science Series: ISSN 1650-7770.

IPCC AR4; Technical Summary of the '*Climate Change 2007: The Physical Science Basis. Contribution of Working Group 1 to the Fourth Assessment Report of the Intergovernmental Panel on Climate Change'* (Source: www.ipcc.ch)

Mitra, A.P., 1996. Global Change: Greenhouse Gas Emissions in India, Scientific Report No. 10, Centre for Global Change, National Physical Laboratory, New Delhi, India.

Molina, M.J. and Rowland, F.S., 1974. Stratospheric sink for chlorofluoromethanes: chlorine atom-catalysed destruction of ozone. *Nature*, **249,** 810-812.

2 Development Pathway

Joyashree Roy, Chhonda Bose, Ranjan Bose, Sarmistha Das, Shobhankar Dhakal, Mitali Dasgupta, Rucha Ghate, Saikat Sinha Roy, Manaswita Konar, Anoja Wickramasinghe, Moumita Roy and Chetana Chaudhuri

Development drivers and their relative role in development process can describe a development pathway as high or low energy, high or low emission, inherently sustainable or unsustainable. The focus is on South Asia because it houses over 1.4 billion people – more than one-fifth of the world's population in 4.6 million square km. which is 3.3 per cent of the world landmass. The region is experiencing rapid energy demand growth and will remain major consumer of energy in next couple of decades. It is already well accepted that in forward looking view what the "non-G8" world thinks and does is of equal if not more importance. Non G-8 is not homogenous group and especially South Asia is a unique region with high poverty, resource rich, cultural, political, environmental diversity with interesting historical past. Following the global trend, regional cooperation to accelerate the process of economic and social development motivated governments of India, Pakistan, Bangladesh, Sri Lanka, Nepal, Bhutan and Maldives to join hands in 1985 to establish the South Asian Association for Regional Cooperation (SAARC). The main policy area includes economic development, poverty alleviation, science and technology, social issues, culture, the environment and combating terrorism. It provides a platform for the people of South Asia to work together in a spirit of friendship, trust and understanding. Added to the existing challenges, new issues emerging from need to climate-proof developmental actions call for locally suited well coordinated global response. Future global scenario will be shaped very much by the direction, drivers, and organization of development pathway in developing nations especially in South Asia. It is important to understand the Business as usual development scenario to identify critical control points for sustainability transition in climate change context. Development scenario is a complex manifestation of the pursuit of multiple goals of multiple actors in a fully networked global socio-economic-political-institutional landscape. In this chapter we focus on certain such manifestations from the past and present motivations which reflect future development scenario to indicate emission commitments.

1. ANALYTICAL FRAMEWORK

A major approach to understand development pathway in current climate change literature is to identify the economic developmental actors from supply side (e.g. industry, agriculture, energy supply etc.), understand their targets, choices for production and service delivery technique and define a reference point or baseline or the Business as usual Scenario (BAU). Globally BAU has been identified as high carbon growth path (an assessment of the complete literature is available in IPCC 2007 and Stern 2007) and it is almost unanimously accepted that it cannot sustain human well-being in the long-run due to its adverse impacts on various life support systems. Since BAU is also not a static scenario so what elements are driving the dynamics of this BAU and are reflected in future goals of these actors in short and medium term are of major interest in the study of control points. However, this analytical framework is biased by exclusive focus on economic actors, few control points and supply side of the economy only. The framework is not wrong but is too simplistic for representing developmental complexity. Historically or otherwise supply side economic actors are neither the sole nor independent drivers of development pathway. This is even more true for South Asian region. So it is important to understand the socio-political-institutional landscape to design economic instruments for any kind of diversion in BAU towards low carbon development pathway.

It is not only hardcore economic production goal oriented developmental actors who decide on developmental pathway independently but they are only one kind of actors who interacts in a complex way with other actors whose primary goal may not be economic development through production per se. The other actors and complex interaction among them is not easy to describe but we try to present a framework which looks at economic development in a much broader perspective that encompasses complex social processes.

Figure 1 is a simple schematic representation of this extended framework that goes beyond supply side economic actors. Some generalization though not full is addressed in Millenium Assessment Report, 2005 where development pathway is considered as a complex reflection of choices made within a set of institutional arrangement to enhance the wealth and well-being of the entire population. We need to understand the inherent interdependence and complexities embodied in a particular development pathway to look for ‘critical control points’, i.e., actions and impediments towards alternative low emission development pathway.

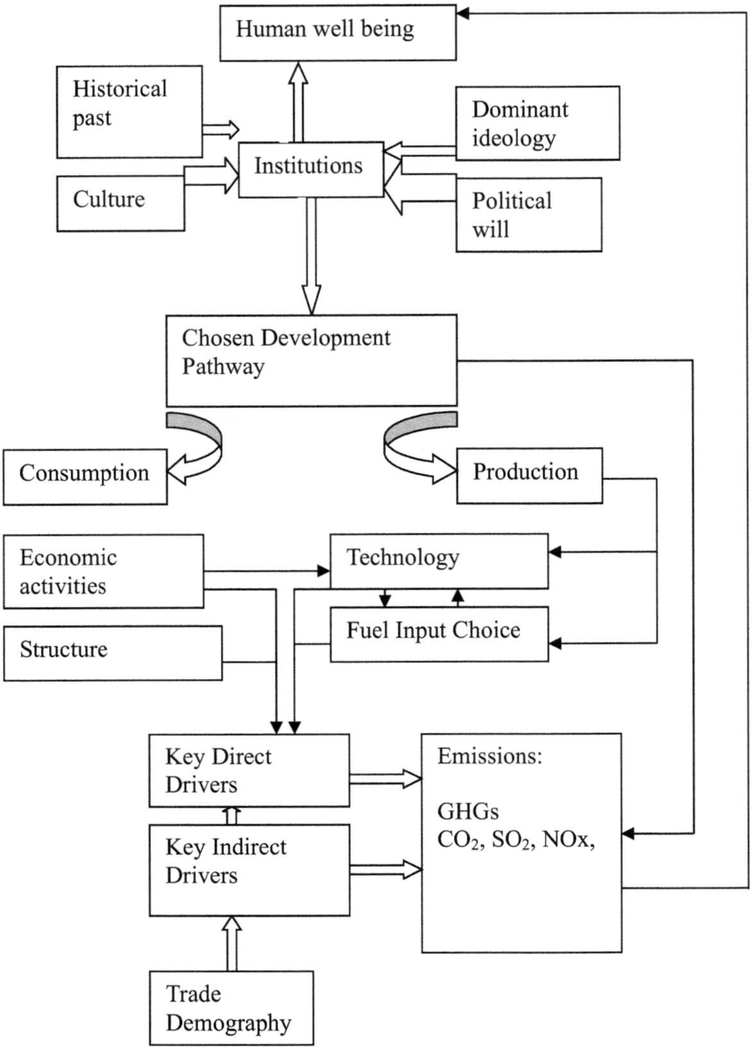

Fig. 1: Complex developmental process.

Any development pathway is chosen and gets defined by the goals, priorities, options and policies. These, in turn, are shaped by the historical past, culture, social structure and dominant ideology, political will – all of which are not equally flexible enough and have various time paths and time horizons. Institutions are basically rule structures regulating the behaviour of population in command area. These institutions are either developed by people seeking to regulate their own interactions or originated with external actors e.g. government, religious organisations, aid agencies etc. (Poteete & Ostrom, 2002). It is now being increasingly recognized that the institutions have material, social, religious and political aspects and for understanding human behaviour, it is essential to expand the range of relevant variables in any analysis. Governance is now a prominent concern in any development discourse. We start from the hypothesis that the chosen development pathway in an institutional configuration (Lunden, 2005) can be observed through the consumption and production behaviour of the participants in the development process. These consumption and production behaviour patterns are the key drivers of GHG emissions. Both demand and supply side actors are critically important who respond to various options, rules, incentives etc.

Development led emission (manifested through consumption and production) commitments will depend on level of activity, and activity composition i.e. structure, technology choice, fuel choice, and consumption pattern. The level of economic activity, structure, technology choice indexed by fuel intensity, and fuel mix (ASIF: **A**ctivity, **S**tructure, **I**ntensity, **F**uel mix) (Schipper and Marie-Lilliu, 1999; Roy et al., 1999; Dasgupta and Roy, 2001; Bose, Mukherji and Gokhale, 2005; Roy et al., 2005) finally determine the emission intensity of a development pathway. Decomposition of energy and emissions drivers into components help in identification of control variables. These control variables are not all of equal strength and at work in same time and space dimensions. This decomposition concept is heavily dependent on the concept of Kaya Identity (IPCC, 2001; Roy, 2007) and no regret policy solutions suggested in the literature. The advantage with this methodology is that it allows for a number of interventions at I, F and A which commensurate with number of multiple scenarios each with a certain energy use and emission levels. Hence the policy makers have a spectrum to choose from and more importantly it makes the framework easily replicable. There are some indirect drivers (MEA, 2005) which determine the emission intensity of a development pathway. These are demographic character of a nation (population growth), trade relation in the process of globalisation and regionalisation, changing notion and development paradigm (SEI/UNER, 1998).

Using the above framework in this chapter we plan to briefly review the literature to highlight the nature of development institutions in South Asia and changing developmental regimes, activity growth pattern, role of technology, fuel intensity, fuel choice in energy consumption pattern, selected demographic characteristics and emerging trade relations. The chapter is not a comprehensive assessment rather an indicative one.

2. PAST DEVELOPMENTAL PROCESS AND INSTITUTIONS IN SOUTH ASIA

Historical analysis will show that the development process in South Asia over past century has been no different than predecessors in global context and are thus converging (unless intervened) towards high energy, fossil fuel intensive, emission loaded pathway. Development regime that emerged in South Asia in 1880 has been state led through investment for public benefit, especially the poor, dominated the scene. Market integration and specialized commercial production emerged in the region around the Indian Ocean. 1930s and 1940s induced nationalists to lay the groundwork for nationally planned economic development, which stressed autonomy, security and national integration under strong central state leadership. In the 1950s, all South Asian countries wrote national plans stressing self-sufficiency and addressing problems of national economic growth, poverty and inequality.

The 25 years between 1950 and 1975 were the heyday of nationally planned development in South Asia. Bangladesh after its emergence pursued progressive, planned national development for all citizens. Funding needs and national pride pushed politicians in South Asia to emphasize economic growth. The policy priorities turned towards larger projects demanding more external finance. Launched in the 1960s, the Green revolution brought an intensive use of pesticides, fertilizer, tractors, tube wells and high yielding hybrid seeds. It raised rice and wheat yields tremendously, secured basic food requirements for national population and spread benefits widely, though unevenly. However, market orientation was never missing. Private enterprise dominated agriculture and industry. Even in India, where national planning had the largest impact, 80 per cent of industrial production remained in the private sector, where public output lowered input prices and state import protection expanded national markets for private enterprise. Sri Lanka, Bangladesh and Nepal led the way in South Asia, starting slowly in the 1970s and accelerating in the 1980s. With declining relative prices for primary product exports, the burden of external debt grew heavier, while raising funds for large development projects became more pressing.

Development regimes in South Asia operate today inside the same national states that managed them in 1975. However today's regimes are fundamentally different, and their transformation has accompanied – if not caused – major shifts in national politics. In India, private capital and state governments have both gained increasing independence. In Nepal, electoral was established in 1991, opening development to wide public debate. Conflict and peace keeping have become cyclical. As fund flow from private and external investors are growing, risk reduction is pushing for more public scrutiny and popular participation to make state regimes more accountable and transparent to citizens. However, national governments are continuing to remain the implementors of development design. Governance has lost its territories. Nation states do not always design and govern development (Lunden, 2005). Millennium development goals are dominating the development agenda of large as well as small countries and institutions. This global common framework for policy thought and action is challenging the nations to adjust national policy to global regimes. Despite this upward movement in institutional convergence, attempt to streamline local micro institutions towards management of local resources is also a predominant trend now in South Asian countries. Box One summarises success and failure stories of participatory approach in local Common Property Resource management (CPRs). Apparently complex combinations of global, regional and local with and without state involvement are coexisting. How far they are mutually reinforcing or contradictory are some of the fundamental burning research questions yet to be resolved. Each country in South Asia now inhabits more than one development regime. National regimes still operate, but each has various local and regional sub-units with distinctive rules of operation, and each must also abide by international rules. The dynamics strongly support very locally efficient systems networked with higher and more global level of growth. This network can be expected to have its mutually strengthening feedback mechanism unless external shock puts constraints.

Box One: Puzzles in Resource Management

The role of community institutions in common property resource management has come to be seen as that of crucial importance to the sustainability of both the resource and welfare of the resource dependent community by giving space to voices of many. Therefore, in recent past there have been concentrated state efforts for involving local communities in CPR management. On the other hand many communities have independently institutionalized their resource management efforts even if there is no backing by the state. In all kinds of community CPR management institutions, cases of both success and failures are reported.

Based on empirical information, scholars have found that certain features of the resource as well as those of the resource-managing group influence the chances of the success or failure of collective efforts. Well-defined boundaries and small size of the resource are factors identified to be behind successful collective action (Wade, 1988; Ostrom, 1990). Group characteristics like small size, clearly defined boundaries, shared norms, existing social capital, homogeneity of the group, etc. are therefore considered to be conducive for collective action as they promote better communication.

The factors considered by most of the authors to be significant for encouraging or discouraging communities' participation are: resource availability/scarcity (Chambers et al., 1989; Hayami et al., 1999; Tang, 1992; Ostrom, 1997), dependence of the community on the resource (Ostrom, 1997; Tang, 1992), alternatives available to private individuals (Balasubramanian and Selvaraj, 2003; Shah et al., 2002; Palanisami and Balasubramanian, 2002; Tang, 1992; Dinar et al., 1997; Sakurai and Palanisami, 2001), availability of exit options (Hayami et al., 1999), proximity to the markets and urban centres (Meinzen-Dick et al., 2002), appropriate leadership (Ghate, 2004) etc.

Regarding availability of the resource, both too much scarcity and too much abundance are found to be damaging for collective action. Hayami, Fujita and Kikuchi (1999) find that while abundance of water obviates the need to organize, severe scarcity may give rise to serious conflicts, making cooperation difficult. While Tang (1992) names "moderate scarcity" as a factor that can bring the farmers together for collective action, Chambers et al. (1989) call it "not too extreme a shortage".

Most scholars have found the group's dependence on the resource to be a factor that constantly creates incentives for individuals to participate in the community resource management efforts. It can be expected that the communities are more likely to get together to protect and manage the resource on which they depend for their livelihood needs. In case of forest dwelling communities, Ostrom (1997) has listed 'resource dependent users' among user attributes that promotes collective action for resource management. Tang (1992) also feels that greater dependence generates more willingness on the part of the farmers to expend labour and resources for the operation and maintenance of the irrigation system.

Even as proximity to the market or urban centres is one source of 'exit options', it is found to be working both against and in favour of the success of collective efforts. Collective action may fail because of demonstration effect, possibilities of making quick money, or influence of foreign values etc. On the other hand proximity can also prove to be conducive if it enhances awareness of the community regarding conservation (Ghate, 2004).

A very typical example from Maharashtra state, however, hints that a set of factors may lead to successful working of an informal institution for one but may not for the other. Saigata and Maral Mendha villages, in the Chandrapur district of Maharashtra, are neighbouring villages. The two villages have much in common e.g. geographical location, terrain, vegetation, exposure to the outside world, resource (forest in this case) dependence etc. Being close together, the contemporary political decisions impacted both the villages in an identical manner. For all communities that depended directly on natural resources, 1955 happens to be a crucial year. In this year the *malgujari* system ended and all forestland and common property water-bodies were transferred to the state for administration and management. This transfer of ownership had similar implications for both the villages. The combined effect of poverty, resulting from low agricultural productivity, small land holdings, growing population etc. caused an unprecedented onslaught on the surrounding forests that had become as good as open access, since abolition of *malgujari*. In both the villages, forestland was encroached upon for agriculture and trees were being cut for immediate income. Lot of felling took place in the forests of both the villages, to satisfy the timber needs of the nearby town of Brahmapuri. The growing population needed bigger amounts of fuel wood and other forest products. In the absence of protection activities, these led to clearing and degradation of large tracts of forest.

Many years later in the early 1970s, when both the villages were engaged in a lot many activities that threatened the sustainability of the forest, some faint traces of anxiety started appearing in both the villages regarding the degradation of the resources.

Saigata, which had destroyed and witnessed the destruction of a major portion of its forest, nurtured these feelings carefully under the leadership of Mr. Suryabhan Khobragade, a villager. Gradually and thoughtfully the community first stalled all forest damaging activities like charcoal making and commercial fuel wood collection and soon graduated over to protection and conservation efforts. In 1979, the community created its "Van Sanrakshan Samiti" (Forest Protection Committee) informally. Different patches of forest were demarcated for different uses and 24-hour patrolling was done by the community members to protect the forest both from outsiders and from wasteful over-harvesting by the villagers themselves. These efforts went on under the unregistered committee till 1993, when the committee came under JFM. After this the committee took up many new activities with support from the Forest Department. The conservation work in Saigata is visible in the healthy and dense forest that the community has managed to develop over the years.

Almost during the same period Maral Mendha also took the initiative in forest protection, mainly due to the fact that neighbouring Saigata had sealed its forest boundaries to outsiders. The community formed its own institution to protect the small patch left of its forest in 1979. This effort however fell apart very soon as the myopic interests of the cultivators saw more potential in converting forestlands into agricultural lands over the unsure benefits of future. The announcement of the Gosikhurd Dam over the River Wainganga, a decade later gave the fatal and final setback to the conservation efforts by the community. In anticipation of the monetary compensation, that was to be paid to farmers with lands falling in the way of one of the canals of the project, every single tree was cut down and all land encroached upon.

While a major share of credit for the success of Saigata's attempt can be given to the able leadership that it received through Mr. Suryabhan Khobragade, Maral Mendha had almost all the attributes required to make collective action successful. Factors like moderate resource scarcity, well-defined boundaries of the resource, small size of the community, resource dependence etc. proved to be ineffective in inducing a collective effort in the Maral Mendha village. The factors that the scholars have found to be working both ways e.g. availability of exit options, exposure to urban societies etc. seem to have worked against Maral Mendha and in favour of Saigata. For scholars of institutions both Maral Mendha and Saigata are puzzles. Why did Maral Mendha not get together to save its forest like Saigata did, and why does Saigata not imitate its neighbour despite the visible prosperity of its neighbour, are some of the questions that prompt further research.

3. MAJOR DEVELOPMENTAL TARGETS

Population and activity growth of South Asian region will dominate for next couple of decades. South Asian region is the home for the largest number of poor in the world. It has the lowest per capita income ($440) of all developing regions. It also has the highest rate of youth illiteracy (23% for males and 40% for females) and the lowest rate of access to sanitation facilities (37%). Growth of basic infrastructure building will be the basic developmental need. The economy, which grew by 5.6% annually in the last decade, depends more heavily on agriculture compared to any other region. In South Asia the per capita availability of food is rising and in general there has been a rise in per capita income affecting food demand. The preferences are shifting from coarse grains to superior grains and from grains to animal and horticultural products. However in spite of the improvement in food trends malnutrition and poverty remains a problem. It is observed especially in the rural areas that there is a lack of access towards food and non-food goods due to lack of infrastructural development. In South Asia 58% of the children are malnourished and by 2020 it is projected to reduce to 48%. Women are the most neglected lot in spheres of health care and literacy. The primary developmental targets that have been set for South Asia to be realised by 2020 are as follows:

- Average annual rate of economic growth: 6 to 7 per cent.
- Average annual rate of population growth: 1.4 per cent.
- Average annual rate of growth of per capita income: 4.6 to 5.6 per cent.
- Annual rate of agricultural growth: 2.5 to 3 per cent.
- Reduce prevalence of child malnutrition: 10 per cent.
- Reduced rate of female illiteracy, universal female enrollment in primary schools, and doubling of secondary- and tertiary-school enrolment.
- Reforestation as percentage of total land area: 25 per cent.

These are aggregate targets for the region as a whole. Country targets are adapted to local conditions. These reflect the huge development led emissions from the demand side drivers for energy demand emerging from life style, preference, income distribution pattern, poverty etc. Energy Demand scenario for South Asia as projected in the literature are:

- South Asia's oil imports are to double by 2020.
- If higher estimates of Bangladesh's undiscovered oil reserve is estimated to be correct then Bangladesh could become a major gas producer and supplier to the potential market in neighbouring India.
- Natural gas usage has increased rapidly over the last decade growing about 59% between 1992 and 2002. Like India, Pakistan plans to increase the use of natural gas for future electric power generation projects, a move that will necessitate a sharp rise in production and/or imports of natural gas.
- If long-term projections of rapidly increasing gas demand for South Asia are correct, the region will require significant increases in production and/or imports. Even with expanded production, however, increased consumption of natural gas in South Asia is constrained by the region's inadequate domestic infrastructure.
- Significant increase in emissions is certain given that demand for coal in India is projected to increase rapidly in the coming decade (from 359 million short tons (Mmst) in 2000 to 430 million short tons (Mmst) by 2010). Indian coal consumption is expected to increase to 510 million short tons (Mmst) by 2020. South Asia's carbon emissions are expected to increase sharply in coming years as a result of increased coal consumption. The net level of coal-fired generation in South Asia is expected to rise. In Bangladesh two coal mining projects are to supply the future coal demand. One plant opened in April 2003, Barapukuria coal mine, and the other, Khalashpir coal mine, is under consideration.

From India's perspective by 2020:

- Life expectancy going up from the present level of 62 years to approximately 70 years, with its implications for public health services.

- GDP going up from present level of US$560 billion to US$ 3 trillion, which is feasible with present trend in growth rates, provided the region takes advantage of the globalising impulse particularly in communications, IT, financial services, etc.
- Tuning with globalization with skill revolution.
- Governance going beyond Government per se. Emerging role of non-government entities.
- Growing regionalism.
- Increasing transformation of the small farms into economically sized mechanized farms, modernization of old cities. The resulting construction boom can be fuelled by the development of liberal construction lending institutions and related lending infrastructure.

Both agriculture and construction must take the full benefits of the latest building technology and construction materials production techniques. A development on these lines will contribute greatly to the industrial and service sectors of economy while maintaining a desirable employment scenario. In fact, the creation of better paying jobs will fuel the consumer sector also and is sure to raise the rate of economic growth so much that a sustained economic growth of above 10% through next couple of decades is in the realm of possibility. By implementing these changes, a heuristically projected structure of the greatly expanded Indian GDP in the year 2020 may consist of 10% share from the agricultural sector, 33% from the industrial sector and 57% from the service sector. Assuming this to be the case, it will necessitate the trimming of the agricultural sector employment from 65 to 15%, a change in the employment pattern to the extent of 50% of the population freed from the agricultural sector.

This scenario may appear scary to the policy makers. It is, however, achievable by implementing bold policy initiatives, especially in the building and construction sector. A 12% sustained annual growth rate translates into doubling of economy in six years. At this rate, the economy in 2020 should increase to six to seven times of its current level. Even a reduction in the share of the agricultural sector from 25% to 10% in the expanded economy may mean two to three fold increase in value of agricultural production in addition to a tremendous increase in the industrial and service sector output. It will be a challenge to higher education institutions in the country to work overtime for training the money managers, town planners and engineers to ensure the reality of Vision 2020. What is more, the process of building these facilities will add considerably to the economic progress by promoting consumerism. Like agriculture small scale sectors play vital role in the growth of the region. They provide the high employment second to agriculture. Small scale sector contributes almost 40 per cent of the gross industrial value added in the Indian economy.

The above high economic growth, high population, high social action led BAU scenarios for South Asia accompanied by high energy and resultant emissions scenario do not include additional technology or policy intervention and alternative development strategies towards transition to low emission pathway.

4. DEVELOPMENT AND EMISSIONS: ACTIVITY, STRUCTURE, TECHNOLOGY AND FUEL MIX (ASIF)

South Asia is in a period of transition as it strives to implement effective economic, political, social, and legal structures to support sustained growth. Multilateral funding agencies have arranged several billion dollars in assistance to the region. Among others, special emphasis is on infrastructure development and energy efficient technology deployment. These being long-term investment projects the choice today will determine the course of the future emission pathway. The nexus between activity growth and emission growth is dependent very much on the structure of the activity, level of efficient energy technology penetration and the fuel mix in use. The traditional route to economic development has been through rapid industrialisation. What the "non-G8" world thinks and does is going to be as important as G-8. The reality of low emission share (Fig. 2) will be changed by 2020 and beyond in BAU scenario unless otherwise induced.

Activity led growth of energy consumption in BAU is going to be a developmental need for the country like India and it can be imagined going to be the same for other SA countries. Example from Indian economy shows, in sectoral energy consumption, industry (Fig. 3) is the major consumer of energy followed by transport. Both account for more than 60% share.

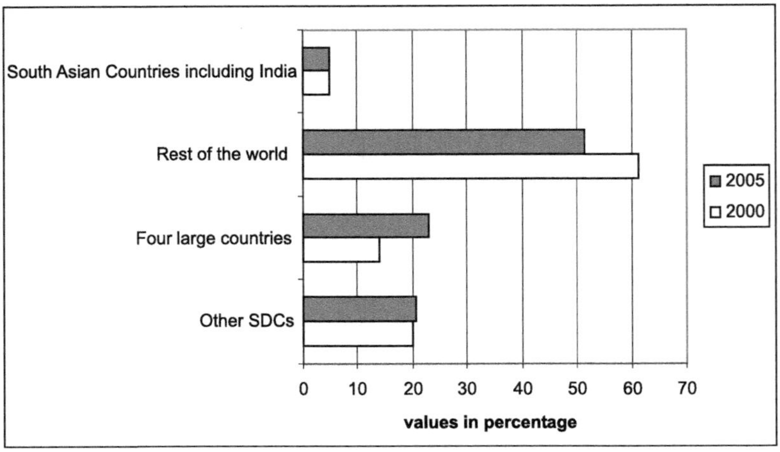

Fig. 2: Share of South Asia in global emissions.
Notes: Five large countries are China, India, Brazil, Mexico and South Africa. Rest of the world: Annex I and small island developing countries excluding Sri Lanka. SDC: small developing countries.

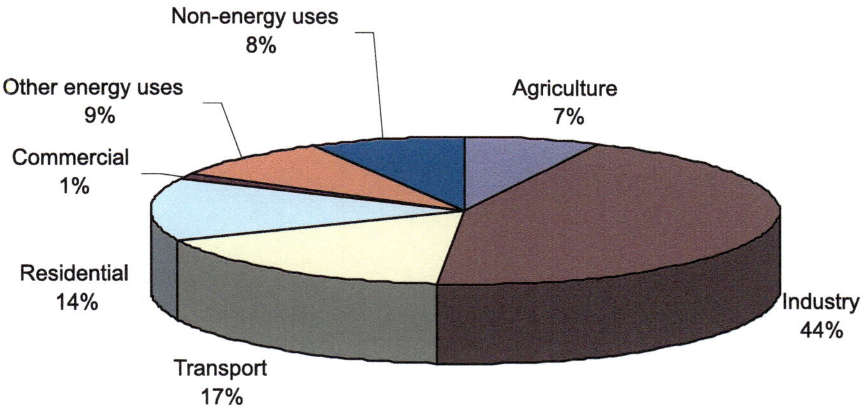

Fig. 3: Final energy consumption by sectors in India in 2005.

Energy use decomposition analysis across countries (for detailed survey please refer to Dasgupta, 2005; Das, 2005; Bose et al., 2005) have shown that rising economic activities – agriculture, industry, transport, commerce, infrastructure – lead to rising energy consumption and emissions. India is by far the largest South Asian country in terms of population, Gross Domestic Product (GDP), land area as well as energy consumption pattern and emissions. South Asian region trying to achieve a very high growth rate in next two decades as mentioned above will inevitably lead to higher emissions level. By sheer size, dynamics of the Indian economy is going to be major trend setter in South Asian Region.

However, whether high energy scenario imply high emission scenario depends on if structural change, fuel mix or efficiency increase (energy intensity ratio declining) can decouple development and emissions. In the past, structural change has been in favour of energy intensive industries and fuel mix has been in favour of fossil fuels. **ASIF analysis** for the Indian economy for past years due to activity growth has led to increase in final commercial energy consumption from a modest figure of 147 mtoe in 1970 to around 438 mtoe by 2001, growing at an average annual rate of 5.2 per cent[1] . However, the activity–emission decoupling could be attained through energy intensity gain. In the eighties intensity grew at 0.86 per cent on an average annual basis, while in the nineties it grew at the rate of –0.95 per cent. The primary commercial energy intensity has declined from 27.2 Btu per thousand US$ in 1989-90 to 25.3 Btu per thousand US$ during 2001-02 (Dasgupta, 2005; Dasgupta and Roy, 2004). It is found that apart from the agricultural sector, energy intensity in all the other sectors (sectoral energy consumption/sectoral GDP) have recorded a negative average annual growth. This can be seen from Fig. 4 which shows the sectoral energy intensity.

[1] http://www.eia.doe.gov

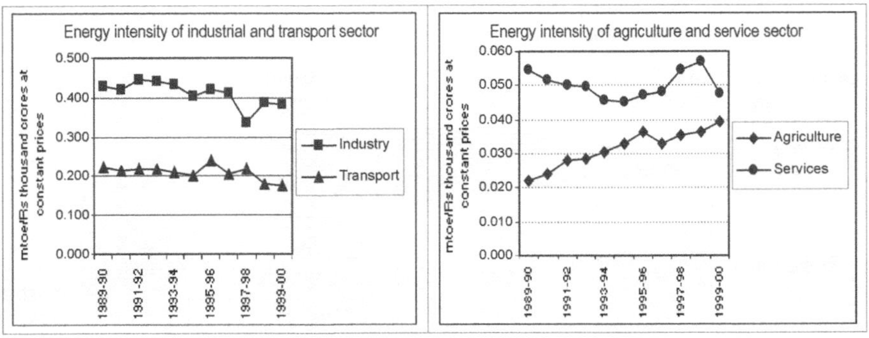

Fig. 4: Sectoral energy intensity in India.
Source: Das Gupta, M and J. Roy (2005)

In percentage terms, in the agricultural sector, the energy intensity increased from 2.2 to 4 per cent between 1989-90 and 1999-00. On the other hand for the other sectors, it had decreased from about 43 per cent to 38 per cent in the industrial sector, from 22 per cent to 18 per cent in the transport sector and from 5.4 per cent to 4.8 per cent in the service sector during the same period. It is important to take a deeper look into BAU of the industry and transport sector as they are the major energy consumers.

4.1 Industry Sector

Despite high share in energy, industrial sector has recorded the maximum energy efficiency improvement during the decade of nineties. Trend shows increasing share of energy intensive industries like iron and steel, pulp and paper, chemical and fertilizer, aluminum, and cement are providing an upward thrust to energy consumption and emissions. But despite activity growth improvement in energy efficiency is enabling India to grow with relatively lower energy consumption despite fast economic growth (Figs 5, 6 and 7). In India energy intensive industries individually have performed well in terms of energy efficiency gain. Intensity effect has been improving over the years (Fig. 8). Fuel Mix has (Fig. 9) also been showing declining bias towards coal in India and increasing bias towards Natural gas. Declining emission from energy intensive industries beyond 1995-96 (Fig. 10) has been the outcome of intensity effect and marginally fuel mix effect (Fig. 11). The main contributors towards a fall in emissions in 2000-01 were primarily the fertilizer, iron and steel industry and cement industries, followed by aluminium and paper industries. For the fertilizer industry, emissions fell drastically by almost 52 per cent in 2000-01 over 1995-96. For each of the iron and steel and cement industries the reduction was to the tune of 23-24 per cent over the same period. In the aluminium and paper industries emissions fell by almost 15 and 10 per cent respectively during this period.

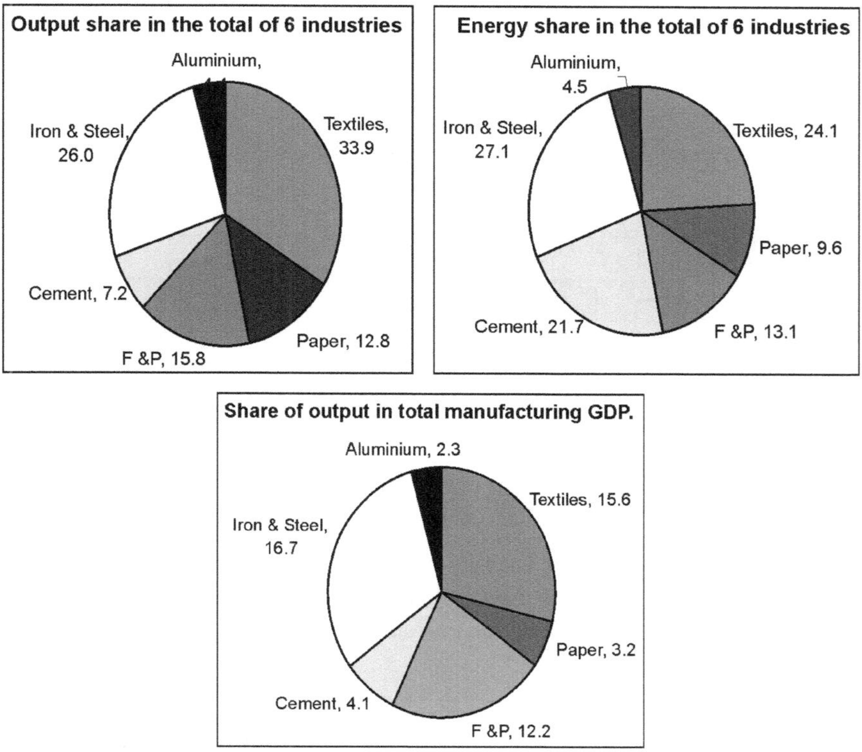

Fig. 5: Role of energy intensive industries in India.
Source: ASI (2000-01), RBI (2002)

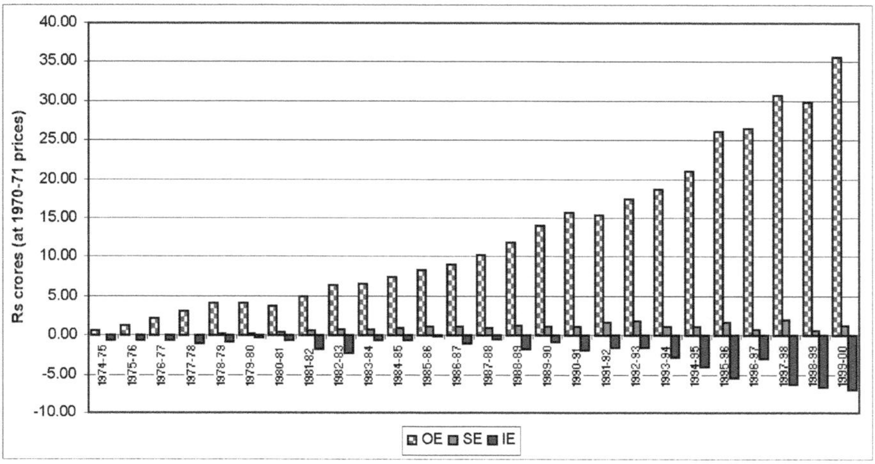

Fig. 6: Intensity decline vis-a-vis activity led increase in energy use.
Note: OE = Output/Activity Effect, SE = Structural Effect, IE = Intensity Effect

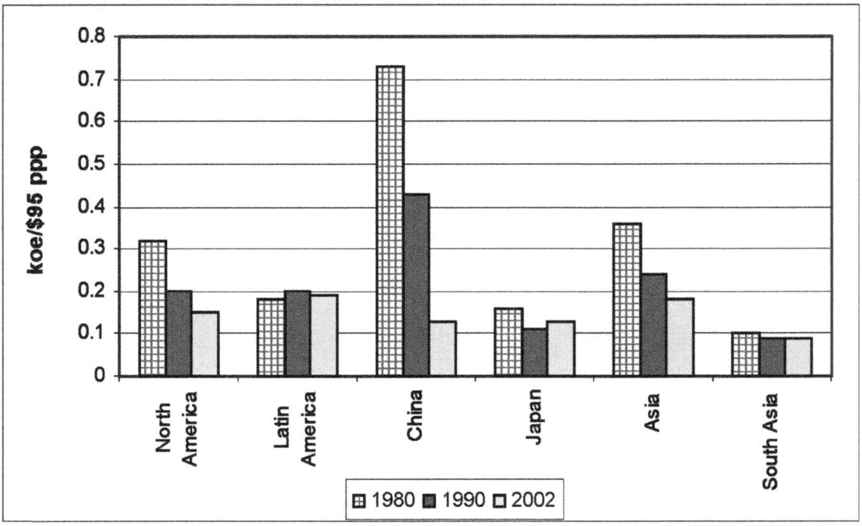

Fig. 7: Energy intensity of industry in select region.
Note: South Asia includes India and Pakistan
Source: Enerdata (2003)

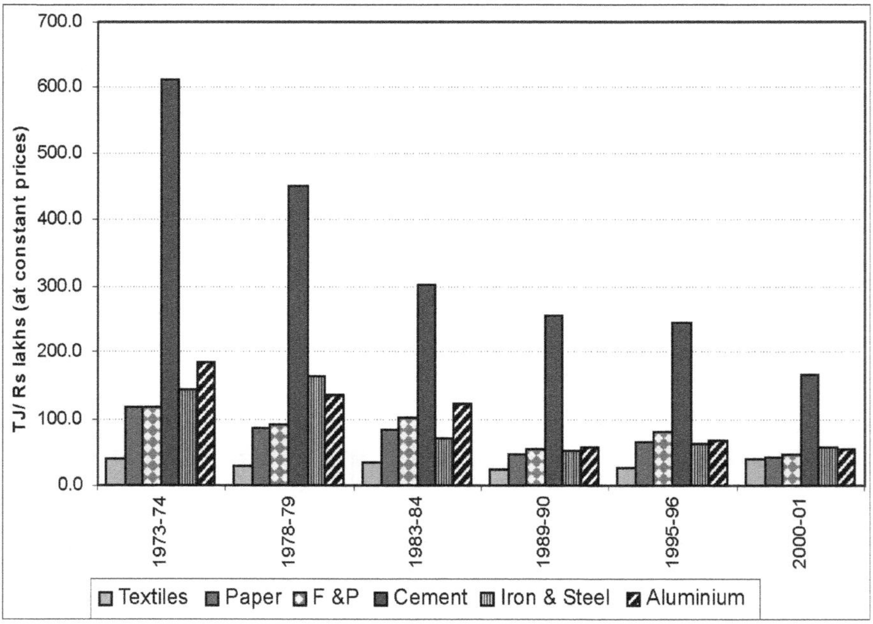

Fig. 8: Energy intensity of the industries.

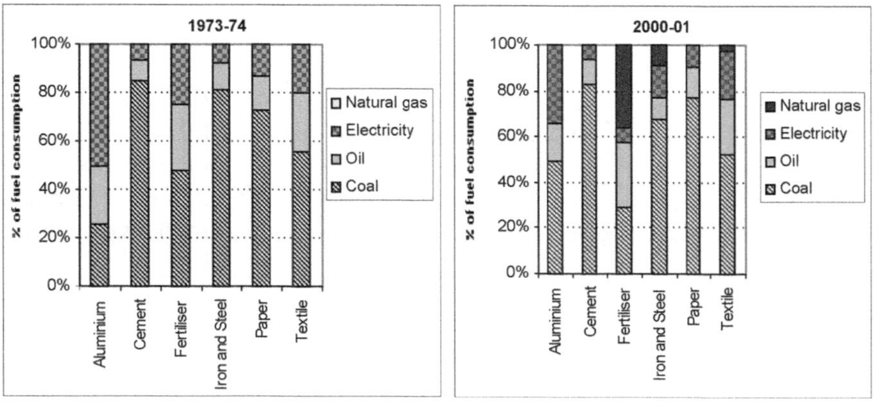

Fig. 9: Fuel mix pattern of industries.

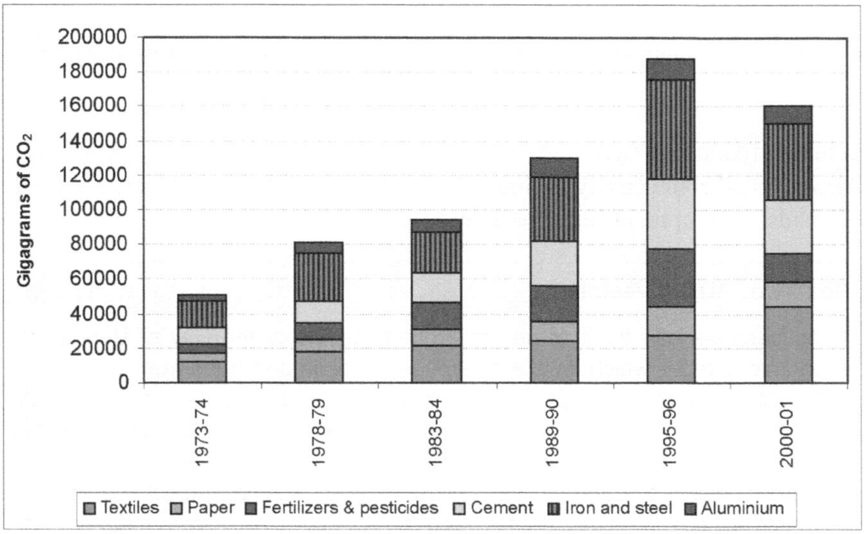

Fig. 10: Emissions trend and industry shares in India.

4.2 Transport Sector

Growing demand for mobility and habits are the major drivers in this sector. Owning a personal motor vehicle is often perceived as the embodiment of development, while other forms of transport, including rail, water transport and non-motorized transport (i.e. bicycles, cycle rickshaws, and walking) are frequently de-emphasized. Traditionally, governments have worked to meet challenge of transport demand by massive investment in road infrastructure accompanied by a sharp increase in road vehicles, both personal motor

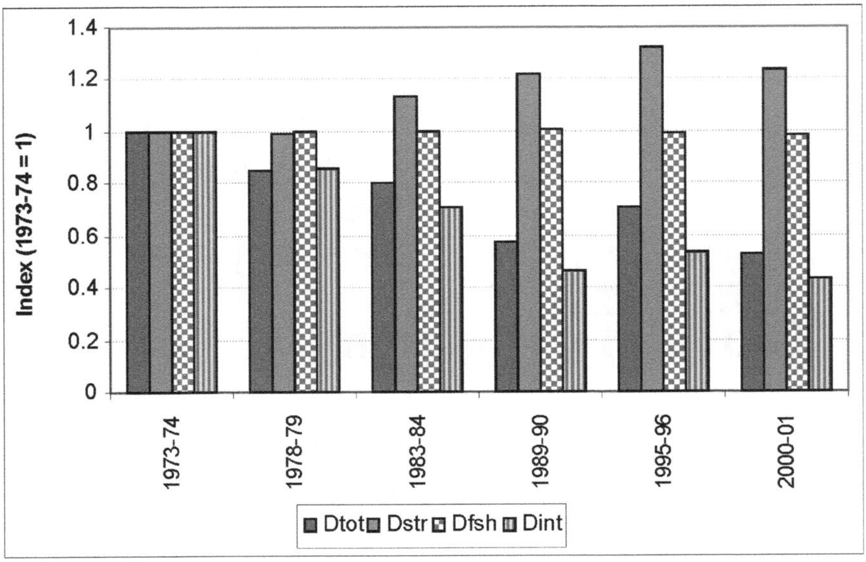

Fig. 11: Decomposition method of total carbon emission intensity from the six energy-intensive industries in India.

vehicles (cars and two wheelers) and trucks. Recent plans by ADB (Box Two) to road connectivity in South Asian Region highlights the large network of roads development and possible emissions loading.

Box Two: ADB Investment

At its base we can think of an arc around the top of the Bay of Bengal, an economic corridor which links a continuum of ports and hubs from Chittagong in the east to Dhaka, Mongla, Kolkata, and Haldia in the west. This base would support a large transport grid. East-west railroads and highways would link the eastern Indian hill states with West Bengal through Bangladesh. North-south transport corridors, such as the Siliguri-Kolkata-Haldia project would give improved access to Nepal, Bhutan, and the hill states of eastern India to ports on the Bay of Bengal. This transportation grid would be linked to the rest of India at Kolkata through the National Highways Development Project, in which ADB is also participating. This so-called Golden Quadrilateral of superhighways linking Delhi, Mumbai, Chennai, and Kolkata, is a top priority in India's road network development programme. Similar grids could be developed for power, hydrocarbons, and information and telecommunications technology. Poverty reduction through growth in this sub region is of the highest priority for ADB.

4.2.1 Urbanization and motorization

The region is characterized by low but generally rising income and rapidly growing young populations. These will raise the requirement of transportation.

Despite the policies and measures that many countries have adopted to promote the use of alternative fuels such as biofuel and compressed natural gas, the share of oil in transport energy demand will remain almost constant over the period 2002-2030 (WBCSD, 2001). Demand for road transport fuels is growing dramatically in many developing countries, in line with rising incomes and infrastructure development. Present supply of Public Transport (PT) mainly in South Asian cities is not uniform across countries and cities and has been inadequate and of poor quality. On the other hand growth of personalized vehicles in the region has resulted in problems of congestion, high costs, higher energy consumption and heavy atmospheric pollution. In contrast to the urbanized areas of the developed world, vehicle-related air pollution in many developing countries is clearly getting worse.

4.2.2 Vehicle fleet and its projections

Estimating the vehicle fleet and its composition in the three cities in South Asian Region (Colombo, Dhaka, Bangalore), the following types of vehicles emerged as commonly identifiable operational vehicle types.

- Personal transport : (1) scooters, mopeds and motorbikes as 2-wheelers, and (2) cars, jeeps and dual/multi-purpose vehicles
- Intermediate public transport (IPT): (1) 3-wheelers, and (2) taxis and maxi cabs
- Public transport: (1) Standard sized buses, and (2) Mini buses
- Goods transport: (1) 3-wheelers, (2) light- and (3) heavy-commercial vehicles

The individual's decision to use 2-wheelers, cars/jeeps, 3-wheelers, taxis and buses is driven by his/her level of income to a very large extent. Rising incomes are expected to lead to a shift up the 'transport ladder' from public transport to 2-wheelers and from 2-wheelers to cars. Hence per capita income is an important economic variable determining the level of motorization. Interestingly, the growth of personal and intermediate public transport in the three cities is strongly influenced by the per capita income. However, the growth of public transport buses and goods transport is strongly influenced by the Gross Domestic Product (GDP). A three city: Bangalore, Dhaka and Colombo, study using **ASIF framework** (**A**ctivity i.e., passenger and freight travel demand expressed in passenger- or tonne-kilometres, modal **S**tructure i.e., share of passenger- or tonne-kilometres occurring on each mode, the modal energy **I**ntensity of each mode i.e., energy burned per passenger- or tonne-kilometre and the emission **F**actors of criteria pollutants – CO, HC, NO_x, PM – and for CO_2 Fuel-to-Carbon ratio i.e., carbon released per unit of energy burned, shows that interventions in I and F (technologies and utilization) have the largest promise of restraint, while policies that affect A and S through broader transport reform will also restrain emissions.

Estimating number of different types of operational vehicles on road in the three cities for a 20-year period from the year 2000 to 2020 at five-yearly intervals, total number of vehicles in Bangalore will be much higher than Dhaka in spite of population size of Dhaka almost double compared to Bangalore. Interestingly, in each of the three cities the total number of vehicles is expected to double in the next fifteen-year period. This is mainly due to the heavy dependence on road-based motorized vehicles in Bangalore and also insignificant share of cycle rickshaws and bi-cycles in the city unlike in Dhaka where a major share of commuter travel demand is met by non-motorized modes. It would be important to note here that this study stresses upon the interventions at the S level of ASIF methodology through reforms achieved by modal shifts towards public transport PT with drop in car and two-wheeler use through traffic restrain measures. The total motorized travel demand for people's mobility is estimated to be the highest in Bangalore, about 63 billion passenger kilometre (bpkm) in 2005 in the business-as-usual or the baseline (BL) case, much higher than Dhaka (6 bpkm) and Colombo (46 bpkm). While in Bangalore the travel demand is expected to increase more than four times between 2005 and 2020, the increase during the same period for Dhaka and Colombo is a little over double and triple respectively. The total motorized freight travel demand is expected to rise from 4.95 billion tonne kilometre (btkm) in 2005 to 14.69 btkm in 2020 in Bangalore (three times); from 3.81 to 9.59 btkm in Dhaka (2.5 times) and from 13.99 to 44.45 btkm in Colombo (over three times) respectively.

Baseline scenario results

Diesel is the pre-dominant auto fuel in the three cities followed by gasoline—ratio of diesel to gasoline is 6.5 in Colombo, 3.4 in Dhaka and 1.3 in Bangalore. Total demand of diesel in Bangalore and Colombo are expected to increase over 3.5 times in 2020 from the current level of 649 ttoe and 963 ttoe respectively. In Dhaka the demand of diesel is quite low but expected to go up three-fold in 2020 from the current level of 179 ttoe. Gasoline demand in Bangalore is significantly higher compared to Colombo. This is mainly because of heavy utilization of two-wheeled vehicles that occupy a significant share of road space. Although CNG cars, taxis and buses have been introduced in Dhaka, but their scale of operation is rather limited due to the supply constraint of CNG infrastructure. Similarly, in Bangalore and Colombo, instead of CNG, LPG is used as an auto fuel. It is expected that CNG demand in Dhaka will increase to almost double in 2020 compared to its present level of consumption (44 ttoe in 2005). While that of LPG demand in Bangalore and Colombo is expected to go up 2.3 times and 1.6 times respectively in 2020 compared to its present level.

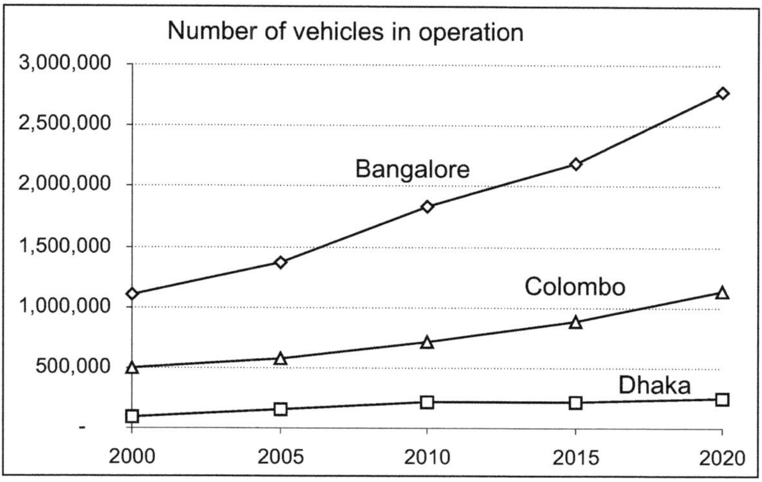

Fig. 12: BAU scenario for three cities.

4.2.3 Alternative transportation

Figure 13 provides energy demand estimates under the alternative scenario where a greater penetration of public transport is assumed with drop in utilization of personal vehicles. The potential for reducing the total transport energy demand differs considerably in the three cities given the different level of penetration assumed in the three cities. For instance, in Bangalore the energy reduction potential is over 21% in 2020 compared to BL. The corresponding figures for Dhaka and Colombo are 15% and 3% respectively.

To minimize the negative impacts of traditional transportation development, especially countries in South Asia require new solutions that enable them to meet transport needs within energy, environmental, and economic constraints. Provisioning of efficient public transport (PT) system, which mainly encompasses bus transport (BT) and rail transport (RT) can bring concrete solutions and put urban transportation on a more sustainable path. PT carries more people in fewer vehicles using less fuel, so it cuts both congestion and pollution. For example, in India, a car consumes nearly five times more fuel than a 52-seater bus and occupies about 38 times more road space to meet a kilometre of passenger travel demand (MoPNG, 2002). Further, replacement of a single bus by an equivalent number of two-wheelers would add to air pollution by 27 per cent. Similarly cars could cause 17 per cent more pollution (TERI, 2001). Given the current fuel availability in the three cities, it is assumed that gasoline, diesel and LPG will be the autofuels in Bangalore and Colombo whereas in Dhaka it will be gasoline, diesel and CNG. In addition, Bangalore and Colombo will have electric trains to meet commuters travel needs. Impact of increasing penetration of public transport purely based on a set of assumptions for a change in modal split taking place in the year 2020 are:

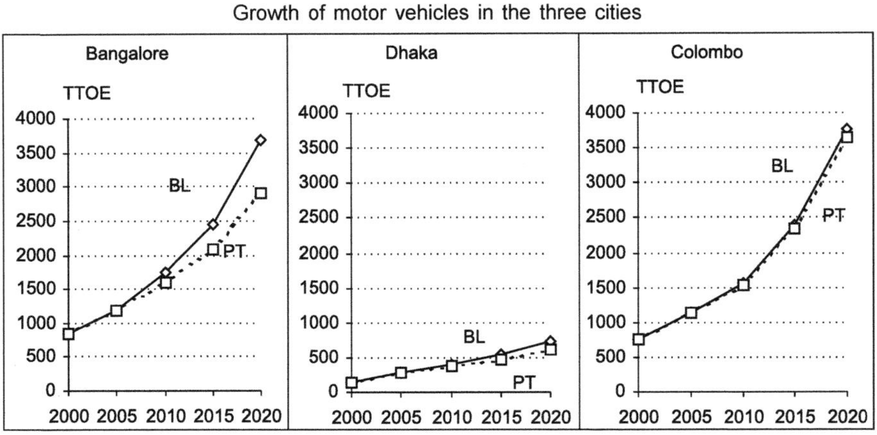

Fig. 13: Estimated total transport energy demand in thousand tonnes of oil equivalent (ttoe).

- An increase in public transport share from 62% to 80% in Bangalore leads to a fuel saving of 765,320 tonnes of oil eqivalent, which is equivalent to about 21% of the fuel consumed in the BL case. The other advantages that ensue are a 23% reduction in total vehicles (642,328) and creating a road space (equivalent to 418,210 cars off the road) and reduce traffic congestion. Air pollution in the city drops significantly: 40% drop in CO, 46% HC, 6% NO_x, and 29% PM. The total CO_2 mitigation potential over the next 15-year period would be 13%.
- An increase in public transport share from 24% to 60% in Dhaka leads to a fuel saving of 106,360 tonnes of oil equivalent, which is equivalent to about 15% of the fuel consumed in the BL case. The other advantages that ensue are a 39% reduction in total vehicles (99,294) and creating a road space (equivalent to 78,718 cars off the road) and reduce traffic congestion. Air pollution in the city drops significantly – 24% drop in CO, 26% HC, <1% NO_x, and 13% PM. The total CO_2 mitigation potential over the next 15-year period would be 9%.
- A marginal increase in public transport share from 76% to 80% in Colombo leads to a fuel saving of 104,720 tonnes of oil equivalent, which is equivalent to about 3% of the fuel consumed in the BL case. The other advantages that ensue are a 5% reduction in total vehicles (47,716) and creating a road space (equivalent to 62,152 cars off the road) and reduce traffic congestion. However, air pollution in the city does not drop much as the city already depends heavily on public transport and the CO_2 mitigation potential is around 2%.

The steep rise in transport energy demand in the three cities is expected to result in rapid growth of CO_2 emissions as shown in Fig. 14. The annual rate of increase of CO_2 emissions will be marginally higher compared to total energy demand in the two scenarios during the period 2005 to 2020.

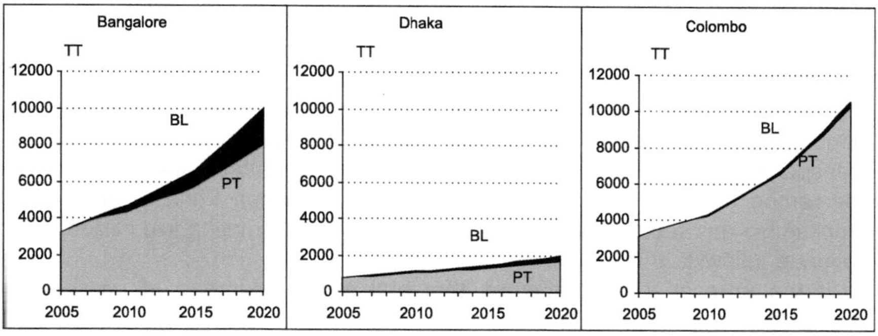

Fig. 14: CO_2 emissions from transport under alternative scenarios in thousand tonnes (TT).

The highest rate of CO_2 growth is observed in Colombo (8.42%), followed by Bangalore (7.97%) and in Dhaka (6.78%).

Box Three: Nepal's National Context

The majority of urban population is concentrated in Kathmandu Valley. About 30% of nation's urban population is concentrated into five municipalities inside the Kathmandu Valley. In recent years in different parts of the country a rapid upsurge of rural-urban migration will add burden to urban areas in view of their limited infrastructure to absorb them. A recent estimate in *Development Plan 2020 of Kathmandu valley* predicts the population to double, to 3.3 million, by 2031. The air quality in major cities in the country is worsening since last decade. Nepal's GHG emissions are relatively smaller. The country's first National Communication to UNFCCC estimates net GHG emissions (CO_2 equivalent) of 39 million-tons in 1994-95. However, transports sector contributes about one third of the total CO_2 emission from fuel combustion in the country[1]. This is primarily due to the low vehicle ownership; for example Nepal has 59 people per vehicle and 348 people per car in mid-2004[2]. Two wheelers, in particular, dominate motorised modes; two wheelers are cheaper and widely used that accounts for 62% of total registered vehicles in the country. In summary, in per capita as well as absolute number, its total GHG emissions and that from road transport sector is rather small and not comparable to its giant neighbours. However, although the scale of motorization is small in Nepal at present, it is embracing rapid motorisation with surging population of cars and two wheelers. Experts suggest that there would be a four-fold increase in GHG emissions by 2030 from transport sector should present trend continue[3].

[1] Share of CO_2 emissions from transport sector is 31% in total fuel combustion in 1994-95.

[2] This figure represents stock of registered vehicles. Cars, here, means light duty vehicles that include car, jeep and vans. As of March 2004, total registered vehicle population is 418,910, car/jeep/van 70,987 and two wheelers 262,067. Source: Economic Survey 2004.

[3] National Communication of Nepal to UNFCCC, July 2004, pp 46.

There are also better future prospects for battery operated electric vehicles or electric transportation such as trolley buses and electric trains (not talking about cost issues at present) as electricity is inherently clean in Nepal and is provided by run-of-river hydropower plants. In this context, Nepal's transport sector may have some prospects to accumulate Certified Emission Reduction Credits (CER) to supplement its rising financial burden of developing urban transportation infrastructure. Recently, the country has entered into agreements for carbon off-setting programme with Prototype Carbon Fund of the World Bank in bio-gas sector. From this alone, experts are estimating five million US$ revenue inflow to this sector.

If the present trend continues the total energy demand of passenger transportation in 2000-2020 will increase by 5% annually, and in 2020, energy demand and environmental emissions are estimated to be about 2.7 times of that of the year 2000. Similarly, environmental emissions for 2020 at present trend are estimated 2.6 times for CO_2, 3.5 times for CO, two times for HC, three times for NO_x, 2.8 times for SO_2, and 2.5 times for TSP from the base year 2000. LDVs and two wheelers are and will be responsible for the majority of emissions. In case of SO_2 and NO_x emissions, contributions of the buses and minibuses are also significant. TSP, in particular, is a major source of health concerns in Kathmandu Valley due to their high concentrations. A 2.5 times increase in TSP will have detrimental implications for the health of the Valley residents, particularly for those in core city areas in 2020. LDVs and two wheelers would be the major sources of TSP emissions in the year 2020.

5. AGRICULTURE AND SERVICES

Section 4 shows that agriculture and services sectors are with relatively low energy consumption but have more potential for energy efficiency. Agriculture sector contributes to emission through non-energy route also through landuse pattern, crop selection, fertilizer use, and agronomic practices. There is very low awareness among the sectoral actors in South Asian Region. The momentum of Green Revolution has waned and future gains will essentially be realized through the generation and adoption of new, appropriate agricultural technologies. The lack of investment, especially in rural infrastructure and in the development of new agricultural technologies, hinders the ability of producers to respond to the market and, therefore, decreases the prospects of overall agricultural growth. The strategies to be undertaken to meet the challenges are:

- Increase investment in agricultural sector
- Enhance access to improved agricultural technologies
- Develop productive, sustainable environment-friendly technologies
- Strengthen efforts to protect environment
- Improve commercialization of agriculture
- Increase investment in human resource
- Improve trade linkages
- Improve government policies

Technologies that promote environmentally sound agricultural practices while increasing productivity, and policies that strengthen property rights, correct tenurial anomalies, discourage fragmentation, and promote land markets that operate more freely, are essential. Because South Asian countries have little comparative advantage in using scarce land and water on cereal production, more attention may need to be paid to growing non-cereals, which yield higher income per hectare and unit of land and water and can be produced for both regional and world markets. Strategies that increase the productivity of rain-fed and marginal land and neglected crops could also help to prevent further land degradation and conserve resources. The scope for wood-fuel and other energy-crop farming in the semi-humid southern parts of South Asia will be gaining momentum in next two decades. ICT's role in agricultural development is assuming larger importance in risk management which is going to increase in changed climate scenario. Subsistence is agriculture getting transformed into agricultural enterprise. But how fast this transition can happen will depend on local institutional transformation. Yield of major crops and livestock in the region is much lower than that in the rest of the world. Considering that the frontiers of expansion of cultivated area are almost closed in the region, the future increase in food production to meet the continuing high demand must come from increase in yield. Integrated nutrient management, arresting deceleration in total factor productivity are the challenges of near future. Investment in irrigation, infrastructure development (road, electricity), research and extension and efficient use of water and plant nutrients are the dominant sources of TFP growth. All the efforts need to be concentrated on accelerating growth in TFP, whilst conserving natural resources and promoting ecological integrity of agricultural system. A recent study (Kumar and Mittal, 2000) has shown that literacy emerged as an important source of growth in adoption of technology, use of modern inputs like machines, fertilisers, and yield. Generation and effective assessment and diffusion of packages of appropriate technologies involving system and programme-based approach, participatory mechanisms, greater congruency between productivity and sustainability through integrated pest management and integrated soil-water-irrigation-nutrient management are going to dominate to bridge the yield gaps in most field crops.

6. ENERGY SUPPLY SECTOR

Section 2 has shown that high growth scenarios will put pressure on energy demand. ASIF analysis (here since analysis is for **F**uel only **ASI** components of ASIF is taken; Figs 15 through 23) for various fuel use in India shows that for oil transport activity, for coal industrial activity and for electricity industrial activities are the major drivers. So higher growth rate projections for transport and industry sectors are bound to drive the oil and coal as well as electricity demands. Power sector is becoming larger in size and pushing the coal demand while agricultural sector with increasing dependence in electricity is pushing the demand upward and transport sector both activity-wise and with expanding road transport is pushing oil demand.

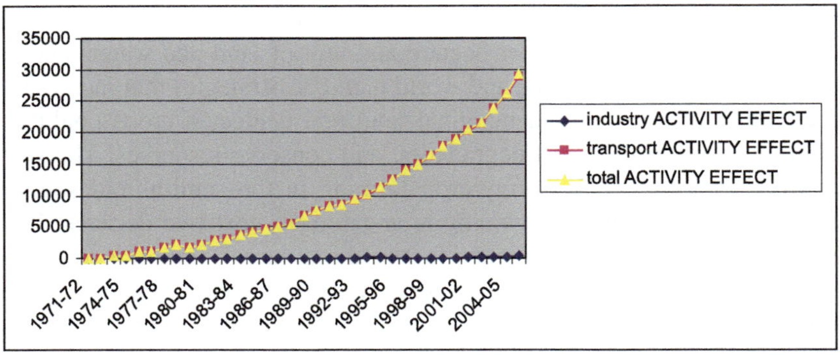

Fig. 15: Activity effect in oil consumption in different sectors.

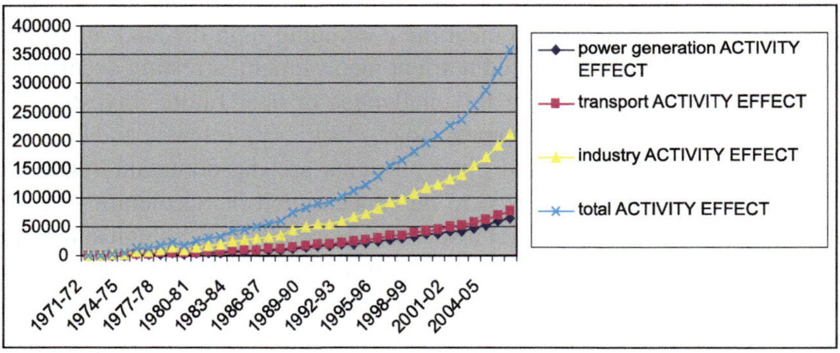

Fig. 16: Activity effect of coal.

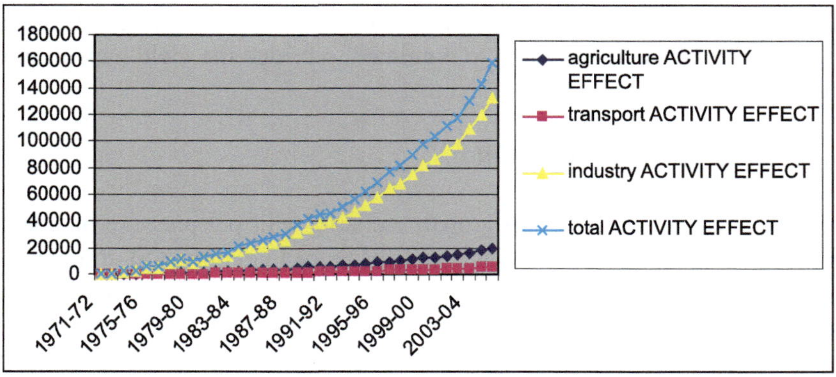

Fig. 17: Output effect of electricity.

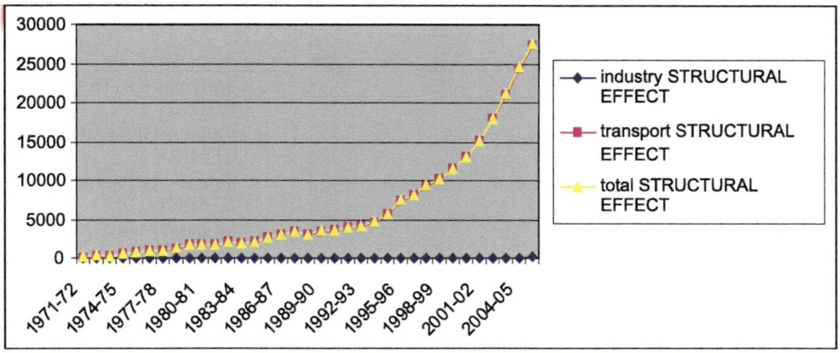

Fig. 18: Structural effect in oil consumption in different sectors.

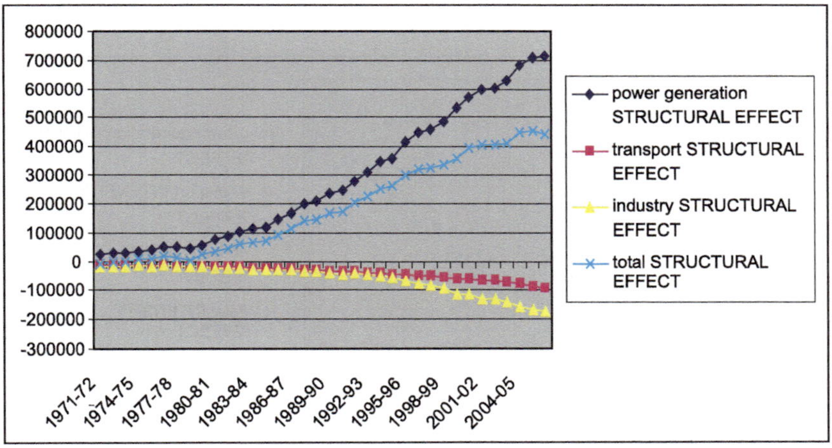

Fig. 19: Structural effect of coal.

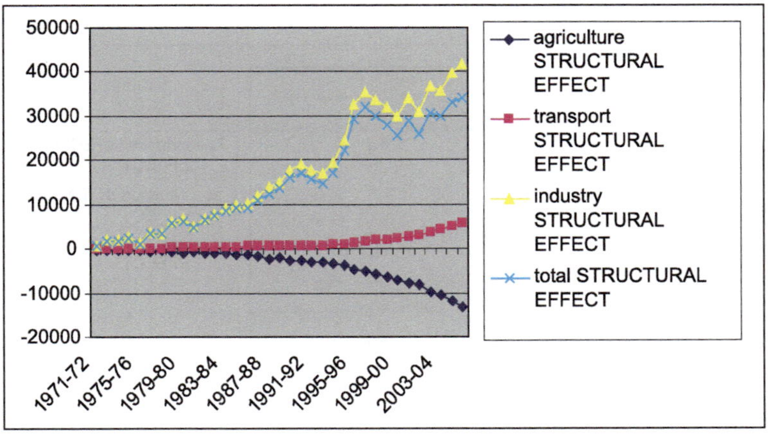

Fig. 20: Structural effect of electricity.

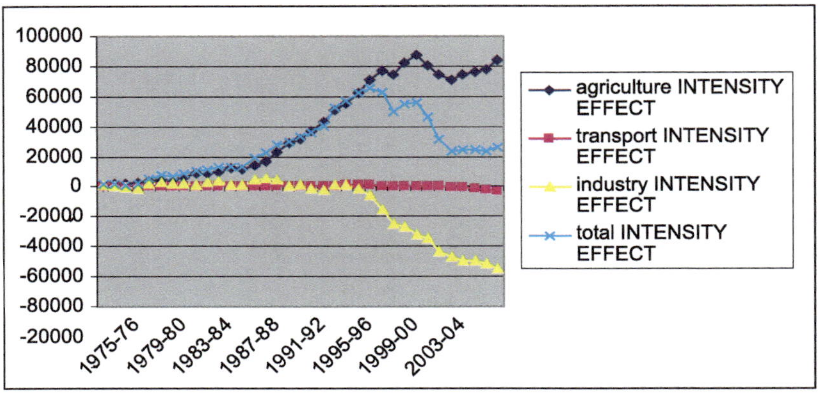

Fig. 21: Intensity effect of electricity.

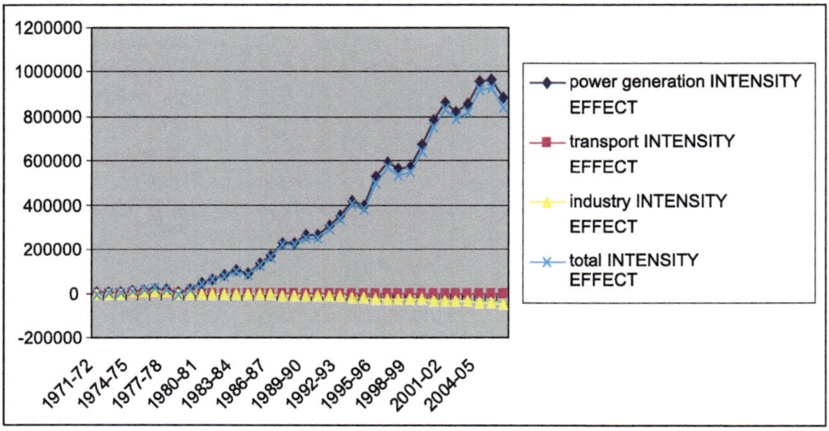

Fig. 22: Intensity effect of coal.

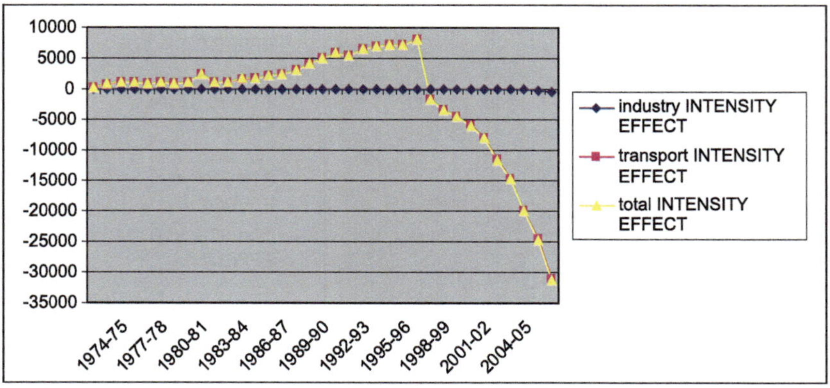

Fig. 23: Intensity effect in oil consumption in different sectors.

How can development led growth in demand for fuels be met through available supply will remain as major challenge for South Asian Region in next two decades. Energy infrastructure development within the region will dominate the regional investment pattern in the energy sector.

6.1 Oil and Natural Gas

The vast majority (around 819,000 bbl/d in 2003) of South Asia's oil production comes from India followed by Pakistan (around 62,000 bbl/d in 2003). South Asia's oil imports are projected to be more than double by 2020. The Middle East has been and is expected to remain the primary source of South Asian oil imports. Natural gas is seen as playing an important part in supplying fuel for new power plants in the region and diversifying from expensive oil imports. As a result, natural gas usage has increased rapidly in South Asia over the last decade, growing about 59% between 1992 and 2002. In 2002 around 42% of natural gas is consumed by India, 39% by Pakistan, and the remaining 19% by Bangladesh. Indian consumption of natural gas has risen faster than that of any other fuel in recent years and accounts for approximately 6.5% of the country's energy demand. Like India, Pakistan plans to increase the use of natural gas for future electric power generation projects, a move that will necessitate a sharp rise in production and/or imports of natural gas. Because natural gas is already Bangladesh's primary source of commercial energy, gas exports are a controversial topic within Bangladesh, as many people feel that Bangladeshi gas resources should be used for domestic purposes rather than exporting. If long-term projections of rapidly increasing gas demand for South Asia are correct, the region will require significant increase in production and/or imports. Even with expanded production, however, increased consumption of natural gas in South Asia is constrained by the region's inadequate domestic infrastructure. Gas imports would require construction of infrastructure – either cross-border pipelines or liquefied natural gas (LNG) facilities – and their success would likewise hinge on the successful construction of domestic gas pipeline infrastructure.

6.2 Coal

South Asia contains coal reserves of approximately 9% of the world's total. Although coal accounts for 43% of South Asia's energy consumption, nearly all of the coal in this region is produced and consumed by India, the only South Asian country with significant coal reserves (93 billion short tonnes) and the world's third largest coal producer after the United States and China. Despite the fact that Indian coal is generally of poor quality – i.e., low in calorific content and high in ash – and primarily located far from major consuming centres, Indian coal consumption is expected to increase to 510 million short tonnes (Mmst) by 2020, up 42% from 360 Mmst in 2000 and 393 Mmst in 2002.

6.3 Biomass

South Asia continues to rely heavily on biomass (i.e., animal waste, wood, etc.) for residential energy consumption, particularly in rural areas. According to the International Energy Agency (IEA), biomass will account for 70% of total residential energy consumption by 2020. Because the primary end uses of biomass are cooking and heating, the expansion of electricity access, used primarily for lighting, is not expected to have a significant effect on biomass use in the near future. But this form of biomass use is inefficient for multiple reasons.

6.4 Electricity

In 2002, South Asia generated 642 billion kilowatt hours (Bkwh) of electricity. Of this, around 81% is from conventional thermal power plants, 16% from hydroelectric plants, 3% from nuclear, and less than 1% from "other renewables" (like wind and solar). Also in 2002, India accounted for the vast majority (85%) of the region's electricity generation, followed by Pakistan (11%), Bangladesh (3%), Sri Lanka (1%), Nepal, Bhutan, and the Maldives (1% total). Regional electricity generation is expected to increase significantly in coming years. Natural gas is expected to displace some coal-fired generation in India, although recently there have been delays in importing natural gas. However, the net level of coal-fired generation in South Asia is expected to rise. Hydroelectricity is expected to fuel new generations, primarily in Nepal and Bhutan. Non-hydroelectric "renewable" capacity (i.e., wind, solar, ocean, biomass, geothermal) is small at present, but it is increasing, with solar and wind power considered most promising. Currently electricity demand in most of South Asia is outstripping supply, and the region is characterized by chronic shortages. In 2002, India generated 547 BkWh of electricity. The International Energy Outlook 2004 projects more than a doubling of Indian power demand from 554 BkWh in 2001 to 1216 BkWh in 2025.

6.5 Prospects of Technology Transfer

Nature of energy supply and end use technology will determine the emission pathway in near future. Under Kyoto mechanism energy supply sector has gained technology transfers in India. CDM projects (Fig. 24) in India are mostly in energy supply sector. But CDM route is only niche sector investments.

The future trends in technology flows within the South Asian region will continue to depend on a large extent on trends in trade and investment and technology basically transferred through FDI or any other new global arrangement. An issue likely to gain importance in years to come will be the need to adopt and adhere to international multilateral agreements in technology transfer including adoption of ISO 9000 and 14000 which will compel the development and transfer of environmentally sound technologies. Failure of

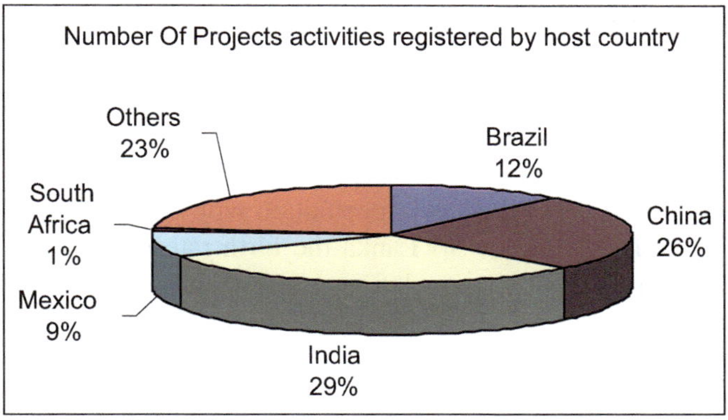

Fig. 24: CDM projects across large developing countries.

adoption will lead to declining exports and restrict the integration of economic activities of developing countries with the developed countries. The adoption of the agreement on TRIPS has important consequences for the transfer and development of technology and integration of industrial activity in the region. Information communication technology (ICT) provides opportunities for industrial development through electronic commerce. For governments it poses the challenge of adjusting the policy environment and undertaking huge investments in the required infrastructure to take advantage of ICT. The government cannot do this alone and will require technology transfer from private sectors abroad. To improve the competitiveness of manufacturing it is essential to generate and adopt new technologies, including ICT. This would require increase in investment in research and development and skill development.

In addition to technology transfer, countries will also have to focus on indigenous technology to strengthen their competitiveness and promote export and integration of their economies. In this context the prospects of LDCs are weak due to their little technological base. However, these countries can benefit from sub-regional and regional cooperation which facilitate the sharing of technologies across borders through company to company cooperation. This in particular is of importance to SMEs which have problems with regard to access of technologies. It is clear that improvement in the quality of scientific and technical education is crucial for overall industrial development and inter-industry complementarities in every country. Unless an environment fostering greater creativity and innovation are provided, the integration of industrial and economic activities will not be possible. This could lead to the marginalization of economies from the global and regional arena. End Use Energy efficiency improvement is another potential area for technology transfer as is evident from ASIF analysis shown above and agriculture and service sectors have the highest potential.

7. DEMOGRAPHY

Demographic trend indicates activity growth in next couple of years. Growth in absolute number of population through next two decades will dominate the South Asian scenario which will have important impact on poverty. SA represents almost 22% of the world population with only approximately 2% of world GNP. It is only for Sri Lanka the birth rate is declining where it is just replacing the present population. In Bangladesh and India both the fertility and mortality rates are declining. However Nepal and Pakistan have still high population growth rates. For India stabilisation in growth rates, which currently stands at 1.3 billion, could plateau at 1.7 billion. Life expectancy going up from the present level of 62 years to approximately 70 years will have its implications for public health services. Globalisation gets reflected in population, capital and raw material movements in the region. Migration Stock represents only a small proportion of the total population in these countries. High levels of migration can cause problems such as increasing unemployment and potential ethnic strife (if people are coming in) or a reduction in the labour force, perhaps in certain key sectors (if people are leaving). For Bangladesh overseas remittances are increasing due to out migration and thereby leading to high consumption growth but however do not add significantly to the productive capacity of the economy. India is getting large number of migrants from neighbouring countries to rural areas. A large number of people continue to migrate to the urban areas from the rural areas of the country. A high population growth will further increase the demand for civic services.

For each of these countries the impact of emigration at the macro level is the remittances it provides for investment in the economy in terms of housing, construction, agriculture as well as economic and social infrastructure. At the microlevel the impact of the overseas migration gets reflected in the change in the consumption pattern of migrant households indicating a change in the lifestyle of migrant households. For example, in Bangladesh migrant households have increased their land ownership, spend more on social ceremonies and religious rituals. The remittances were an important source of revenue for these countries. It is interesting to note from table that India and Bangladesh feature among the top fifteen countries with the highest total remittances received in 2001. The proportion of urban population in Bangladesh increased to 24.5% from 4.2% in 1951. It is evident that Bhutan ranked first followed by Nepal and Bangladesh in terms of annual urban population growth rate. Annual growth rate of urban population is much higher than the national average for all countries with the only exception of Sri Lanka where it is found to be very close to the national average after 1965-70. In 1995, however, the urban growth rate far exceeds the average growth rate and will continue to be so for next decades unless investment in peri-urban and rural areas increase.

8. GLOBALISATION, REGIONALISM AND SOUTH ASIAN TRADE

Globalisation and regionalisation have simultaneously moved at a fast pace the world over during the 1990s and thereafter. South Asia is not an exception. The South Asian countries have intensified trade reforms during the 1990s leading to a greater integration of the region with the world market. Along with multilateral trade and investment policy reforms, the South Asian countries have moved forward with trade policy reforms through regional and bilateral trading arrangements.

The South Asian countries embarked on trade reforms after a prolonged phase of import substitution with a view that liberal trade policies would serve as an engine of growth. Apart from reforms in exchange rate and investment regimes, trade reforms were undertaken by removing distortions arising from quantitative restrictions (QRs), other non-tariff barriers, and tariff protection. While Sri Lanka initiated trade liberalization in the late 1970s, most other South Asian countries underwent trade reforms during the 1990s. In Bangladesh effective protection might have increased since the mid-1990s.

India initiated economic deregulation in the mid-1980s when the import restrictions on a number of goods were relaxed by expanding the positive OGL list and tariff rates on capital goods were brought down (Sinha Roy, 2001, 2004). The reforms agenda was made more comprehensive covering trade, industrial and exchange rate policy regimes since 1991. QRs on most tradeables except consumer goods were phased out in early years of 1990s (Pursell, 1996) and that on the remaining items were phased out by 2000-01. On the tariff front, duty rates were bound on a large spectrum of agriculture and non-agriculture items under the WTO Agreement, and subsequent trade reforms were in accordance with such multilateral commitments. However, the process of tariff reform relatively slowed down thereafter during 1997-98 to 2001-02. Average (unweighted) tariffs, thus, declined from 41 per cent in 1994-95 to 32.3 per cent in 2001-02, but they (mostly peak tariffs) continued to remain high on major agricultural products and agro-based industries. Imports also continued to be protected through specific tariffs, anti-dumping duties, technical standards and regulations and sanitary and phytosanitary rules, especially for agricultural and food products, apart from tariff rate quotas in the agricultural sector. Towards further trade reforms, special economic zones were set up and trade procedures were substantially simplified in order to reduce trade transactions cost. On the whole, the pace of India's trade reforms was gradual.

In Nepal, the import licensing system and quantitative restrictions were eliminated and tariff rates were reduced and the duty structure rationalized since the early 1990s. The monopoly granted to fertilizer imports was abolished in 1997 and there remained no traditional import QRs or government import monopolies, except for petroleum products. Additional measures initiated to promote international trade include the introduction of a bonded warehouse, duty-drawback scheme, initiation of the multi-modal facility (dry port) and an export-processing zone. In 2002 a security tax was imposed on imports, the rate varying across commodities, thus increasing the protection levels of Nepalese domestic industries.

Pakistan introduced trade policy reforms in 1988 aiming at removing the various anti-export biases. The second package of trade reforms announced in 1993 included a phased reduction of maximum tariff levels to 50 per cent within three years subject to exemptions and concessions to a large number of products. Trade reforms became comprehensive in Pakistan since 1997-98 to include agricultural commodities. By 2003, barring few exceptions such as ban on imports from India, use of local component requirements in auto industry and technical standards in some industries, Pakistan had eliminated all remaining traditional QRs and parastatal import monopolies. Average tariffs were reduced. Pakistan also increasingly resorted to anti-dumping measures to protect some of its industries from cheaper imports. On the whole, the trade regime in Pakistan underwent significant reforms.

On the whole, South Asian countries have undergone substantial trade policy reforms. Except for India, the average tariff rate for most South Asian countries has been low. With substantial liberalization of trade regimes, tariff regimes in particular, the South Asian countries have increasingly become trade oriented. Leaving aside the Maldives, Sri Lanka has the highest trade-orientation among South Asian nations, while it is the lowest for India. With substantial trade reforms, the decline in trade-GDP ratio in Pakistan observed for the 1990s have been arrested. The ratio increased from 33 per cent in 1995-97 to 34.1 per cent during 2000-02. Growing trade with rest of the world as well as rising intra-regional trade at the margin explain the phenomenon.

South Asian Trade Performance: Growth, Composition and Markets

Merchandise trade in South Asia grew at high rates during 1995-2003. While merchandise exports grew at 8.52 per cent, imports grew at an even higher average rate of 9.62 per cent during the period. These were indeed significantly higher than the observed growth in world trade. Such high average growth for the region, however, masks the inter-country variations in trade performance. Export growth during the period ranged from 5.76 per cent in Pakistan to 12.21 per cent in Bangladesh. The range of import growth was also quite similar between 4.26 per cent in Sri Lanka and 11.74 per cent in India.

Underlying the robust, but varying, South Asian trade performance is the changing commodity composition and markets. On the import front, these countries sourced mostly manufactured items from rest of the world. While Bangladesh and Sri Lanka imported 'other manufactured goods in relatively large proportions', the import basket for most countries was spread across manufacturing product groups including machinery and capital goods. Apart from manufactures, the South Asian countries also imported crude and processed fuels in growing proportions, while food-deficient South Asian countries continued to import food items.

Manufactures increasingly accounted for exports from the South Asian countries, while the share of agricultural items in total merchandise exports declined after 1995. Agricultural commodities, food items in particular, accounted for more than two-thirds of merchandise exports from the Maldives in 2003. Among manufactures, the export composition in these countries comprised labour-intensive products such as textiles and garments, leather and manufactures, non-metallic mineral products, wood and paper products, etc. The predominance of textiles and garments in South Asia's exports assume importance especially when the sector in these economies is expected to gain substantially in the post-MFA regime. It can also be observed that rising proportions of chemicals and allied products featured in the export basket of these countries, the share being significantly large in India and Nepal. However, the South Asian countries except India were not large exporters of non-traditional manufactures such as machinery and transport equipment. On the whole, South Asian export structures continue to be predominated by labour intensive manufactures. The structural mobility towards more skill and technology intensive goods was restricted; the only exception is perhaps India's exports. While structural immobility of South Asian exports can be due to price competition and world demand conditions; these low skill and technology intensive merchandise exports are also low value adding (RIS, 2002; 2004). On the whole, such limited mobility in export structure of South Asian countries provided only limited push for growth in South Asian trade.

A large proportion of this growing trade from the South Asian countries was with the USA and the EU countries. Some of these countries such as India and the Maldives also increasingly exported to the ASEAN+3 (APT) countries. However, the proportion of merchandise exports to ASEAN+3 from Bangladesh, Nepal, Pakistan and Sri Lanka either remained low or declined between 1997 and 2003. The share of South Asian exports to other regions in the world such as Africa, Europe, Middle East or Western Hemisphere countries have remained relatively insignificant except Pakistan's increasing exports to the Middle East. However, on the import front, the pattern that has emerged by 2003 is significantly different from that of South Asia's merchandise exports. While imports from EU countries are proportionately large, the ASEAN+3 countries have emerged as the most important source of imports in South Asia. The evidence also shows that most South Asian countries, except for India and Pakistan, have significant dependence on other countries within the region for merchandise imports. Merchandise from Middle East have accounted for relatively large proportions of total imports in South Asia, especially Pakistan.

Intra-regional trade in South Asia have also continued to remain small in proportions despite increasing regional engagements of these countries. However, smaller countries in South Asia, namely the Maldives and Nepal, have larger trade dependence on the region. For Nepalese goods, India was the market and exports from the Maldives were directed to Sri Lanka. The intensity of intra-South Asian trade was largely on account of limited complementarity in labour-intensive products. However, trade within the region could have been high but for the existence of large unrecorded border trade.

High growth in South Asian trade has resulted from a significant diversification towards growing trade partners. Among other trading partners of South Asia, the broad based recovery in the region from the 1997 crisis assisted the expansion of the global economy and world trade. The accession of China to WTO and the consequent MFN status that it received from other WTO member countries may have some implications for the competitiveness of the South Asian countries' exports. This is because South Asia and China compete in third country markets for a number of goods including textiles and garments, leather goods, light engineering products, chemicals and pharmaceuticals, among others. The alternate possibility is that the Chinese market may also become more accessible for South Asian exports. Evidence also shows that the gains would be most from liberalization of primary products by China (Wook and Hongyul, 2002). Recent data show that trade between China and some South Asian countries, especially India, increased manifold after China's accession to WTO.

On the whole, it is evident that on the upswing of the global growth cycle, expanding demand from the world economy had a favourable impact on South Asian trade, while the trade outlook was adversely affected on the downturn. While East Asian crises of 1997 triggered the slow down in growth of South Asian trade in 1997, the revival since 2002 is a result of the strong impact of growth in large trading partners such as the US and the EU and the observed turnaround in Japan. The step up in growth in Asian developing countries, and robust growth in China had significant impact on South Asia's trade prospects at the margin. The observed dependence of growth in South Asian trade on the fluctuations in global economy is perhaps an outcome of deeper global integration of South Asian economies since the 1990s. Regional trade integration in South Asia operationalised through various rounds of South Asian Preferential Trade Agreement (SAPTA) since 1995 is being carried forward towards implementing South Asian Free Trade Agreement (SAFTA) in 2006. The South Asian countries committed incremental degrees of tariff liberalization in these different rounds of SAPTA, even though the number of commodities under SAPTA-I, SAPTA-II, and SAPTA-III are not comparable and the level of disaggregation of products differs significantly across countries.

Intra-regional trade of individual South Asian countries increased manifold during the period since 1995. With regard to trade balance, there are variations in country-wise pattern. While it is a favourable regional trade situation for India since 1995, the balance turned positive for Nepal during the period. The regional trade balance position for Pakistan varied over the years. However, regional trade deficit persists for Sri Lanka, Maldives and Bangladesh during this period. The difference in regional balance of trade status between India and other trade partners in South Asia is on account of a larger production and export base in India as against lack of production and adverse trade balance in other South Asian countries.

South Asia has embraced globalization and regionalism almost simultaneously. While unilateral trade liberalization in most of these countries started since the early 1990s, the WTO agreements increasingly directed process of trade reforms in South Asian countries since the mid-1990s. These concurrent processes of globalization and regionalization led to significant liberalization of the trade in individual South Asian economies by eliminating quantitative restrictions and resorting to large tariff cuts. These reforms aimed at reducing distortions in relative prices and reallocate resources accordingly. In addition, trade liberalisation would provide access to cheap inputs and allow competition at the margin. These changes in trade policy would necessarily result in higher volume of trade in South Asia.With multilateral and regional trade liberalization, trade orientation of South Asian countries improved. Despite fluctuations, trade grew at high rates across countries since the mid-1990s.

9. RESEARCH GAPS

Data gaps, rigorous analysis of energy implication of envisaged development pathway and possible alternative pathways are major research gaps. Such study can help in understanding energy cost and emission intensity of the development pathway and potential cost of decarbonisation in the region. Regional study based on macro models to understand development strategies and possible impacts can help in formulation of low carbon development pathway. How developmental goals get translated into emissions scenario is important to be assessed to understand control points for sustainability transition. How short-term development goals can be integrated with long-term global emission pathway scenarios through technology choice, transfer, development and deployment, sectoral investment need assessment; policy priorities are some of the studies which need rigorous analysis. Indian data base and analytical capability is much ahead of other countries in the region. How can Indian leadership in the region generate cooperation, and sustainable development in future need to be understood in bigger developmental framework. There are fragmented disciplinary studies but comprehensive holistic approach to regional study can generate information about how integration with global deal in climate change can emerge and be beneficial for resilience building for the region as a whole. How developmental decisions get managed needs comprehensive analysis.

10. CONCLUDING REMARKS

South Asian region is emerging with multiplicity of institutional mix where governance space of nationalist governments are getting contested not only by political opponents but also by INGOs, NGOs, religious institutions, experts, socially conscious globally enlightened citizens' forums and variety of institutions with variety of objectives cutting across social, economic and environmental agenda. But despite all these dynamics high economic growth still dominate the future South Asian scenario. High growth scenario is a developmental need and cannot be slowed down if poverty reduction is the goal. However, what can be controlled is emission intensity of the growth path. In the past whatever technology development and deployment has been achieved to reduce fossil fuel and resultant emissions intensity have happened in the industry sector and in the transport sector. But it is important to understand that BAU regulatory mechanism could achieve reversal of intensity growth only from early nineties in industry sector and late nineties in transport sector. But the concern got voiced from first oil crisis in early seventies. This indicates that if sustainability transition has to happen in South Asian region in next one decade then not only the alternative development pathway needs to be depicted now but should be acted upon from now itself with more well concerted institutional mechanism, regulations and technological deployment in all sectors if time lag in adjustment is considered.

Industry may be ready for efficient technology deployment due to existing success story but major effort needs to be initiated in all other sectors before major lock in happens, e.g. rural construction projects and infrastructure development. There is a need for technological improvement in coal use efficiency in power sector, oil use efficiency in transport sector, and electricity use efficiency in agriculture. Trade liberalization and FDI along with new financial mechanism and demand management need to be followed with supply management mechanisms in the energy sector. Besides low emission pathway resilience of the economy needs to be planned within the development pathway. Alternative transport, alternative fuels, and implementation of policy alternatives have much scope to be implemented through market and regulatory mechanisms to go beyond BAU development pathway in a scenario over next two decades where economic activity growth, population growth, energy consumption, land use pattern change, technology change, and financial flow are all going to be very high. Institutional changes need to favour high growth to enhance resilience. Loose regional integration will need to be tightened further to ensure stability in the region with new challenges from climate change with likely impact on intraregional migration of poor and vulnerable population.

With increasing democratization of the political process, globalization facilitated through ICT, trade liberalization making human and physical capital footloose, rise in demand for best practice is inevitable. This can have positive effect on drivers of development pathway. South Asia being late comer in global development process with huge development goals in near future and almost on that pathway can play an important role in shaping global sustainable development pathway through choice of right kind of drivers, institutions and goals. Integrating local and global issues is the major challenge now and in years to follow to ensure smooth transition for one fifth of the global population and lead the future by not following high emission commitment through choice of alternative development pathway. Question is how far the decision makers engaged in managing development process in the region are capable of delivering the leadership role for the region in directing development in an alternative pathway. How 'developmental decision making process' is going to make use of this opportunity for South Asia to be realized is much bigger question than what technologies can define. Policy and institutions that can create enabling environment for technology deployment in South Asian region is of primary importance now as new opportunities are emerging in GHG constrained world through second industrial revolution.

REFERENCES

ADB, 2001. Asian Environment Outlook 2001. Asian Development Bank, Manila.

Annual Survey of Industries (Various issues): Factory Sector. Volume 1. Central Statistical Organization (Industrial Statistics Wing), Department of Statistics, Ministry of Planning and Programme Implementation, Government of India.

Annual Survey of Industries (Various issues): Summary Results for Factory Sector. Central Statistical Organization (Industrial Statistics Wing), Department of Statistics, Ministry of Planning and Programme Implementation, Government of India.

Atkins, W.S., 2001. Bangalore Central Traffic Area Management Project. pp. 36.

Balasubramaniam, R., 2008. Community Tanks vs Private Wells: Coping Strategies and Sustainability Issues in South India. *In*: Rucha Ghate, Narpat Jodha and Pranab Mukhopadhyay (Eds), Promise, Trust and Evolution: Managing the Commons of South Asia. Oxford University Press.

Bose, R.K., Mukherji, S. and Gokhale, S., 2005. Energy, Environment and Mobility in Three South Asian Cities – Making Way for Public Rapid Transit: Benefits and Opportunities. Research report submitted to the International START Secretariat, Washington DC.

Chambers, Robert, Saxena, N.C. and Tushaar Shah, 1989. To the Hands of the Poor: Water and Trees. Oxford & IBH Publishing Company, New Delhi.

Das Gupta, M. and Joyashree Roy, 2001. Understanding Changes in Energy Consumption by Energy Intensive Manufacturing Industries in India. Staff paper, CSO. Government of India.

Das Gupta, M. and Joyashree Roy, 2001. Estimation and Analysis of Carbon dioxide Emissions from Energy Intensive Manufacturing Industries in India. *International Journal of Energy, Environment and Economics*, **11(3)**, 165-179.

Das Gupta, M. and Joyashree Roy, 2002. Energy Consumption in India: An Indicator Analysis. *Development Alternatives Newsletter* (COP-8, Climate Change Special), **12(10)**.

Dinar, Ariel, Rosegrant, M.W. and Ruth Meinzen-Dick, 1997. Water Allocation Mechanisms – Principles and Examples. CEPIS Policy Review Working Paper no. 1779.

Enerdata, 2003. Energy Efficiency/CO_2 Trends in the World. www.enerdata.fr

Ghate, Rucha, 2004. Uncommons in the Commons: Community Initiated Forest Resource Management. Concept Publishing Company, New Delhi.

Hayami, Yujiro, Masako Fujita and Masao Kikuchi, 1999. The Conditions of Collective Action for Local Commons Management: A Case of Irrigation in the Philippines. Workshop in Political Theory and Policy Analysis, Colloquium Presentation – V691, September 13, 1999.

IEA, 2002. Bus Systems for the Future – Achieving Sustainable Transport Worldwide. International Energy Agency, Paris.

IPCC, 2007. Climate Change 2007: Mitigation of Climate Change, Working Group III contribution to the Fourth Assessment Report of the Intergovernmental Panel on Climate Change. Summary for Policymakers and Technical Summary. UNEP, WMO.

IPCC, 2001. Third Assessment Report, Climate Change 2001. UNEP, WMO.

Kumar, Praduman and Surabhi Mittal, 2000. Final report of ICAR-ACIAR Collaborative. Project Equity Driven Trade and Marketing Policies and Strategies.

Ludden, David, 2005. Development Regimes in South Asia History and the Governance Conundrum. *Economic and Political Weekly*, Sept 10, 4042-51.

Millenium Ecosystem Assessment, 2005. http://www.millenniumassessment.org/en/Reports.aspx.

Meinzen-Dick, Ruth, Raju, K.V. and Gulati, Ashok, 2002. What Affects Organisation and Collective Action for Managing Resources? Evidence from Canal Irrigation Systems in India. *World Development*, **30(4)**, 649-666.

National Communication of Nepal to UNFCCC, July 2004.

Ostrom, Elinor, 1990. Governing the Commons: The Evolution of Institutions for Collective Action. Cambridge: Cambridge University Press.

Ostrom, Elinor, 1997. Self Governance in Forest Resources. Conference on Local Institutions for forest management: How can Research make a Difference. CIFOR, Bogor, Indonesia, November 19-21. http://www.cgiar.org/cifor.

Palanisami, K. and Balasubramanian, R., 2002. Management of Common Pool Irrigation Tanks of Southern India. *In*: Dinesh K. Marothia (ed.) Institutionalizing Common Pool Resources. Concept Publishing Company, New Delhi.

Poteete, A. and Ostrom, E., 2004. An Institutional Approach to the Study of Forest Resources. *In*: J. Poulsen (ed) Human Impacts on Tropical Forest Biodiversity and Genetic Resources. Centre for International Forestry Research (CIFOR), New York: CABI Publishing.

Pursell, Garry, 1996. Indian Trade Policies since the 1991-92 Reforms. Washington, DC: World Bank.

RBI, 2001, 2002. Handbook of Statistics on Indian Economy. Reserve Bank of India. Mumbai.

Roy, Joyashree, Sathaye, Jayant, Alan Sanstad, Katja Schumacher, and Mongia, Puran, 1999. Productivity Trends in Indian Energy Intensive Manufacturing Industries. *Energy Journal*, **3**, 33-61.

Roy, Joyashree, Katja Schumacher, Sathaye, Jayant, and Sanstad, Alan 2005. Analysis of Fuel Input Use and Technical Change in Selected Energy-Intensive Industries in India. *In*: Gautam Gupta, (ed) *Resource Revisited*, Allied Publishers Ltd., Kolkata.

Roy, Joyashree, 2007. De-Linking Economic Growth from GHG Emissions through Energy Efficiency Route – How far are we in India? *The Bulletin on Energy Efficiency*, **7** (Annual Issue), 52-57.

Sakurai, T. and Palanisami, K., 2001. Tank irrigation management as a local Common Property: The Case of Tamil Nadu, India. *Agricultural Economics*, **25**, 479-498.

Schipper, L. and Marie-Lilliu, C., 1999. Transportation and CO_2 Emissions: Flexing the Link – A path for the World Bank. Environment Department Papers of the World Bank, Paper No. 69.

SEI, 1993. Long Range Energy Alternatives Planning System: Overview for version 94.0. Boston, Massachusetts, USA: Stockholm Environment Institute.

Shah, Tushaar, Seenivasan, R., Shanmugam, C.R. and Vasimalai, M.P., 2002. Sustaining Tamil Nadu's Tanks: A Review of issues and Approaches. *In*: Dinesh K. Marothia (ed.) Institutionalizing Common Pool Resources. Concept Publishing Company, New Delhi.

Sinha Roy, Saikat, 2001. Post Reform Export Performance in India: An Exploratory Analysis. Discussion paper no 13, New Delhi, RIS.

Sinha Roy, Saikat, 2004. Factors in the determination of India's Exports. Unpublished Ph.D Thesis submitted to Trivandrum Centre for Development Studies, India.

Stern, N., 2007. The Economics of Climate Change – The Stern Review. Cambridge University Press, UK, London.

Tang, Shui Yan, 1992. Institutions and collective action: Self-governance in Irrigation. ICS Press, San Francisco.

TERI, 2000. Developing a Sustainable and Environmentally Sound Transport System for Delhi. Prepared for GTZ, Germany; Department of Transport, Govt. of Delhi; and Ministry of Urban Development, Govt. of India. The Energy and Resources Institute, New Delhi. Project code: 98UT42.

TERI, 2000a. Transport and Sustainable Development in Developing Countries. The Energy and Resources Institute, New Delhi. Project code: 2000UT41.

Wade, Robert, 1988, 1994. Village Republics: Economic Conditions for Collective Action in South India. ICS Press, Oakland.

WBCSD, 2001. Mobility 2001 – World Mobility at the end of the twentieth century and its sustainability. Working Group of the World Business Council for Sustainable Development by the Massachusetts Institute of Technology and Charles River Associated Incorporated, USA.

Wook Chae and Han Hongyul, 2002. Searching for a better regional surveillance mechanism in East Asia, 2002, Working Papers, **No. 02-01**, pp. 79.

3 Instrumental, Terrestrial and Marine Records of the Climate of South Asia during the Holocene: Present Status, Unresolved Problems and Societal Aspects

A.K. Singhvi, K. Rupakumar, M. Thamban, A.K. Gupta, V.S. Kale, R.R. Yadav, A. Bhattacharyya, N.R. Phadtare, P.D. Roy, M.S. Chauhan, O.S. Chauhan, S. Chakravorty, M.M. Sheikh, N. Manzoor, M. Adnan, J. Ashraf, Arshad M. Khan, D.A. Quadir, L.P. Devkota and A.B. Shrestha

1. INTRODUCTION AND OBJECTIVES

In South Asia, monsoon is almost a synonym to climate and constitutes a critical resource for the region's largely agrarian economies. Region's water needs are largely dependent on summer monsoon rainfall. In some regions however, of the total water budget, glacier melt water can measure up to ~10-15%, the rest being the rainfall. The past century has seen a marked expansion of global/regional observational networks for accumulation of data on climate state parameters. However, so far, these have found only a limited use in policy planning in the socio-economic sector and for strategies toward sustainable development, particularly in the context of anticipated global warming. A critical synthesis of all the climatic information is prerequisite for a scientifically sound policy planning. Consequently, a regional scale description of the present state of our understanding of climate state parameters and an estimation of the likely amplitude of their variability are crucial for this purpose.

The climate of the region is largely dominated by seasonally reversing wet and dry monsoon winds – called as the Indian or the South Asian monsoon system. During summers (June-August), the southwesterly winds (the summer monsoon) pick up moisture from the oceans, travel over land and subsequently drop their moisture over the region. During winters (December-February), however, the monsoon winds are dry and variable, blowing from the northeast out to sea. The summer or southwest (SW) monsoon affects societies and agriculture-based economy of large parts of South Asia (Fig. 1). It also plays an important role in modulating the global climate through its control on hydrological and carbon cycles. The effects of the monsoon are preserved in proxies such as the tree rings, soils, ice, lake sediments, peat deposits, cave deposits and marine sediments. In general, such paleo-proxies inform about both the low and high frequency changes in the monsoon rainfall and in some cases temperatures. Evidence of abrupt, decadal, and multi-decadal to century and millennial scale changes in the monsoon during the Holocene exist. However, centennial to millennial scale changes in the monsoon are largely driven by solar activity (Fleitmann et al., 2003; Gupta et al., 2003, 2005). Overall, both the land and marine records indicate that the monsoon during the Holocene was not static; rather it underwent repeated occurrences of large amplitude variations throughout this period.

An important aspect that has been largely overlooked so far, relates to the response times of proxies. Different sedimentary archives and proxies have their own response times for the same climatic perturbation. Thus, for example, creation and preservation of sediment record in lakes, oceans and land is determined by three factors, viz. the sediment supply, the transport potential and sediment preservation. This implies that there is a "*window of opportunity*" for the formation of a sedimentary record on land and consequently, temporal differences in spatially distributed archives are to be

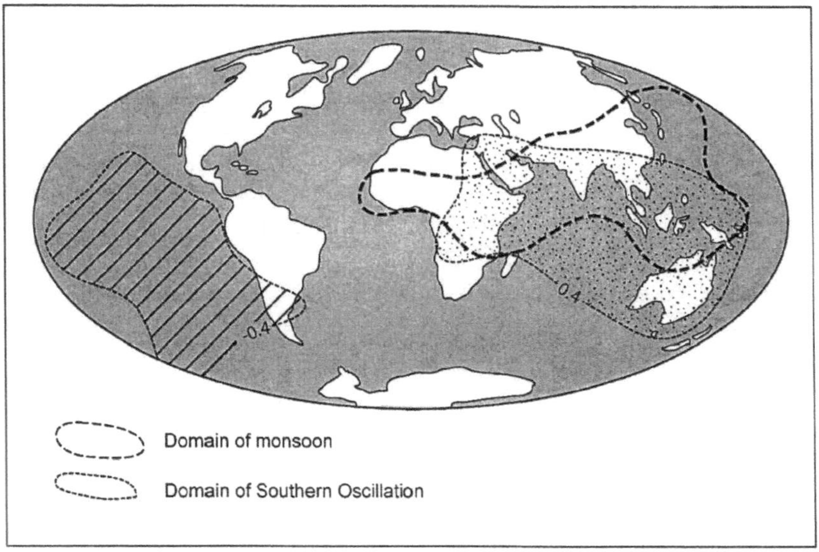

Fig. 1: The range of the monsoon in Asia and its reach. The region affected by the Southern Oscillation is also indicated (modified from Williams et al., 2006). Correlation >0.4 of annual surface pressure anomalies with those of Jakarta, Indonesia. Stippled area is a positive correlation and hachure is a negative correlation.

expected and need to be constrained. In the marine records, preservation is not an issue but bioturbation can disturb the climate signal. It is then implicit that a comparison of marine and land records requires due caution. In general, with respect to a forcing function, both the marine and land reconstruction should respond differently. And between the two, the terrestrial reconstruction, by definition should be lagged to variable extent (i.e. site specific). The terrestrial records are in general truncated as the very agencies that create these records, erode them as well. The duration of hiatuses is variable and is determined by the response of sediment supply, transport and preservation to the forcing function. Thus, the correlation of land records (with hiatuses) with marine records (that are continuous but partially time averaged) becomes non-trivial and it seems reasonable to suggest that conventional approaches of associating landforms with climates (that engendered them) are not valid.

Another important question that requires attention is – do the marine proxies provide data that can be translated into rainfall reconstruction on the land? This is particularly of interest in the context of the fact that agriculture and the migration of people depend on rainfall and its distribution in a season, on land. Also the question as to, how reliably ocean parameters, like the sea surface temperatures and wind speeds, can be used to predict the monsoon intensity and its spatial distribution, has not been addressed to adequately. Paleo-reconstruction of monsoon rainfall on land can play an important role in addressing to such issues.

It is considered that the Indian monsoon evolved some time in the late Oligocene-Miocene (~30-6 Ma) and underwent a major shift ~8.5 Ma (e.g., Quade et al., 1989; Kroon et al., 1991; Gupta et al., 2004). Rapid changes in the monsoon are documented in numerous recent publications that are based on proxies such as:

1. marine sediments that record the summer monsoon winds (Schulz et al., 1998; Anderson et al., 2002; Gupta et al., 2003),
2. cave deposits that record precipitation in Oman (Neff et al., 2001; Fleitmann et al., 2003),
3. peat deposits that indicate humidity and temperature (Hong et al., 2003) and monsoon history (Phadtare and Pant, 2006),
4. river runoff and sediment fluxes in the Bay of Bengal (Kudrass et al., 2001; Chauhan, 2003; Chauhan et al., 2004),
5. fluvial sediments indicating the relative changes in the amount of precipitation (Sharma et al., 2004a; Juyal et al., 2006; Williams et al., 2006),
6. desert dunes (Singhvi and Kar, 2004), and
7. organic matter in the Arabian Sea sediment cores and the $^{15}N/^{14}N$ ratio that is related to the upwelling and monsoon winds (Altabet et al., 2002).

Strong monsoon events had a potentially dramatic effect on the fluvial systems (Goodbred and Kuehl, 2000), the fauna and flora, and human populations in South Asia during the Holocene (Gupta, 2004; Gupta et al., 2006). The present contribution aims at summarizing and synthesizing the present understanding of the instrumental and paleorecords with a view to examine both the spatial and temporal variability and the validity/domain/information content of the proxies. A survey of the meteorological aspects of monsoon, its variability and modelling is provided by Gadgil (2003).

A key societal aspect is to recognize the factors controlling the spatial variability and the distribution patterns of rainfall in a season. The agricultural output in South Asia depends on monsoon. It is therefore important to understand the causes of agricultural droughts (i.e. distribution of rainfall in a season) rather than the mean seasonal rainfall. At the outset, it must be admitted that we are far away from achieving this. Occasional insights are emerging, particularly from securely-dated tree ring reconstructions and well-dated sedimentary records. In the following text, an account of climate variability over Bangladesh, India, Nepal and Pakistan is presented. Both the instrumental records and chronologically constrained paleorecords, wherever available, are examined.

2. INSTRUMENTAL DATA

2.1 India

Historical Data

Systematic analysis of historical documents from western India by Abhayankar (1987) indicated an increase in droughts/famines since the sixteenth century with maximum drought intensity during 18th and 19th centuries. A synthesis of historical and archeological records of rainwater harvesting practices by Pandey et al. (2003) suggests that 1500 BC, 800 BC, 300 BC, 550 AD and 1148-50 AD were dry or arid periods (Fig. 2). Though the duration of these events can not be ascribed, it is likely that they represent a span of ~30 years. The dating in general is indicative and not definitive. Numerous historical documents describing climate and revenue records that can provide data on agriculture revenue (hence rainfall performance) in these regions do exist and await a proper synthesis. These can provide quantitative inferences on monsoon performance for the past several hundred years. Similarly, the rich archaeological record has not been fully exploited for paleoclimate reconstruction.

Rainfall Variability

The Indian summer monsoon circulation dominates rainfall over South Asia. All-India summer monsoon rainfall (AISMR) displays predominant interannual variability (Fig. 3), marked by recurrent large-scale droughts and

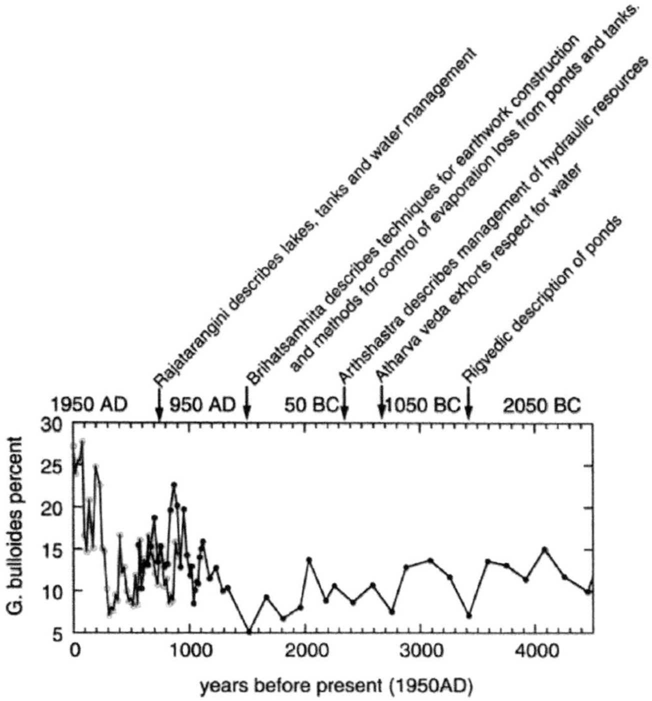

Fig. 2: Monsoon winds reconstructed for the past 4500 years using *Globigerina bulloides* abundances in boxcores from the Arabian Sea. Arid episodes correlate well with ancient Indian records containing description of early work that suggest contemporary aridity, Rigveda, 1500 BC (dating is not certain), Atharva Veda (800 BC) Arthashastra (3000 BC) Brihatsamhita (550 AD), Rajtarangini (1150 AD) (From Pandey et al., 2003).

floods. Years of large-scale deficient and excess monsoon rainfall are usually identified with the criteria of the AISMR being below and above 10% of the long-term mean, respectively. A remarkable feature of anomalous monsoon situations is the spatial coherence of seasonal rainfall anomalies over large areas of the country. The effect of droughts is accentuated by the higher coefficient of variability over regions of lower seasonal rainfall (Parthasarathy, 1984) and their occurrence in 2 or 3 consecutive years on several occasions (Chowdhury et al., 1989).

Studies during the last four decades clearly demonstrate that monsoon rainfall lacks any trend and, on an all-India scale, it randomly fluctuates around a mean value (Mooley and Parthasarathy, 1984). However, on a smaller spatial scale, notable trends do emerge. Rupa Kumar et al. (1992) found that the west coast of India, northern Andhra Pradesh and northwestern India registered an increasing trend in monsoon seasonal rainfall, whereas east Madhya Pradesh and adjoining areas, northeastern India and parts of Gujarat and Kerala experienced a decreasing trend. The long-term trends

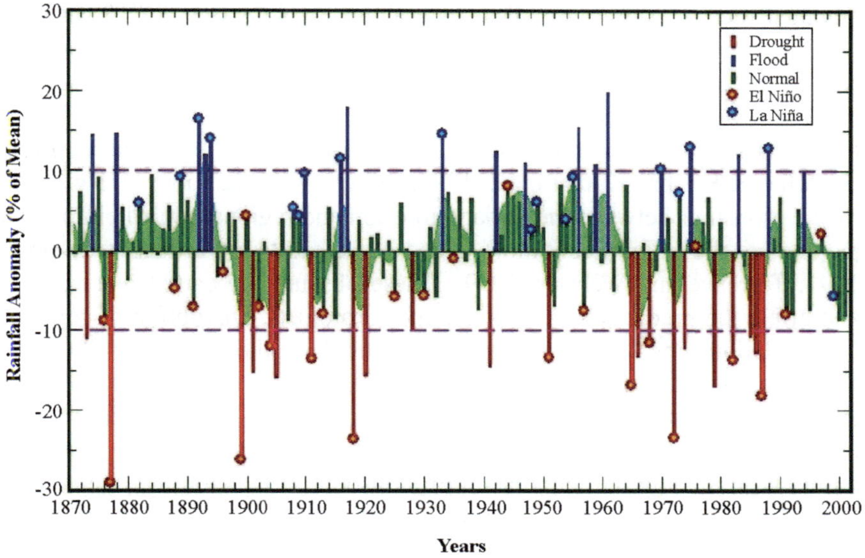

Fig. 3: Variation of all-India summer monsoon rainfall anomalies during 1871-2004 and its association with El Niño/La Niña occurrences. The horizontal dashed lines delineate anomalies crossing the standard deviation. The data is from IMD.

account for only a small part of the total variance in the rainfall on an annual scale. The interannual variability dominates the rainfall fluctuations on all space scales. A reassuring feature of the Indian rainfall variability is that, unlike the Sahelian region, the Indian summer monsoon shows stable, long-term characteristics with extremes such as droughts and floods being only a part of its natural variability, i.e. the monsoon in India never failed and, will not fail in the future as well.

On a decadal scale, AISMR displays an epochal pattern, with alternating multi-decadal periods (3-4 decades each) of more frequent droughts and of higher rainfall (Mooley and Parthasarathy, 1984). However, this does not appear to be part of any regular periodicity in monsoon rainfall. This low frequency signal is common to the parameters generally used to describe the behaviour of the monsoon system like onset, break monsoon days and number of storms and depressions (Pant et al., 1988). Numerous attempts have been made to identify the significant periodicities in monsoon rainfall series, which suggest quasi-biennial periodicities in the range of 2 to 3 years being significant (e.g. Mooley and Parthasarathy, 1984).

The monsoon displays significant global/regional teleconnections. Observational data show evidence of the association between weak monsoon, a large negative Southern Oscillation Index (SOI) and El Niño events and also with a strong monsoon, large positive SOI and absence of El Niño events (Pant and Parthasarathy, 1981; Pant and Rupa Kumar, 1997). During 1871-2003, 11 out of 23 drought years were El Niño years, whereas none amongst the 19 excess rainfall years was an El Niño year. In 15 cases of normal monsoon years, when El Niño occurred, only four were in the upper half of the normal range of monsoon rainfall. Besides these, several global and regional parameters are seen to contribute to interannual variability of the monsoon. These form the basis for seasonal forecasting of the monsoon rainfall (see Krishna Kumar et al., 1995; for a review, see Table 1). However, the relationships between the Indian monsoon and regional/global circulation parameters show significant multi-decadal changes (e.g. Parthasarathy et al., 1991; Torrence and Webster, 1999). These obscure the causal mechanisms and affect the predictive skills of the meteorologists. Disruption of some of these relationships by global warming, particularly those with ENSO, have also been reported (Krishnakumar et al., 1999). More recent work by Goswami and Xavier (2006) suggests the linkages of the length of rainy season with ENSO, such that the stronger ENSO reduced the duration by delaying the onset and hastening the withdrawal of monsoon. This is through the control of SST on troposphere temperatures and, if confirmed, it can provide data of immense value in predictions. We would also like to mention the projection of increasing droughts by the year 2050 based on simulation of enhanced greenhouse gases and aerosols (Ramanathan et al., 2005).

Table 1: Parameters used for prediction of monsoon (Rajeevan et al., 2004)

No.	*Parameter*	*Month*	*Correlation Coefficient 1983-2002*
P1	Arabian Sea SST	Jan-Feb	0.55
P2	Eurasian Snow Cover	Dec.	-0.46
P3	NW Europe Temperature	Jan	0.46
P4	Nino 3 SST anomaly	July-Sept.	0.42
P5	South Indian Ocean SST Index	Mar	0.47
P6	East Asia Pressure	Feb.-Mar	0.61
P7	N. Hemisphere 50hPA wind pattern	Jan, Feb.	-0.51
P8	European Pressure gradient	Jan	0.42
P9	South Indian Ocean 850hPa zonal wind	Jun	0.45
P10	Nino 3,4 SST tendency	Apr-Jun, Jan-Mar	0.46

Temperature Variability

Monitoring and analysis of atmospheric temperatures on global, as well as regional, scales have acquired special importance during the last few decades due to the clear indications of global warming in the post-industrial era. The Fourth Assessment Report (AR 4) of the Intergovernmental Panel on Climate Change (IPCC) concluded that the global mean surface air temperature has increased by 0.76°C during the period 1850-1899 to 2001-2005 (IPCC, 2007). Considerable emphasis has also been placed on the manifestation of regional warming/cooling in terms of day and night temperatures, because of their links to changes in cloudiness, humidity, atmospheric circulation patterns, wind and soil moisture (Karl et al., 1993). Easterling et al. (1997) reported that the amplitude of variation between daytime high and nighttime low temperature has decreased for most part of the world during the period 1950-1993.

In one of the early studies for the Indian region in the context of contemporary global warming, Hingane et al. (1985) reported that the mean annual temperature increased by about 0.4°C/100 years in India during the period 1901-82. The rise in the annual mean temperature is mainly contributed by the post-monsoon and winter seasons. Temperatures during the monsoon season do not show any significant trend over a major part of the country. The spatial distribution of temperature trends indicates large areas of a significant warming trend along the west coast, the interior peninsula and over northeastern India. Northwestern India is conspicuous by its significant cooling trend. Rupa Kumar et al. (1994) pointed out that, while the mean temperature trends over India were similar to the global and hemispheric trends, the diurnal asymmetry of surface temperature trends observed over India is quite different from that in the other parts of the world (Karl et al., 1993). The increase in the mean temperature over India was almost solely contributed by maximum temperatures, with the minimum temperature remaining practically trendless. Rupa Kumar et al. (1994) have also shown that the all-India temperature trend does not have a significant urban/non-urban bias. The AIR (all India rainfall) also does not exhibit any trend that could be distinctly attributed to global warming.

Krishnan and Ramanathan (2002) observed that the all-India surface air temperature during the drier part of the year (January-May) has been subject to a relative cooling by ~0.3°C during the last three decades, when the global effects of greenhouse gases and natural variability are filtered out from the temperature series. However, it must be pointed out that this is rather a perceived cooling, and the overall temperature trends still indicate significant warming, presumably due to greenhouse forcing. Kothawale and Rupa Kumar (2002) have reported significant post-1970 warming in the lower troposphere over India (Fig. 4). An interesting new approach to monitoring surface ground temperatures is the bore hole temperature profile that suggests warming of ~0.9°C over the past 150 years and indicates that this warming began before the widespread changes in surface air temperatures. It is further suggested that total warming for 1980 base line and 1990 is 1.2°C (Roy et al., 2002).

Greenhouse gas changes in the atmosphere appear to be affecting not only the surface and near-surface conditions but also those in the upper troposphere, the mesosphere and the stratosphere (Tyson et al., 2001). While there has been a significant surface warming over India, there are indications that it is associated with a cooling in the upper-troposphere levels (Rupa Kumar et al., 1987; Kothawale and Rupa Kumar, 2002). Using updated data sets up to 2003, Kothawale and Rupa Kumar (2005) reported substantial recent changes in temperature trends over the last three decades (Fig. 4). While all-India mean annual temperature has shown significant warming trend of 0.05°C/10 yr during the period 1901-2003, the period 1971-2003 has seen a relatively accelerated warming of 0.22°C/decade, which is mostly

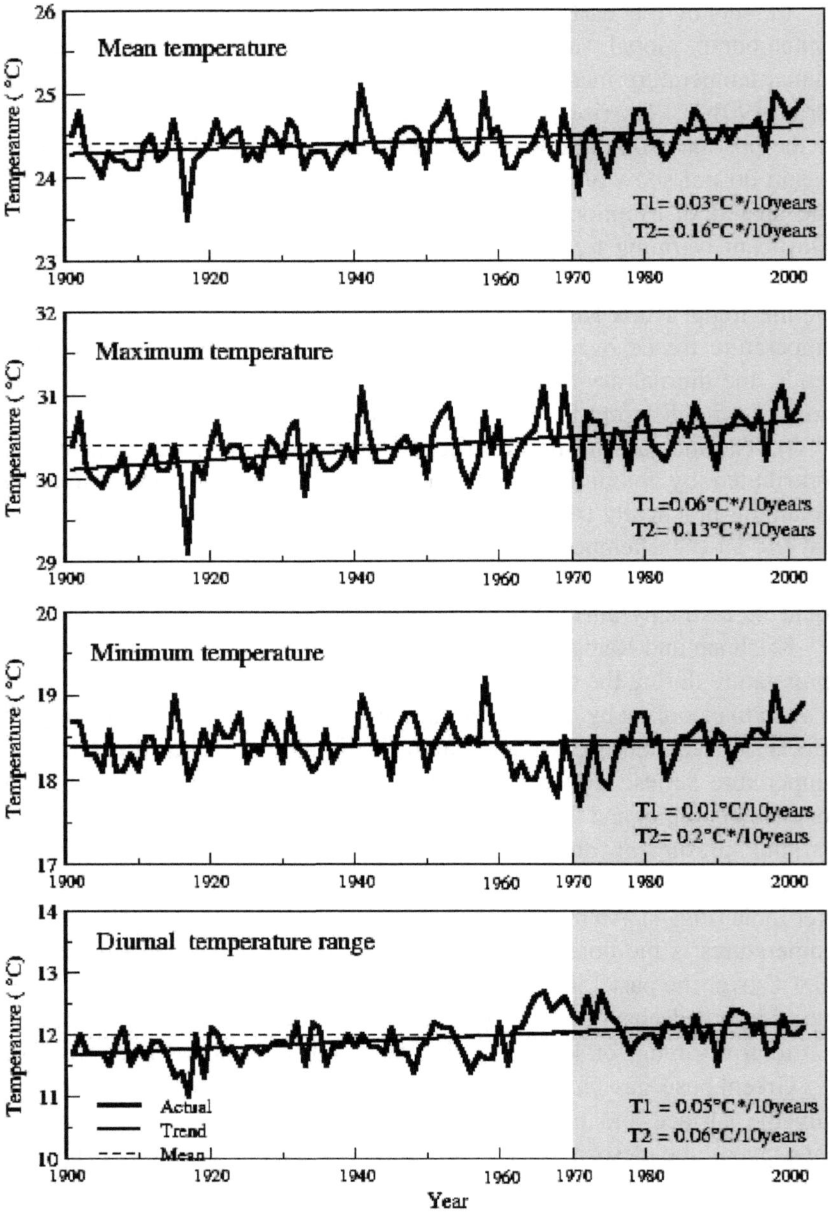

Fig. 4: Variations of all-India mean, maximum and minimum temperatures during 1901-2003. The data is from IMD.

during the last decade. Further, in a major shift, the recent period is marked by rising temperatures during the monsoon season. This change results in a weakened seasonal asymmetry of temperature trends reported earlier. The minimum temperatures also show an increase, indicating that the warming over India is manifested equally in the day-time and the night-time temperatures.

Extreme Climatic Events – Large-scale Droughts/Floods

During anomalous monsoon years, the deficient/excess rainfall conditions are widespread over the continent. However, the spatial variability of the monsoon rainfall is large and there are occasions when some parts of the country experience floods due to heavy rains while at the same time other parts have serious rainfall deficiency leading to drought. The recent 2005 rainfall event over Mumbai was one such event. Therefore, the aerial extent of the extremes in rainfall has to be incorporated in the procedures that identify a large-scale drought or a flood. Instrumental records over the past 130 years do not indicate any marked long-term trend in the frequencies of large-scale droughts or floods in the summer monsoon season (Goswami et al., 2006). Low frequency change that is clearly discernible is the alternating sequence of multi-decadal periods of more frequent droughts followed by periods of less frequent droughts. This feature is part of the well-known epochal behaviour of the summer monsoon. Some cyclicity has also been noticed in the occurrence of large-scale floods, which was linked to sunspot activity (Bhalme and Mooley, 1981).

Arid-area Extent

Large tracts in northwestern India and interiors of peninsular India experience arid conditions. Though desertification is a complex environmental process involving geomorphologic, atmospheric and human-induced processes, it is observed that the rainfall regimes closely demarcate the arid region boundaries. The "arid area" of India can be demarcated on the basis of a simple isohyetal criterion, i.e., the area of the country with annual rainfall less than or equal to 560 mm (Singh et al., 1992). The fluctuations of such meteorologically defined arid area are extensive. In general, during extremely deficient years of SW monsoon over the Indian subcontinent, aridity expands over the semiarid areas in the north and even extends deep down south to the peninsular India. A cursory look at annual changes in spatial patterns brings out the clear contrasts in the aerial extents of aridity in the region. On an average, about 19% of the area of the country experiences arid conditions every year, of which 15% is in northern India and 4% in Peninsular India.

Over northern India, the interannual variations of the arid-area extent are relatively small and the arid area expands and contracts in a somewhat organized manner. Its eastward expansion and westward contraction nearly follow the average annual isohyets. It has been reported that the total arid area shows decreasing trend from the beginning of the 20th century over India, possibly due to a westward shift in the monsoon rainfall activities (Singh and Sontakke, 2002) and/or changes in the land use pattern over northwestern India (Pant and Hingane, 1988).

Heavy Rainstorms

Rainfall on a daily scale over India is highly variable and has very little or no relationship with the monthly or annual averages. There are many places in India, which have recorded 40 to 100% or more of their mean annual rainfall in a single day (Dhar and Mandal, 1981). Heavy rainfall events have profound local-scale impacts, and information about such extreme events is vital to infrastructural development. Climatologically, the heaviest falls are mostly recorded at coastal and hill stations. Extreme short-duration rainfall events show significant increasing trend over the northern parts of the west coast of India and decreasing trend over the southern parts of the west coast, while no spatially coherent trends could be noticed in other parts of India (Krishna Kumar et al., 1997).

The statistical characteristics of the daily rainfall over India also have significant spatial variations. The spatial distribution of the mean amount of rainfall per rain-day (a rain-day is defined as a day with 1 mm or more rainfall) during summer monsoon over India indicates that the amount of rainfall varies from 10 mm/day to >30 mm/day (Soman and Krishna Kumar, 1990). The lowest values occur over the rain shadow regions east of the Western Ghats and the highest along the west coast and some parts of northeastern India. The average number of rain-days during the southwest monsoon season over the Indian stations varies from <10 to >100 out of the maximum possible 122 days, i.e., every day of the season can be a rain-day. Further, at most of the individual stations, the summer monsoon rainfall in just 10 to 20% of the total number of rain-days accounts for as much as 50% of the seasonal rainfall (Soman and Krishna Kumar, 1990).

2.2 PAKISTAN

Pakistan, located between 24°-37°N and 61°-76°E, is an area of marked changes in relief ranging from the Arabian Sea in the south and the Himalayan mountains in the north. This small space of 13° in latitude contains diverse physiography ranging from marine, coastal, desert, riverine plains and mountains. Consequently the climate is highly variable and marginal changes in these are likely to affect the local eco-systems on amplified scales.

Physiographically, the region north of 35°N is dominated by the winter (December-March) rains due to western disturbances mostly active in winter and in the transition period of pre-monsoon. The western disturbances are active more or less throughout the year by the passage of eastward moving depressions, secondary lows and waves. The region abounds in sizable glaciers particularly the Karakoram Range. Snow and ice melt constitute an important water resource for the country. The region south of 35° to 31.5°N, including the outer Himalayas or the sub-montane region within almost the eastern half of these latitudes, constitutes the summer monsoon dominated region in the country. Monsoon depressions from the Bay of Bengal or monsoonal systems from the Arabian Sea contribute significantly to the rainfall over this region and over the eastern plains of the country. The major parts of the southern plains of Pakistan are basically arid with some hyper-arid parts in southwestern parts of Balochistan Plateau.

M.M. Sheikh and N. Manzoor (2005) provide a detailed account of a synthesis of regional temperature and precipitation trends for 1951-2000 and the discussion below is based on this review. Tables 2-4 respectively provide region-wise trends in the mean, maximum and minimum temperature. Table 5 provides trends in the annual and seasonal precipitation over Pakistan. Figures 5a (i-ii) provide the 100-year trend changes in the mean temperature and precipitation based on Climate Research Unit (CRU) data. These tables and figures given are part of the Global Change Impact Studies Centre (GCISC) Research Report (GCISC-RR-01, June, 2009). A synthesis of the 1951-2000 data enables the following salient inferences:

1. The mean annual temperatures exhibit a mixed trend from no change in the coastal region to 1.17 °C in Balochistan to 0.72 °C in the Western Highlands. Mean summer (April-May) temperatures over all the regions indicate an increase ranging from 0.03°C to 2.17 °C. Mean maximum temperatures generally show higher increasing trend than the minimum temperatures.
2. In the Greater Himalayan region, both the winter (December-March) and subsequent summer (April-May) maximum temperatures have increased. This may decrease the accumulation of snowfall over the mountains and enhance the glacier melt.
3. In Balochistan region, both the maximum and minimum temperatures have increased and this implies the desert regions becoming warmer.
4. The monsoon (June-September) rainfall has increased in all the regions with the exception of the Balochistan Plateau and coastal areas where it has decreased.
5. Winter (December-March) precipitation has slightly decreased in the Greater Himalayan region whereas it is increased in the Sub-montane, Central and Southern Punjab and the eastern parts of Balochistan Plateau.

Table 2: Mean temperature trends (1951-2000)

Regions/Seasons	*Annual*	*Monsoon (Jun-Sep)*	*Winter (Dec-Mar)*	*Apr-May*	*Oct-Nov*
Region I(a): Greater Himalayas (Winter dominated)	0.04	-0.80	0.32	1.09	-0.06
Region I(b): Sub-montane Region and Monsoon dominated	-0.19	-0.57	0.00	0.13	0.12
Region II: Western Highlands	-0.72	-1.48	-0.65	0.17	-0.47
Region III: Central & Southern Punjab	0.11	-0.25	0.03	0.83	0.31
Region IV: Lower Indus Plains	-0.08	-0.55	-0.07	0.35	0.15
Region V(a): Balochistan Province (Sulaiman & Kirthar Ranges)	0.44	0.11	0.36	0.63	0.86
Region V(b): Balochistan Plateau (Western)	1.17	1.3	0.43	2.17	1.80
Region VI: Coastal Belt	0.00	-0.18	0.05	0.03	0.30

Table 3: Mean maximum temperature trends (1951-2000)

Regions/Seasons	*Annual*	*Monsoon (Jun-Sep)*	*Winter (Dec-Mar)*	*Apr-May*	*Oct-Nov*
Region I(a): Greater Himalayas (Winter dominated)	0.63	-0.16	0.73	1.91	0.98
Region I(b): Sub-montane Region and Monsoon dominated	0.04	-0.46	0.08	0.55	0.29
Region II: Western Highlands	-0.42	-1.10	-0.55	0.78	-0.25
Region III: Central & Southern Punjab	-0.14	-0.20	-0.54	0.78	-0.06
Region IV: Lower Indus Plains	-0.02	-0.17	-0.33	0.63	0.08
Region V(a): Balochistan Province (Sulaiman & Kirthar Ranges)	0.54	0.36	0.53	0.86	0.59
Region V(b): Balochistan Plateau (Western)	0.83	1.23	0.10	1.97	1.17
Region VI: Coastal Belt	-0.08	-0.08	-0.20	-0.25	0.43

Table 4: Mean minimum temperature trends (1951-2000)

Regions/Seasons	*Annual*	*Monsoon (Jun-Sep)*	*Winter (Dec-Mar)*	*Apr-May*	*Oct-Nov*
Region I(a): Greater Himalayas (Winter dominated)	0.80	-1.58	-0.23	-0.10	-1.23
Region I(b): Sub-montane Region and Monsoon dominated	-0.32	-0.68	-0.14	-0.19	-0.08
Region II: Western Highlands	-1.45	-1.82	-1.10	-0.60	-0.78
Region III: Central & Southern Punjab	0.41	-0.35	0.77	0.76	0.99
Region IV: Lower Indus Plains	-0.2	-1.18	0.12	-0.02	0.00
Region V(a): Balochistan Province (Sulaiman & Kirthar Ranges)	0.36	0.10	0.27	0.53	0.96
Region V(b): Balochistan Plateau (Western)	1.33	1.40	0.67	2.20	2.50
Region VI: Coastal Belt	0.13	-0.23	0.25	0.43	0.23

Table 5: Percentage precipitation trends (1951-2000)

Regions/Seasons	*Annual*	*Monsoon (Jun-Sep)*	*Winter (Dec-Mar)*
Region I(a): Greater Himalayas (Winter dominated)	24.5	86.5	-2.0
Region I(b): Sub-montane Region and Monsoon dominated	15.0	19.0	26.5
Region II: Western Highlands	-1.0	11.0	0.0
Region III: Central & Southern Punjab	31.5	28.5	49.5
Region IV: Lower Indus Plains	11.0	22.5	-13.5
Region V(a): Balochistan Province (Sulaiman & Kirthar Ranges)	59.5	58.0	57.0
Region V(b): Balochistan Plateau (Western)	5.0	-10.0	-20.0
Region VI: Coastal Belt	-41.0	-67.0	0.0

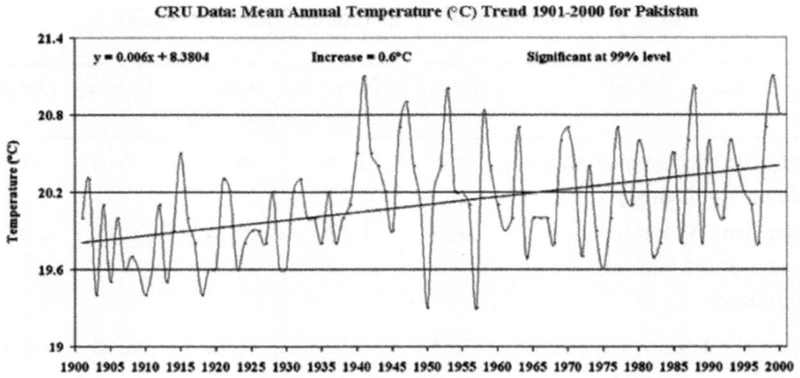

Fig. 5 a (i): Mean annual temperature trend changes over Pakistan based on CRU Data (1901-2000).

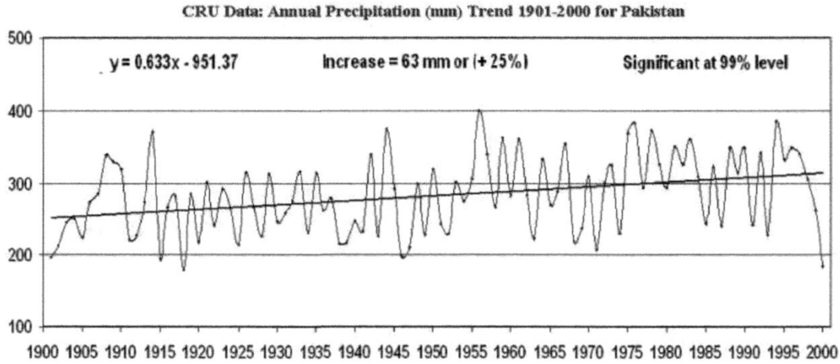

Fig. 5 a (ii): Annual percentage precipitation trend changes over Pakistan based on CRU Data (1901-2000).

Climate Extreme Indices

A suite of climate extreme indices using all the 27 ETCCDMI (Expert Team on Climate Change Detection and Monitoring Indices) core indices, 16 for temperature and 11 for precipitation (Table 6) are derived from daily temperature (both maximum and minimum) for the period 1971-2000 and daily precipitation totals over the period 1961-2000 for 32 meteorological stations distributed all over Pakistan. The index Rnn (Annual count of days when precipitation>nn), a user defined/dependent threshold is not chosen for analysis because of varying threshold values. Even the index ID0 (Ice days) i.e. Annual count of days when TX (daily maximum temperature) < 0°C rarely happens in tropical, subtropical and even temperate parts of the South Asia region, is not considered for analysis. However, two additional indices i.e. Extreme Temperature Range (ETR) and Annual Contribution from very wet days (R95PT), not directly calculated by RClimDex are included in the analysis with the definitions:

$$\text{ETR} = \text{TXx} - \text{TNn} \text{ and } \text{R95PT} = (\text{R95P}/\text{PRCPTOT}) \times 100$$

Table 6: List of ETCCDMI core climate indices used in the project

ID	*Indicator name*	*Definitions*	*Units*
FD0	Frost days	Annual count when TN (daily minimum)<0°C	Days
SU25	Summer days	Annual count when TX (daily maximum)>25°C	Days
TR20	Tropical nights	Annual count when TN (daily minimum)>20°C	Days
TXx	Max Tmax	Monthly maximum value of daily maximum temp	°C
TNx	Max Tmin	Monthly maximum value of daily minimum temp	°C
TXn	Min Tmax	Monthly minimum value of daily maximum temp	°C
TNn	Min Tmin	Monthly minimum value of daily minimum temp	°C
TN10p	Cool nights	Percentage of days when TN<10th percentile	Days
TX10p	Cool days	Percentage of days when TX<10th percentile	Days
TN90p	Warm nights	Percentage of days when TN>90th percentile	Days
TX90p	Warm days	Percentage of days when TX>90th percentile	Days
WSDI	Warm spell duration indicator	Annual count of days with at least 6 consecutive days when TX>90th percentile	Days
CSDI	Cold spell duration indicator	Annual count of days with at least 6 consecutive days when TN<10th percentile	Days
DTR	Diurnal temperature range	Monthly mean difference between TX and TN	°C
RX1day	Max 1-day precipitation amount	Monthly maximum 1-day precipitation	Mm
Rx5day	Max 5-day precipitation amount	Monthly maximum consecutive 5-day precipitation	Mm
SDII	Simple daily intensity index	Annual total precipitation divided by the number of wet days (defined as PRCP>=1.0mm) in the year	Mm/day
R10	Number of heavy precipitation days	Annual count of days when PRCP>=10mm	Days
R20	Number of very heavy precipitation days	Annual count of days when PRCP>=20mm	Days
Rnn	Number of days above nn mm	Annual count of days when PRCP>=nn mm, nn is user defined threshold	Days
CDD	Consecutive dry days	Maximum number of consecutive days with RR<1mm	Days

(*Contd.*)

(Contd.)

CWD	Consecutive wet days	Maximum number of consecutive days with RR>=1mm	Days
R95p	Very wet days	Annual total PRCP when RR>95th percentile	Mm
R99p	Extremely wet days	Annual total PRCP when RR>99th percentile	mm
PRCPTOT	Annual total wet-day precipitation	Annual total PRCP in wet days (RR>=1mm)	mm

Trend changes in these climate extreme indices for South Asian countries viz. Bangladesh, India, Nepal, Pakistan and Sri Lanka using the software RClimDex Ver. 1.3 and RH Test are consolidated in Table 7. The above work

Table 7: Number of stations showing trend changes in climate extreme indices in South Asian countries for the period 1971-2000 for temperature and 1961-2000 for precipitation indices

S. No.	*Indices*	*Bangladesh*		*India*		*Nepal*		*Pakistan*		*Sri Lanka*	
		+	-	+	-	+	-	+	-	+	-
1	TXx	13	12	88	32	18	2	18	14	10	1
2	TXn	12	13	68	49	13	6	15	16	8	1
3	TNx	13	12	59	60	13	6	5	24	10	1
4	TNn	12	12	75	42	13	7	24	7	10	1
5	TX10p	0	6	30	90	2	17	12	18	2	9
6	TX90p	6	1	80	39	17	3	17	15	10	1
7	TN10p	2	5	24	97	6	14	15	16	1	10
8	TN90p	4	3	83	37	16	4	15	16	10	1
9	CSDI	3	4	26	94	9	11	18	12	2	9
10	WSDI	5	2	76	43	16	4	15	17	9	2
11	SU25	22	2	78	43	17	3	20	11	9	2
12	TR20	12	13	68	49	13	6	15	16	8	1
13	DTR	18	7	53	68	15	4	17	15	4	7
14	FD0	0	1	8	5	1	4	10	16	1	0
15	ETR	13	11	61	60	11	8	12	19	5	6
16	CDD	2	17	54	89	13	42	10	21	8	3
17	CWD	9	10	86	57	25	30	19	13	2	9
18	RX1day	10	8	77	66	22	32	20	11	4	7
19	RX5day	10	10	77	63	28	27	22	8	5	6
20	R95p	12	8	78	64	32	23	21	11	3	7
21	R99p	11	9	81	61	25	30	18	12	5	6
22	R10mm	16	4	74	66	29	23	23	8	1	10
23	R20mm	14	6	64	80	28	27	23	8	2	9
24	PRCPTOT	14	5	73	69	33	23	22	10	2	9
25	SDII	10	8	60	80	24	31	19	13	8	3
26	R95PT	10	10	81	62	30	25	19	13	7	4

Note: Results shown in the above section are taken from GCISC Technical Report: Trends in Extreme Rainfall and Temperature Indices over South Asia (under publication).

was done under the APN project: "Development and Application of Climate Extreme Indices and Indicators for monitoring Trends in Climate Extremes and their Socio-economic Impacts in South Asian Countries" awarded to GCISC in 2007. Results drawn for Pakistan reveal that:

Temperature Indices

1. TXx (Warmest day temperature) and TXn (Coldest day temperature) have increased comparatively at more stations whereas TNn (Coldest night temperature) has increased at more than 70% of the stations. TNn has increased at 4 out of 5 stations in the Karakoram Range, the abode of glaciers.
2. TNx (Warmest night temperature) have dropped at a greater number of stations across the country, in particular, in the mountain north of Pakistan.
3. Warm days (TX90p) and Warm nights (TN90p) have increased at about 60-70% of the stations; however, warm nights have dropped at 6 out of 7 stations in the greater Himalayan region.
4. Cool nights (TN10P) and cool days (TX10P) have decreased at 60% of the stations across the country; however, cool nights have increased in the greater Himalayan region at around 70% of the stations.
5. CSDI (Cold Spell Duration Index) and WSDI (Warm Spell Duration Index) both have an overall mixed trend yet CSDI has an increasing trend at 5 out of 7 stations in the greater Himalayan region.
6. DTR (Diurnal Temperature Range) overall has a mixed trend; however, it has decreased in the monsoon rains receiving regions (1b, III and IV). In Zone-II, the winter dominated region, the DTR has increased at all the stations.
7. TR20 (Tropical nights) has an overall increasing trend at comparatively more stations, but the decreasing trend is seen at 6 out of 7 stations in the mountainous north. It has a general increasing trend in other parts of the country.
8. SU25 (Summer days) shows increasing trend at around 80% of the stations.

Precipitation Indices

1. The precipitation indices, RX1day, RX5day, R10mm, R20mm, R95p and R99p have a general increasing trend at more stations.
2. PRCPTOT has increased at around 70% of the stations. Around 78% of the stations, however, show increasing trend in the mountainous north i.e. Zone I (a) and the monsoon rainfall receiving parts i.e. Zones (1b, II and IV).
3. SDII has a mixed trend across the country, but the Greater Himalayas and the sub-montane region has 10 out of 15 stations where this index has increased.

4. CDD (Consecutive Dry Days) have decreased at around 65% of the stations, in almost all the zones. Almost 60% of the stations show an increasing trend in CWD (Consecutive Wet Days).

Frequency of Temperature and Precipitation Extremes

Analysis carried out for 52 meteorological stations in Pakistan for highest temperature and heaviest rainfall events recorded during the 24 hours over the entire 40-year period (1961-2000) are shown with decadal frequency of occurrence in Tables 8 and 9.

Table 8: Frequency of daily temperature extremes during different decades (1961-2000)

Period	1961-70	1971-80	1981-90	1991-2000
No. of stations with highest daily temperature	4	12	16	20

Table 9: Frequency of heaviest 1-day precipitation during different decades (1961-2000)

Period	1961-70	1971-80	1981-90	1991-2000
No. of stations with highest daily precipitation	6	18	11	17

ENSO Tele-connections

El-Nino and La-Nina episodes based on Tahiti minus Darwin pressure and on the Nino 3.4 temperature index are related to the rainfall departures from the mean of past 50 years (1951-2000) for the monsoon region. Some six out of eight events before 1990 are well associated with deficient rainfall and the two that occurred during 1994 and 1997 showed a positive relationship (Table 10, Fig. 5b). The available records for the period 1971-2000 of monsoon depressions from the Bay of Bengal indicate almost no depressions reaching

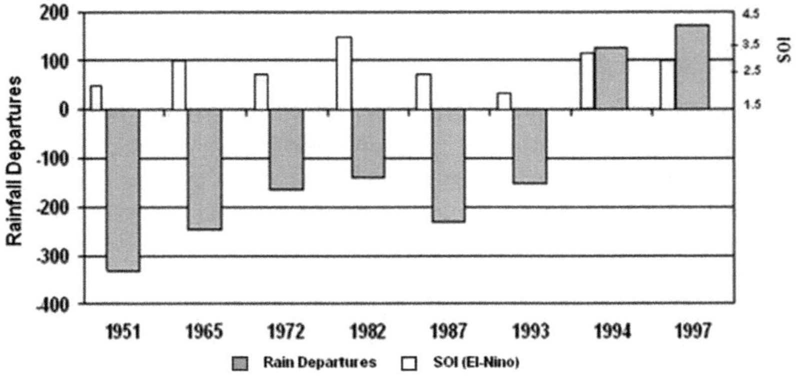

Fig. 5(b): Association of El Nino events with rainfall departures.

Pakistan during the warm ENSO phase (Table 11). La-Nina episodes show a mixed relationship with the rainfall departures (Table 12, Fig. 5c).

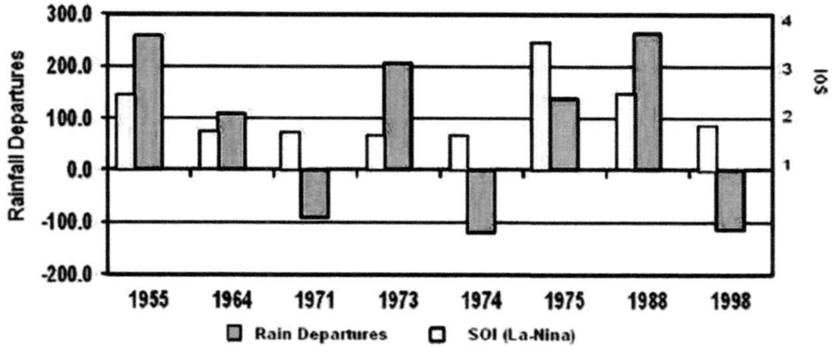

Fig. 5(c): Association of La-Nina events with rainfall departures.

Table 10: Association of El-Nino events with rainfall departures (monsoon region)

Year	*SOI (El-Nino)*	*Rainfall departure*
1951	1.9	-331.3
1965	2.7	-244.1
1972	2.4	-162.8
1982	3.5	-139.0
1987	2.4	-229.8
1993	1.8	-151.5
1994	3.0	126.7
1997	2.6	173.1

Table 11: Monsoon depressions reaching Pakistan during the El-Nino years

Year	*Depressions from Bay of Bengal*	*Depressions from Arabian Sea*	*Depressions reaching Pakistan*
1972	5	-	Nil
1982	7	-	Nil
1987	1	-	Nil
1993	1	-	Nil
1994	-	-	Nil
1997	6	-	2

Table 12: Association of La-Nina events with rainfall departures (monsoon region)

Year	*SOI (La-Nina)*	*Rainfall Departure*
1955	2.4	258.0
1964	1.7	108.6
1971	1.7	-91.7
1973	1.6	206.1
1974	1.6	-119.8
1975	3.5	135.5
1988	2.4	261.4
1998	1.9	-112.8

2.3 Nepal

Out of 147,181 km^2 total area of Nepal, about 86% area comprises hills and mountains and remaining 14% are flat lands. Within south-north range of 145 km to 241 km, the altitudinal variations ranged from ~60 m above mean sea level in the southern plain (called Tarai) to Mount Everest (8848 m) in the northeast. Such variation in altitude results in a spectrum of climatic conditions ranging from subtropical to alpine/arctic. These are manifested in the presence of diverse habitats, vegetation and fauna.

Numerous studies on the climate of Nepal have been carried out over the last four decades (Chalise, 1994; Domroes, 1979; Hormann, 1994; Jha, 1996; Kripalani et al., 1996; Malla, 1968; Mani, 1981; Nayava, 1974, 1980; Parthasarthy, 1958; Ramaswamy, 1962; Rawson, 1963; Ramage, 1971; Rao and Desai, 1973; Robinson, 1976; Shrestha et al., 1999, 2000; Shrestha, 2000; Yoshino, 1984). The southwest monsoon contributes to over 80 percent of the annual rainfall in Nepal during June-September. The lower troposphere wind from east-south-easterly direction over entire Nepal is the most common circulation feature during this season. However, during the break monsoon in India, when the monsoon trough shifts to foot-hills of northern mountains, the westerlies dominate the lower circulation pattern over Nepal and release significant quantities of rainfall. The remaining amount of the annual precipitation is predominantly contributed by the passage of the extra-tropical circulation, called western disturbances. The climate of Nepal is, thus, controlled by altitude, topography, and the seasonal atmospheric circulations.

Temperature Variability

Spatial variation of surface air temperature, in general, follows the topography of Nepal and shows a variety of trends in seasonal patterns. The western part of Nepal has a slightly larger temperature range between summer and winter than the eastern part, due to the combined effects of northward tilt of the topography and more continental climate in the west. The annual cycles of all-Nepal maximum and minimum temperatures indicate that the monsoon circulation reduces the difference between maximum and minimum temperatures. The minimum surface temperature rises steadily from the months of January to July/August followed by a trend reversal from September to December each year. However, the maximum surface temperature is modified by the monsoon system and is consistent with the South Asian modification (Pant and Rupa Kumar, 1997).

The time series of all-Nepal maximum and minimum temperatures including a linear regression of data are presented in Fig. 6a. Figure 6b provides the Kathmandu temperatures, all Nepal trends and global reconstructions (Shrestha et al., 1999). Besides interannual variations, an increasing trend in both the maximum and minimum temperatures is also

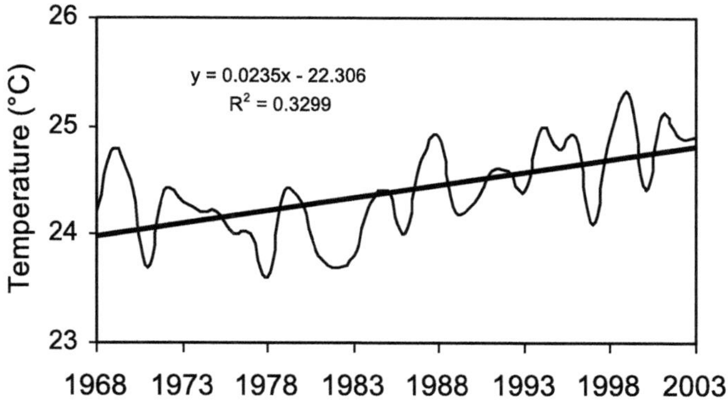

(i)

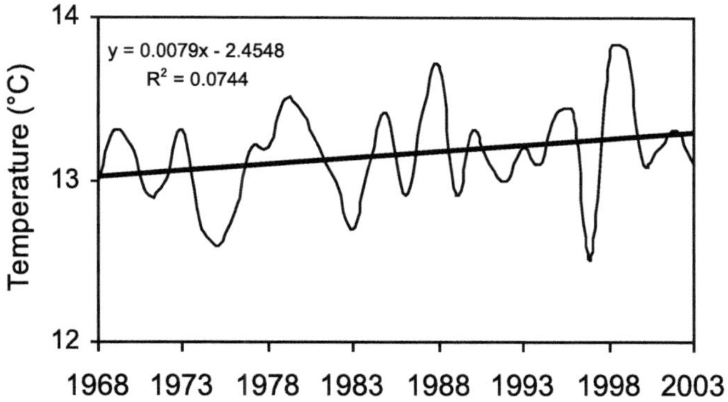

(ii)

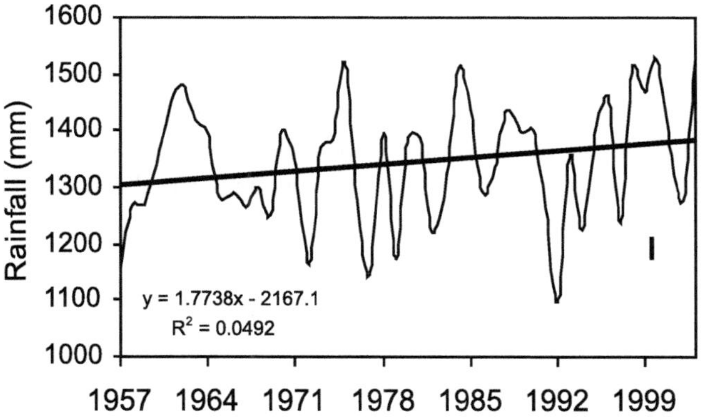

(iii)

Fig. 6(a): All Nepal time series of (i) maximum temperature, (ii) minimum temperature and, (iii) precipitation.

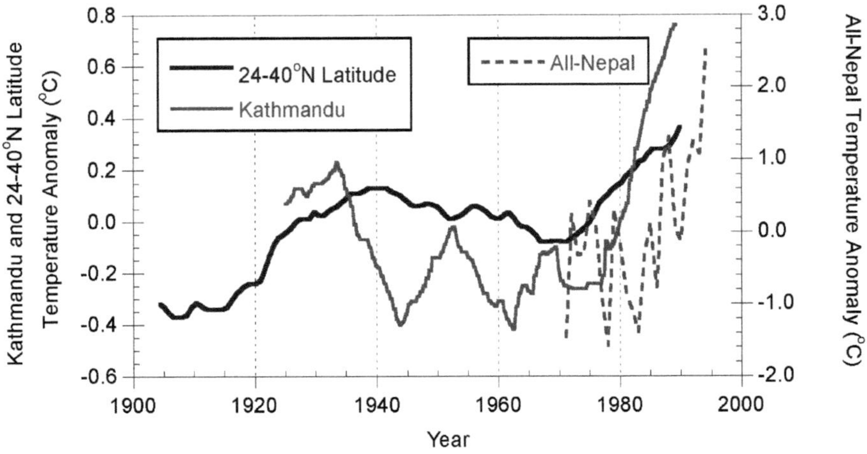

Fig. 6(b): Comparison of Kathmandu temperature with all Nepal temperature and global trends. Despite significant sub decadal fluctuations, an overall correspondence with the global trends is seen (Shrestha et al., 1999).

observed. The average warming in annual temperature between 1977 and 1988 was 0.06°C/year. The warming was more pronounced in the higher altitude regions and lower in the lower altitude Terai region, where also a trend is seen. Also, the increase is a little higher in maximum than in minimum temperature series. The temperature data for Kathmandu, when compared with the global data in the latitude belt 24-40°N, show a general similarity between the two series i.e. an overall decreasing trend from 1940-1970 and a monotonous increase thereafter. Given that the Tibetan plateau also shows similar trends (Liu and Chen, 2002), it seems that the warming has a more pronounced effect on the higher altitude regions compared to lower altitude ones. Since Nepal is a mountainous country, the temperature variation with altitude also plays an important role in vegetation and other aspects of social life. The lapse rate of the maximum temperature is always higher than that of the minimum temperature. The difference between the highest and the lowest lapse rates for maximum and minimum temperatures are observed during monsoon and pre-monsoon seasons, respectively. The highest lapse rate is during pre-monsoon season with the temperature being the maximum. However, the lowest lapse rate is seen during winter season with the minimum temperature.

Rainfall Variability

Strong spatial and temporal variations exist in the rainfall distributions of Nepal (Shrestha et al., 1999, 2000; Shrestha, 2000). The seasonal mean rainfall is highest during summer monsoon season and lowest during winter. Pre- and post-monsoon thunder activities and the occasional passage of the western disturbances make rainfall during these periods a little higher than in winter. However, the variability is highest during post-monsoon and lowest during monsoon seasons. Although, the variability of the monsoon rainfall is small, the anomalies in either sides of this rainfall have severe impacts on the socio-economics of the country. All-Nepal summer monsoon rainfall time series (1957-2003) is presented in Fig. 6a and this illustrates that the monsoon rainfall has both, interannual variation and a increasing trend of about 20% of the average per decade. The extreme monsoon rainfall events are also analyzed and the results show that, over a period of 47 years, seven droughts and eight flood conditions associated with intraseasonal variation of monsoon rainfall occurred, which had a direct impact on both agriculture and water resources. Table 13 provides an assessment of the temperature and rainfall changes in Nepal based on an assessment of 12 recent general circulation models (OCED, 2003).

Table 13: GCM estimates of temperature and precipitation changes in Nepal

	Temperature change (°C) mean (standard deviation)			*Precipitation change (%) mean (standard deviation)*		
Year	*Annual*	*DJF**	*JJA***	*Annual*	*DJF**	*JJA***
Baseline average				1433 mm	73 mm	894 mm
2030	1.2 (0.27)	1.3 (0.40)	1.1 (0.20)	5.0 (3.85)	0.8 (9.95)	9.1 (7.11)
2050	1.7 (0.39)	1.8 (0.58)	1.6 (0.29)	7.3 (5.56)	1.2 (14.37)	13.1 (10.28)
2100	3.0 (0.67)	3.2 (1.00)	2.9 (0.51)	12.6 (9.67)	2.1 (25.02)	22.9 (17.89)

* DJF refers to December, January and February.
** JJA implies June, July and August.

Glacier Fluctuations and Water Discharge

There is an overwhelming evidence of shrinking glaciers in the Himalaya. As the glaciers are an important water resource for Nepal and India, their widespread shrinkage has a large impact. Analysis of river flow data, however, does not show such a trend (Fig. 6c). Figure 6d clearly establishes that since 1976 the glaciers have been retreating and this has accentuated significantly since the late 1990's (Fig. 6d). Further, Asahi et al. (2000) studied glacier fluctuation in Ghunsa-Khala region of the Kanchenjunga area. Using aerial photo interpretations they compared the glaciers of 1958 with those of 1992. In this region out of 57 glaciers, 50% retreated, 38% were static and 12% were advancing. The state of a glacier is determined by precipitation and temperature and changes in them. The fact that a major fraction does show a retreat implies that, either of temperature increase (estimated as ~0.02°C/year) or more possibly a reduced precipitation has played a significant role in glacier retreats.

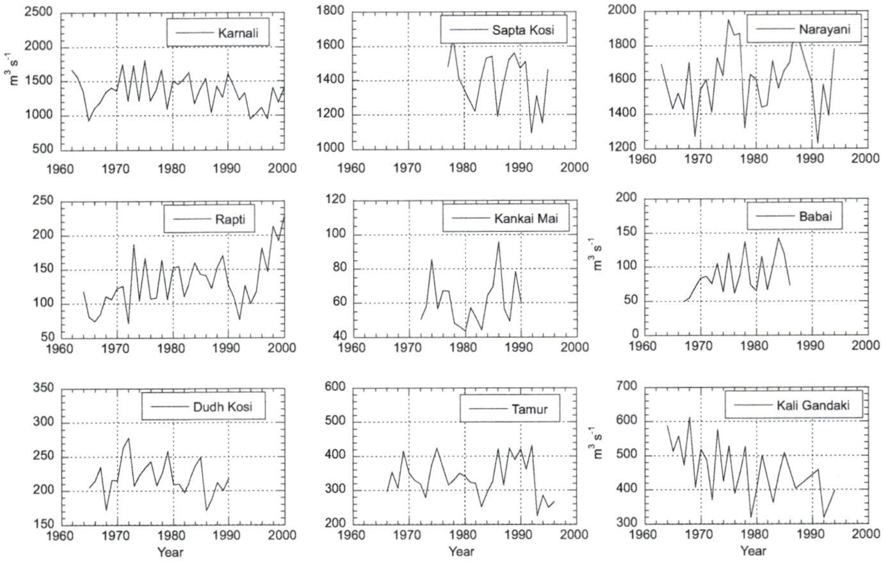

Fig. 6(c): Discharge data of selected rivers: Top row, Large Rivers; middle row, southern rivers; and bottom row, snowfed rivers. *Source:* DHM (1996).

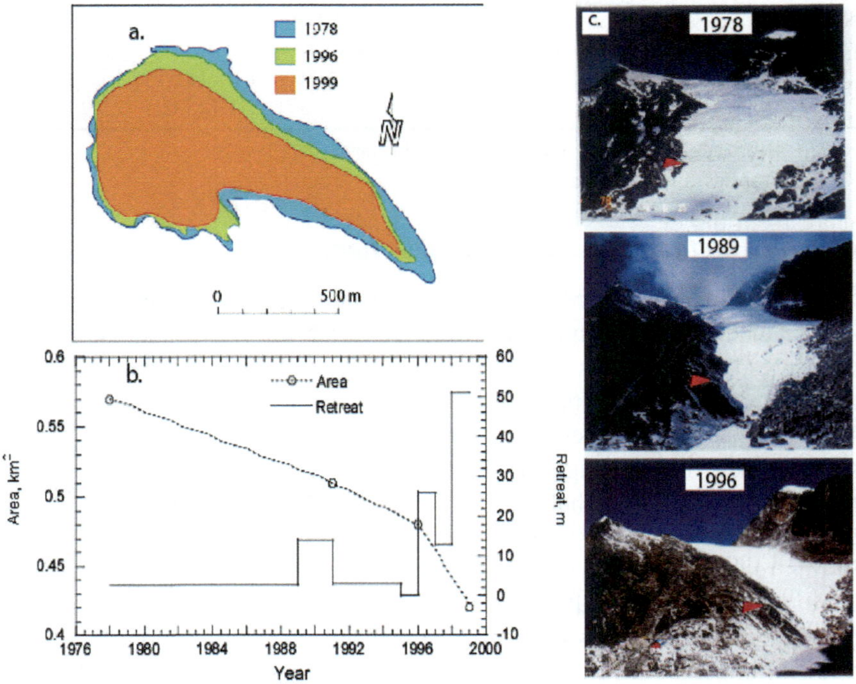

Fig. 6(d): The changes in the glacier area and terminus retreat for Shornng Himal glacier. The retreat from 1978 to 1989 was 30 m, which is equivalent to 12 m thinning of the glacier surface.

Over 6000 rivers and rivulets flow over Nepal. A comprehensive analysis of trends in river flow has not been performed yet. Nevertheless, a preliminary analysis of river discharge was carried and this shows trends in rivers of three categories: large outlet rivers, southern rivers and snow fed rivers (Fig. 6c). Among the large rivers, Karnali and Sapta Koshi show a decreasing trend although the record of Sapta Koshi is of a limited duration. In contrast another large river, Narayani shows increasing trend. The southern rivers do not show any trend. All of the three snow fed rivers examined here exhibited a declining trend in discharge. It is clear that trends observed in the river discharge are neither consistent nor significant in magnitude. It has to be noted that the ambiguity is due to limited duration of the record and high inter-annual variability in discharge data. Nevertheless, another independent study suggests that the number of flood days and consecutive days of flood events are increasing (Shrestha et al., 2003).

2.4 Bangladesh

Bangladesh is located within the 20°34'-26°38' N and 88°01'-92°42'E and occupies an area of 147,570 km^2. All the major rivers of Bangladesh originate in the Himalaya and flow over thousands of kilometres over several countries. 92% of the catchments of the rivers flowing through Bangladesh lie outside Bangladesh. Thus, the rainfall that occurs over the 92% areas outside the country ultimately drains through 8% of the catchments in Bangladesh. Consequently, the floods over Bangladesh are caused by excess rainfall both inside and outside Bangladesh. Bangladesh has a monsoonal climate with its principal rainfall in the southwest monsoon season (June-September).

Temperature Variability

The temperature and rainfall trends discussed below are based on a statistical analysis of the grided dataset for the period 1961-2004 (44 years). Data from 30 meteorological stations was used for generating the grided data. The climate variability in Bangladesh has been analyzed by Choudhury et al. (1997, 2003), Karmaker and Shrestha (2000), Kripalani et al. (1996) and Quadir et al. (2002, 2003, 2004). The variability and trends of temperature and rainfall of the region indicate that the maximum temperature has bimodal characteristics with maximum in April and secondary maximum in September. The relatively low temperature between April and September indicates the impact of monsoon rainfall on the thermal regime. Minimum temperature shows a unimodal distribution, with a maximum in August. The amplitude of annual variation of the minimum temperature is higher than that of maximum temperature. Besides, the diurnal variation of temperature is higher in the winter than in the summer. The time series analysis of seasonal and annual mean minimum and maximum temperatures shows strong inter-annual fluctuations over 2-4, 5-7 years and decadal time scales (Fig. 7a). The late eighties and the nineties were warmer and 1999 was the warmest. The annual mean minimum and maximum temperatures exhibit an increase at the rate of 0.09°C/decade and 0.07°C/decade which is significant at 1% level (Table 14).

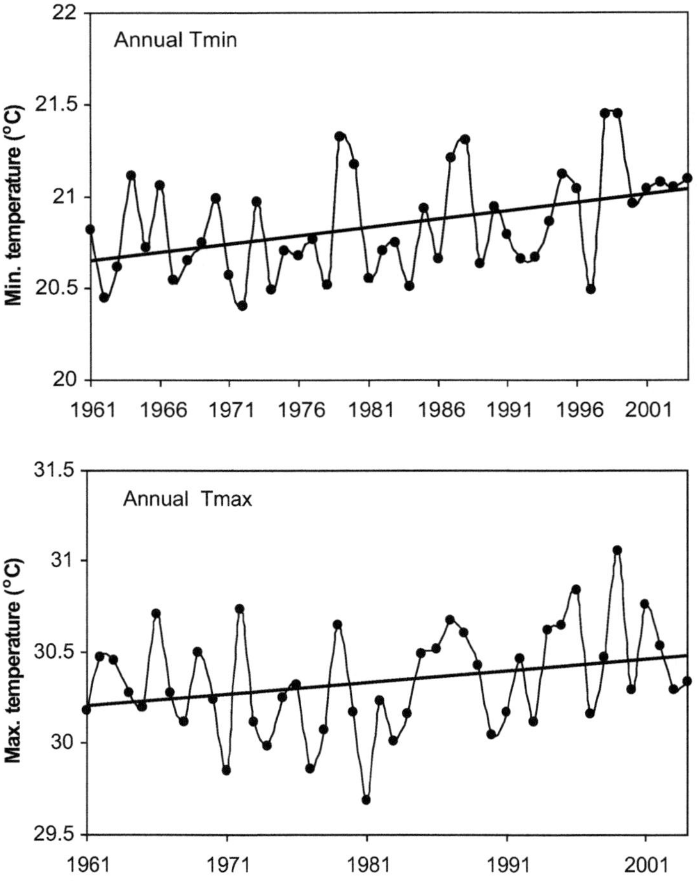

Fig. 7(a): All Bangladesh time series of minimum and maximum temperature from 1961-2004 (44 years). The trends are shown by solid straight lines.

Table 14: Trends of minimum and maximum temperature (°C/decade) and rainfall (% of normal per decade) during the period 1961-2004 in Bangladesh

Parameter	*Winter*	R^2	*Pre-monsoon*	R^2	*Monsoon*	R^2	*Post-monsoon*	R^2	*Annual*	R^2
Tmin	0.13*	0.10	0.07	0.03	0.07*	0.13	0.10	0.04	0.09**	0.18
Tmax	-0.03	0.10	-0.10	0.03	0.18**	0.44	0.23**	0.34	0.07*	0.09
Rainfall (%)	7.65	0.02	7.02*	0.08	0.73	0.01	3.41	0.01	2.26*	0.06

* 5% significance level ** 1% significance level

Sea Level Variations

A linked aspect is the sea-level variations in the Bangladesh coast which shows a rise at the rate of 4.0 mm/year in its western side (at Hiron Point), 6.0 mm/year at the Meghna estuary (at Char Changa) and 7.8 mm/year in the southeastern side (at Cox's Bazar) (Khan et al., 1999; Singh et al., 2000). These rates are 3-5 times higher than the 100-year global rate of sea level rise of 1-1.5 mm/year. These are substantive and worrisome.

Rainfall Variability

The winters are dry. The rainfall gradually increases in the pre-monsoon season. The pre-monsoon rainfall is from frequent thunderstorms/local severe storms, locally termed as 'Kalbaisakhi'. Numerous tornadoes with high destructive force also occur in this season. The convective activities associated with thunderstorms are due to the strong atmospheric instability caused by the passage of the subtropical westerly troughs carrying relatively cool air masses above the low level hot and moist air masses carried from the Indian Ocean by the low level southerlies. Additionally, tropical disturbances in the form of depressions occur during monsoon season and tropical cyclones originate in the Bay of Bengal occur during April, and May and October-December. The seasonal statistics indicate that over 70.62% of annual rainfall occurs during the monsoon and over 18% during pre-monsoon seasons.

The rainfall over Bangladesh shows a large spatial variation. The eastern and northeastern parts experience highest rainfall and the lowest is in the central-west. The average annual rainfall varies from 1429 mm in the central-west to 4338 mm in the northeast (Quadir et al., 2006). Temporal variability of the country average monsoonal rainfall is shown in Fig. 7b. The dashed lines show the upper and lower limits of standard deviation and the thick straight lines represent the linear trends. The trend analysis shows that the pre-monsoon and annual rainfall exhibit increasing trends of 7.02% and 2.26% per decade respectively. The monsoon rainfall (ABMR) shows a

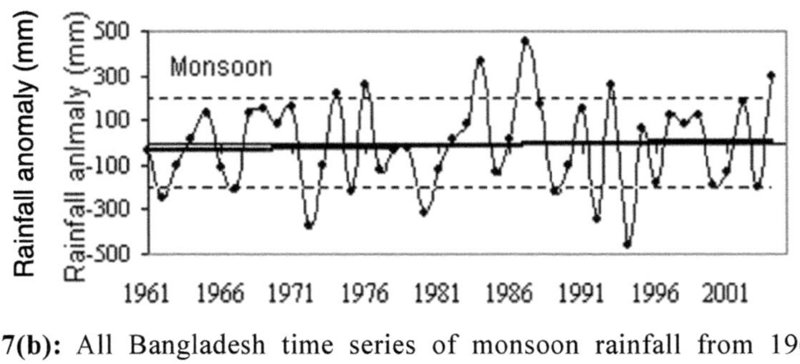

Fig. 7(b): All Bangladesh time series of monsoon rainfall from 1961-2004 (44 years). The dotted lines show the limits of standard deviation, the thin solid straight lines present the trend and the thick solid lines give five years moving average.

trend of 11.9 mm/decade i.e. 0.73% per decade. Spatially, the rainfall trends are not homogeneous (see Choudhury et al., 2003 for more details).

Analysis of the rainfall data from 1939-1977 of 14 stations by Kripalani et al. (1996) did not exhibit any correlation with: (i) Northern Hemispheric Surface Temperature (NHST), (ii) Darwin Pressure Departure (DPD) and (iii) position of the 500 hPa subtropical ridge along 75°E. Neither the ABMR nor its regional subsets exhibited any correlation. ABMR has a correlation of 0.5 with the average rainfall over NE India. A negative correlation (–0.3) has been observed with the rainfall in the northern part of the east coast (20°N and 80°E) which extends to the northwest up to northwestern India. These results indicate some non-homogeneity among the monsoons of India and Bangladesh.

Cyclonic Storms

Well-documented data on cyclonic storms, monsoon depressions and low-pressure systems over the Indian region, extending to more than a century in the past are available (Quadir, 2006). Table 15 provides the classification of tropical disturbances in the region. It distinguishes between cyclonic disturbances by their speeds; cyclonic storms (CS ~ speed 62-88 km/hr), severe cyclonic storms (speeds > 88 km per hour) and storms with hurricane intensity (speeds of 118 km/hr or more). Mitra et al. (2003) reported that annual average of cyclonic disturbances in the north Indian Ocean is 15.7 with a standard deviation of 3.1. The annual average of tropical cyclones is 5.6 with a standard deviation of 1.85. The annual frequency of severe cyclonic storms with a core of hurricane winds is about 1.3. Figure 8 (a-c) plots the five-year running average of annual frequency of cyclonic disturbances along with all-India mean annual surface temperature which since 1970 indicates a significant decreasing trend of tropical disturbances (Lal, 2001; Patwardhan and Bhalme, 2001; Quadir, 2006). The annual frequency of CS+SCS and individual trends for CS and SCS of the Bay of Bengal are indicated in Fig. 8 (a-c), which indicates that substantial year-to-year variation occur and no clear long-term trend is discernible in the frequency of CS + SCS.

Table 15: Classification of cyclonic storms

Country	*Types of disturbances*	*Wind speed km/hr)*
Low and	1. Low	<31
Depression	2. Depression	31-50
	3. Deep Depression	51-61
Tropical Cyclone	4. Cyclonic Storm	62-88
	5. Severe Cyclonic Storm	89-117
	6. Severe Cyclonic Storm with hurricane intensity	118-221
	7. Super Cyclone	>222

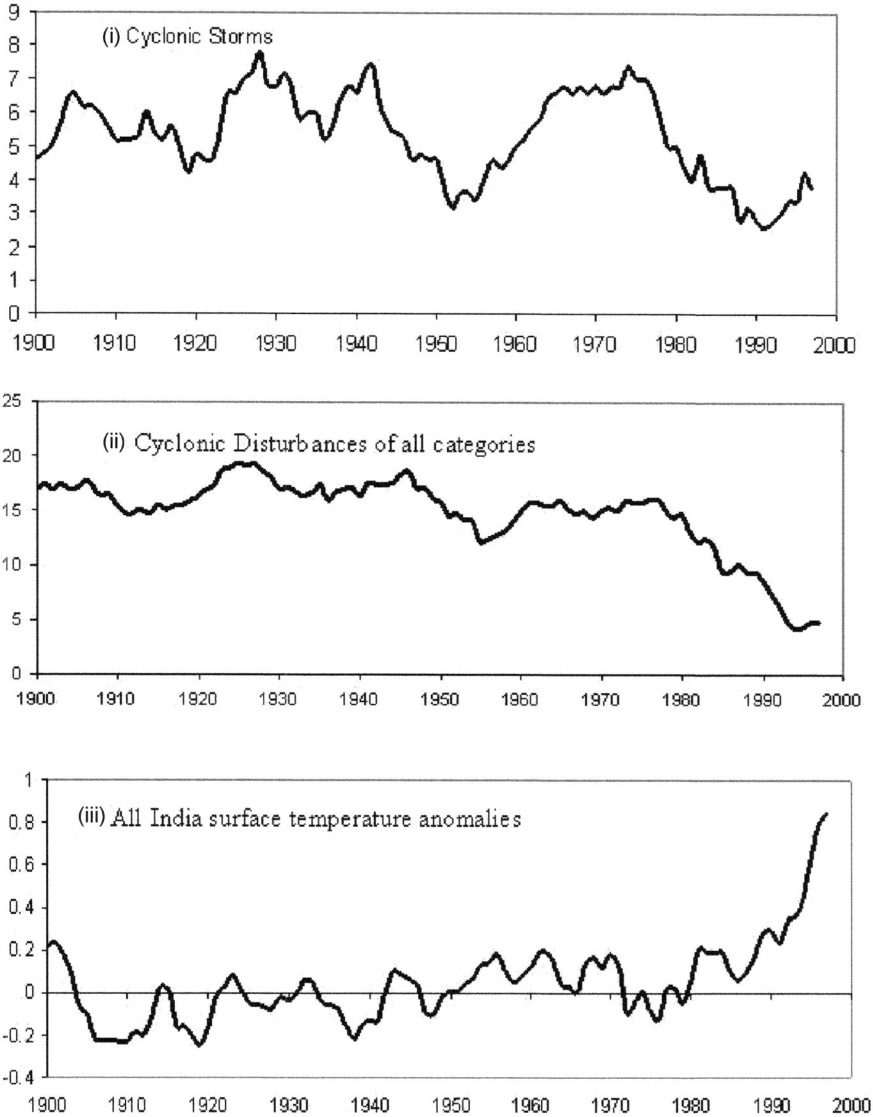

Fig. 8(a): The five-year moving average of cyclonic storms; (i) cyclonic disturbances (cyclones and depressions); (ii) of the north Indian ocean (Bay of Bengal and Arabian Sea) and (iii) all-India mean annual surface temperature [the data has been extracted from Lal (2001) and plotted by Quadir (2006)].

During the recent 3-4 decades however, the frequency of severe cyclonic storms around November has consistently increased and has nearly doubled to a value 2.3 from 1.3 during November. Thus, while the total frequency of cyclonic storms that formed over the Bay of Bengal has nearly remained constant during 1877-1997, there appears to be a conspicuous increase in the frequency of 'severe' cyclonic storms during November. It is equally possible that this apparent increase could be a manifestation of the better cyclone monitoring technology in the recent period rather than a real change in the frequency of severe cyclonic storms. However in the absence of any abrupt change in the frequency since the beginning of satellite era in mid-1960, this factor can be discounted. In fact the increase predates the beginning of observational era and was also seen in the 1920 (Dvoark, 1984; Singh et al., 2001). The time series of the frequency of CS and SCS of the Bay of Bengal (Fig. 8(b; c)) indicates both short- and long-term variability. The decrease of CS and almost simultaneous increase of SCS in the Bay of Bengal implies that a larger number of disturbances intensified from the stage of CS to SCS from around 1937. The cause of such increase of SCS is not understood, but may have some link with the increase of SST as well as surface air temperature. It is almost certain that an increase of SST is accompanied by

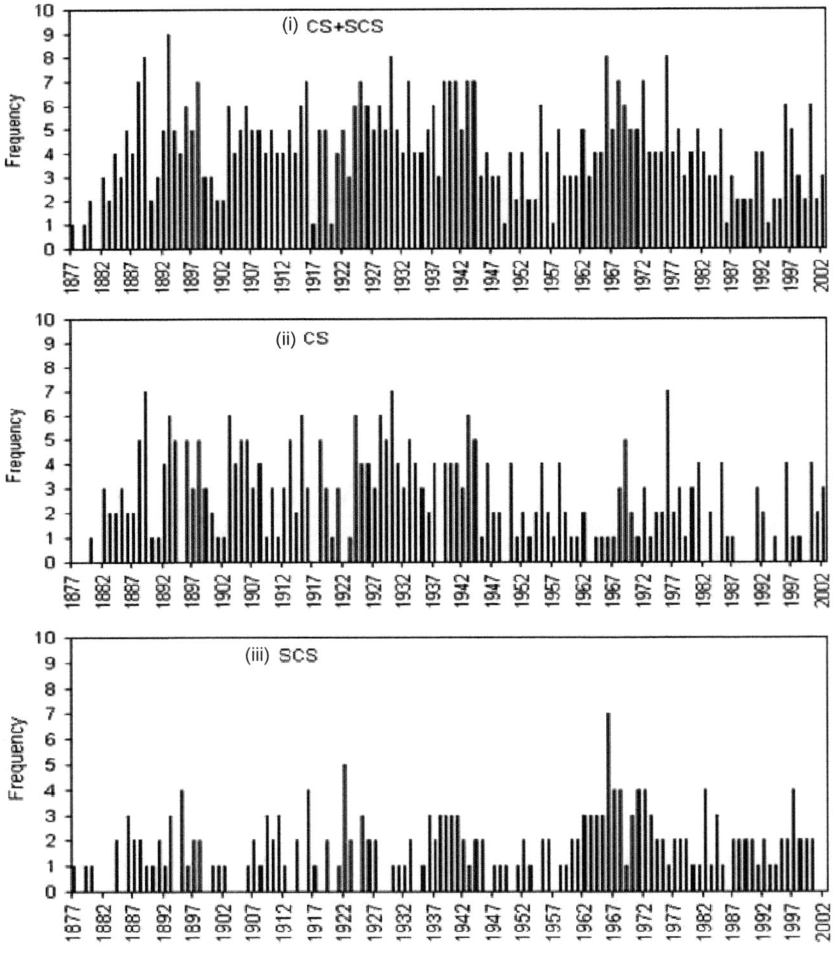

Fig. 8(b): The inter-annual distribution of the tropical cyclones of Bay of Bengal: (i) CS+SCS, (ii) CS and (iii) SCS (Quadir, 2006).

corresponding increase in cyclone intensity. Higher SST enhances the evaporation causing higher supply of energy in the form of latent heat through the boundary layer. Emanuel (1987, 2005) has developed a relationship between maximum sustained wind speed (MSWS) and SST which suggests that a 1° C rise of SST will increase the MSWS by 4%, 2°C by 10% and 4°C rise by 22%. The wind stress on the water surface increases as the square of the wind speed. Thus the impact of enhanced SST on wind speed vis-à-vis on storm surge height would be quite large. Considering 2°C and 4°C rise of SST, Ali (1996) developed scenarios of storm surge heights with no sea level rise, and it was found that storm surge would increase by 21% and 49% with respect to the present.

Based on the data of tropical disturbances from 1977-2003, the vulnerability of the coastal zones of the Bay of Bengal to tropical cyclones has been compiled (Karmaker, 1998; Alam et al., 2002, 2003; Quadir, 2006). It is seen that out of 532 tropical cyclones, formed during the past 123 years, 16.2% hit Bangladesh, 58.8% went to the east coast of India, 10.9% hit Mayanmar and 3.4% dissipated. The storm surge information used in the present study has been obtained from Karmaker (1998), Dube et al. (2004), Dube (2005) and Prasad (2005). Karmaker (1998) has produced the vulnerability of the coastal zone of the Arabian Sea (the west coast of India and coast of Pakistan) using the landfall data of 39 major cyclones during 1618-1995 for which complete data were available. The results show that the Gujarat coast is most vulnerable to tropical cyclone followed by the Maharashtra coast. Vulnerability map of storm surges in the coastal zone are provided by Prasad (2005). Based on peak storm surge, the entire coast line of India and Bangladesh has been divided into four zones, i.e., very high risk, high risk, moderate risk and minimal risk corresponding to surge height of >5 m, 5-3 m, 3-1.5 m and <1.5 m. Detailed reviews of numerical modeling of storm surges in the Indian oceans is given by Ghosh (1977), Ali (1979), Murty (1984), Murty et al. (1986), Dube et al. (1997) and Dube (2005). The model of Dube is widely used for prediction of storm surges associated with a land falling cyclone.

Two interesting results were reported by Mandal (1989, 1990) regarding the recurving and landfall properties of the tropical cyclones of the Bay of Bengal with respect to ENSO activity. In the north Indian Ocean, about 35% of the cyclones are of recurving type and 65% of the cyclones are non-recurving type. The analysis of the tracks of the cyclones for seven severe El Nino years during 1901 to 1990 showed that in the El Nino years 39% and 61% are found to be recurving and non-recurving types. Based on the analysis of cyclone tracks during the period 1966-1988, it was observed that tropical cyclones forming in the Bay of Bengal mostly cross the Indian coast south of 20°N in the El Nino years. In La-Nina years most of the cyclones crossed coastal regions north of 20°N thereby affecting West Bengal and Bangladesh. Singh et al. (1987) have shown that there is an increase of the frequency of tropical cyclones in the low latitude during the negative phase of southern oscillation. Mandal (1989) and Mukherjee (1990) studied the relation of Quasi Biannual Oscillational (QBO) phases with tropical cyclones of north Indian Ocean, and showed that the storm genesis would be more favourable during westerly phase of QBO. The relationship of QBO with the tropical cyclone behaviour in different ocean areas has been investigated by various workers.

Although significant advances have been made and models developed to understand recent cyclone behaviour in the Bay of Bengal, little has been done to examine a relation between global warming trend and cyclone intensities and frequencies. It is believed that the increased surface air temperatures due to greenhouse warming will influence both the SW monsoon and cyclone behaviour in the years to come. Thus, there is a need to develop models, not only for monsoon rains but also for cyclones.

3. PALEORECONSTRUCTIONS

3.1 The Land Records

Tree Rings

Himalaya influences the regional and extra regional circulation system. High-elevation mountain heating, snow/glacier albedo and glacier melt fresh water flux into the ocean provide strong land-ocean-atmosphere interactive system. Consequently, long-term, high-resolution climate records for this data-scarce region are crucial for an improved insight into the climate, both on regional and global scale. This region displays a succession of tropical to sub-alpine vegetation from low to high elevations. Many conifer species, important constituent of subtropical to sub-alpine vegetation, some of which are known to grow for a millennium or more, provide valuable proxy data for long-term climatic reconstructions. Since the late seventies (Pant, 1979), several precisely-dated tree-ring chronologies from western, central and eastern Himalaya for climatic studies have been developed and are discussed below.

The Western Himalaya

Several century long chronologies of *Abies pindrow*, *A. spectabilis*, *Cedrus deodara*, *Pinus roxburghii*, *P. wallichiana*, *Picea smithiana*, *Taxus baccata* etc. from different geographic regions in the western Himalaya have been prepared e.g. Bhattacharya et al., 1988; Hughes and Davies, 1987; Borgaonkar et al., 1994, 1996; Yadav and Singh, 2002; and Singh et al., 2004. Singh et al. (2004) prepared a 1198 years tree-ring chronology (AD 805-2002) of *Cedrus deodara* and later extended to AD 747-2003.

These studies show that for most of the species, the tree growth is favoured by cool and wet conditions during the pre-monsoon period, prior to the onset of growing season. The climate reconstructions so far are for mean temperature of the pre-monsoon season (Borgaonkar et al., 1996;

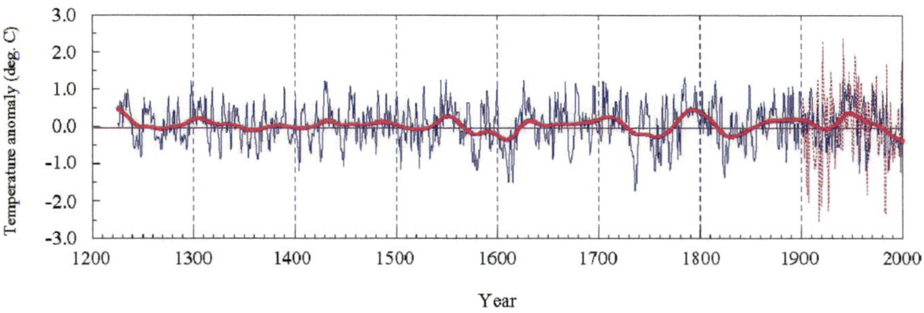

Fig. 9(a): Pre-monsoon summer temperature reconstruction (AD 1226-2000) for western Himalayas. Anomalies are relative to 1961-1990 mean. The dotted curve is instrumental data superimposed over the reconstructed one. The thick smooth line represents the 50-year spline curve with 50% frequency response cut off (From Yadav et al., 2004).

Hughes, 1992; Yadav et al., 1997, 1999; Yadav and Singh, 2002; Yadav et al., 2004). Further, irrespective of the length of record, the climatic reconstructions consistently show a periodicity on decadal to inter decadal scale. The longest mean pre-monsoon temperature reconstruction is from a network of 16 homogeneous sites in the western Himalaya, each >500 years, and the maximum extending to 1226 AD (Fig. 9a) (Yadav et al., 2004). This mean premonsoon temperature reconstruction (AD 1226-2000) shows variability during the Little Ice Age (LIA). Cool episodes around 1573-1622, 1731-1780, and 1817-1846 and the LIA cooling in the western Himalaya can be detected. The temperature decline began since 1560s and reached their lowest values between 1590 and 1610. This cool episode coincides with the extension of glaciers in the Himalaya and plausibly represents the onset of little ice age (LIA) in the region. The 18th century has been reported to be relatively warmer in high-latitude northern hemispheric regions. This period witnessed a cooler regime in the western Himalaya. The 18th and early 19th century cooling episodes in the western Himalaya are contemporaneous with cooling in Nepal, Tibet, Central Asia and Karakoram (Briffa et al., 2001; Esper et al., 2002; Cook et al., 2003). Such a consistent pattern in temperature variability in the Asian mountain region suggests the relevance of regional forcing factors.

The premonsoon temperature reconstruction does not show warming towards the late 20th century. This is similar to other tree-ring based temperature reconstruction from the Asian region. Analyses of century long meteorological data of stations from the western Himalaya show steady warming in maximum temperature throughout the 20th century. However, contrary to this the minimum temperature shows a warming trend up to 1950's and a decrease, thereafter. Reduction in minimum temperatures since 1960's outbalances the warming trend in maximum temperature and, this leads to slight cooling trend in mean temperatures after the 1950's. As the

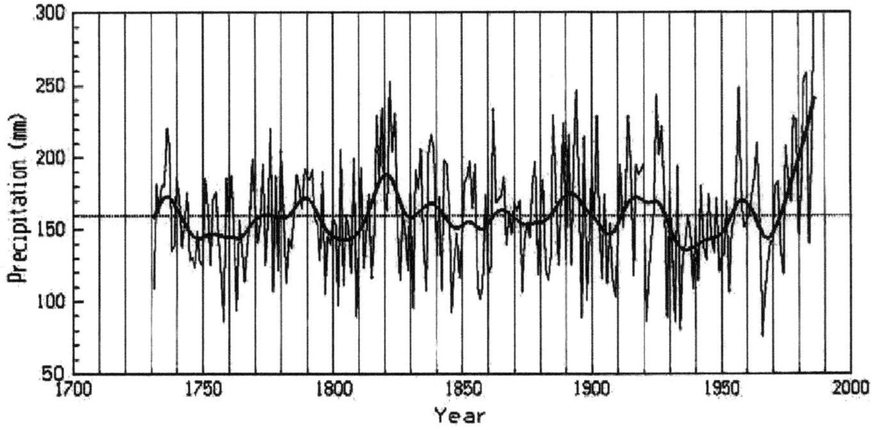

Fig. 9(b): Mean pre-monsoon precipitation reconstruction in western Himalayas. The smooth line represents the cubic spline designed to remove 50% of the variance in a sine function with wavelength of 20 years (From Singh and Yadav, 2005).

tree rings respond to mean temperature only, the maximum temperature has not been reconstructed. This is to be contrasted with the recession of glaciers and extension of tree line to upper elevations (Dubey et al., 2003), which respond to both the maximum temperature and precipitation.

Reconstructions for the paleo-precipitation for the Himalayan region are few (Hughes, 1992; Borgaonkar et al., 1994; Yadav and Park, 2000; Singh and Yadav 2005). The large topographic relief of the region induces corresponding spatial variability in precipitation. The precipitation records of low elevation weather stations usually do not adequately represent the precipitation over high elevation tree-ring sites. For this reason, tree-ring data from sites far from the weather stations leads to poor calibrations. However, using network of tree-ring data from 15 moisture stressed, homogeneous sites in the region, Singh and Yadav (2005) reconstructed premonsoon precipitation for the period AD 1731-1986. The reconstruction showed strong decadal to inter-decadal scale variability in precipitation (Fig. 9b). This reconstruction revealed lowest and highest 20-year mean precipitation during the 20th century. An increasing mean premonsoon precipitation trend since 1970s coincides with a decrease in summer monsoon rainfall over the Indian region. Winter and premonsoon precipitation over the Himalayan region have inverse relationship with subsequent summer monsoon rainfall over India. This implies that such long-term information on premonsoon precipitation could be useful in understanding the intricacies of monsoon system and its predictions.

Central Himalaya

Tree rings of several conifers viz., *Abies spectabilis, Pinus wallichiana, Cupressus dumosa,* and *Pinus roxburghii* reflecting diversified ecological conditions of Nepal were used to develop tree-ring chronologies of these taxa (Bhattacharya et al., 1992) and regional temperature of past several centuries for two seasons, viz. for February-June (AD1546-1991) and October-February (AD1605-1991). This climatic reconstruction reflects temperature variability associated with the LIA and subsequent warming. The October–February reconstruction does not show any evidence for late 20th century warming, while February–June temperature shows a cooling trend since 1960. Recent temperature changes are consistent with the instrumental data from Kathmandu and show an unusual cold period during 1815-1822. This seems to be coeval with the Tambora volcanic eruption in Indonesia (Cook et al., 2003).

Eastern Himalaya

The conifers in the eastern Himalayas have been found suitable for dendroclimatology (Chaudhary et al., 1999; Chaudhary and Bhattacharyya, 2002). Tree-ring analysis of *Larix grffithiana* (the only deciduous conifer of this region) shows that May temperature limits its growth (Fig. 9c; Chaudhary and Bhattacharyya, 2000). Similarly growth of *Abies densa* correlates to July-September temperature. The dendroclimatic reconstruction from the eastern Himalaya extending back to 1507 AD has been made using composite tree-ring chronologies of *Abies densa* from two sites, T-Gompa (Arunachal Pradesh) and Yumthang (Sikkim). Reconstructed temperatures show:

(a) No significant change in temperature during the past five centuries (see Figs 9c and 9d).
(b) Decadal scale fluctuations. Thus the 1760s, 1780s, 1800s, 1830s, 1850s and 1890s are recorded as cool decades with the minimum occurring in 1801-1810 (-0.31°C). The period 1978-1987 (+0.25°C) was the warmest one (Bhattacharyya and Chaudhary, 2003).
(c) No long-term cooling corresponding to the LIA was seen. Although, a reduction in late-summer temperature variability from the late 1700s to 1900 was seen by these authors, significant negative anomalies over the longer duration temperature variability were not seen.

Recently, *Pinus* ring-width chronologies from Mishmi Hills, Arunachal Pradesh, Bhutan and Thailand have been compared to global surface temperatures for the past 150 years. The shifting correlations through three seasonal averages, two seasons preceding the monsoon (Dec–Feb and Mar-May) and the monsoon season (Jun–Sep), show clear patterns for each of the chronologies that highlight links to areas of known influence on the Asian monsoon: the Indian Ocean, the tropical eastern Pacific Ocean, and the high-latitude Asian landmass. The correlations are strongest for seasons preceding the summer monsoon. Chronologies from Arunachal Pradesh and Thailand using *Pinus merkusii* show a strong correlation with tropical Indian and Pacific Ocean climate. The *Pinus wallichiana* record from Bhutan correlates to climate over the north Pacific and Asian landmass. These results indicate

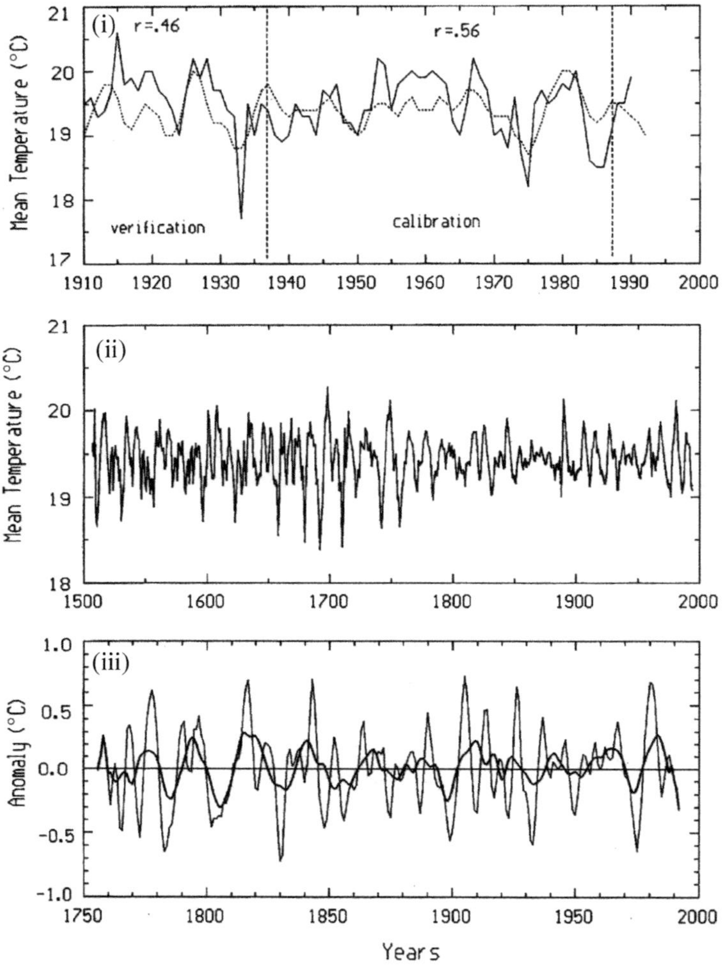

Fig. 9(c): (i) Observed (solid line) and reconstructed (dotted line) mean late summer temperature from 1910-1987. The reconstruction was calibrated using 1938-1987 temperature data; (ii) Reconstructed mean July-September temperature for the Eastern Himalaya since A.D. 1507; and (iii) The reconstructed late summer temperature (A.D. 1757-1994) plotted as anomalies relative to the mean of the fully reconstructed series. The thick line superimposed over the reconstruction is a 10-yr smoothing spline (after Bhattacharya and Chaudhary, 2003).

that network of pine chronology from South-East Asia would be useful in understanding past climate variability with respect to large scale features related to monsoon and ENSO (Buckley et al., 2005).

Karakoram

Limited analysis of *Juniperus* tree-ring widths from the upper timber line (4000 masl) in the Hunza Valley in Karakoram, shows a climatic connection

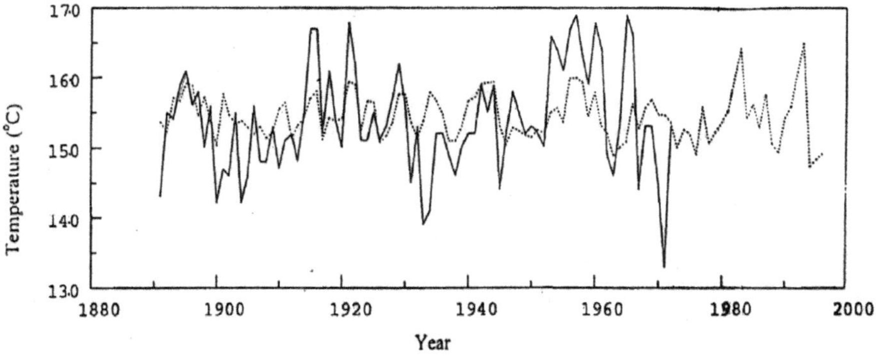

Fig. 9(d): Observed (solid line) and reconstructed (dotted line) May temperatures. The reconstruction was calibrated 1921-1978 temperature data at Darjeeling (after Chaudhary and Bhattacharya, 2000).

(Esper, 2000). Smaller ring widths suggest unfavourable cooler conditions and larger widths indicate a warmer condition. In a five hundred-year reconstruction extending back to 1494, periods of favourable (optimum) condition during AD 1560-1603, 1792-1824 and 1933-1964 and unfavourable ones during, 1601-1661, 1825-1850, 1905-1932 and 1964-1984 were identified. Overall this data does not show any systematic trend of a consistent improvement of favourable condition associated with the global warming (Esper et al., 2002).

Corals

Seasonal reversal of wind pattern results in upwelling – both in the open ocean and the coastal areas of the Arabian Sea. As a result, sea surface temperature drops in areas and this is recorded in the oxygen-isotope ratio of annual bands of coral. In the Lakshadweep Island (10°N, 73°E), the oxygen-isotope signature of *Porites* corals was used to reconstruct the sea-surface temperature for 1974-1990. A rise in $\delta^{18}O$ by 0.7-0.8‰ implied a decline of sea-surface temperature by 3 to 4°C (Chakraborty and Ramesh, 1993, 1997). Comparison of the reconstructed temperature and the instrumental data however indicated that reconstructed SST had an uncertainty of ±1.9°C (op cit). A coral *Favia speciosa* from the northeast region of the Arabian Sea was analyzed by Chakraborty and Ramesh (1998) that produced 40-year records of coralline oxygen and carbon isotopes.

A major hurdle in using corals for monsoon reconstruction is a poor understanding of the relationship between Indian Ocean SST and monsoon rainfall, particularly at lead times greater than 1-2 months before the summer monsoon season (Clark et al., 2000). Analysis of the instrumental data and modeling simulations confirm that the Arabian Sea SST influences subsequent monsoon rainfall on timescales less than a month. Cooler Arabian Sea SST implies reduced Indian rainfall and the converse also holds (Shukla, 1975).

The oxygen isotope of the Gulf of Kutch and *Porites coral* from Marbat (Oman coast; 16°N 48E, 54°42'E) indicates such correlation (Tudhope et al., 1996). The NE monsoon (i.e. Dec-Jan-Feb) coral of Dahlak Archipelago (16°N, 40°E) $\delta^{18}O$ have a significant inverse correlation with annual Indian Ocean and annual southern Red Sea SST (Klien et al., 1997). These observations suggest a relation between coral oxygen isotopes and monsoon rainfall in the northern Arabian Sea. Although, both SW monsoon and NE monsoon precipitation affect coral oxygen isotope, NE rainfall, however appears to play a greater role in controlling coral $\delta^{18}O$. Other than NE monsoon variability coral oxygen-isotopes also appear to co-vary with the southern oscillation index (SOI) (Klein et al., 1997). Studies by Pfeiffer et al. (2004) indicated that Porite corals from Chagos Archipelago (7°S, 72°E) reliably records temporal variations in precipitation associated with intertropical Convergence Zone.

Charles et al. (1997) have analyzed a 3-m long coral (*Porites lutea*) from the Mahe Island (4°37'N, 55°49'E). $\delta^{18}O$ was measured at 1 mm interval to reproduce a 150 year record of equatorial Indian Ocean climate variability on decadal scale. The correlation between monthly $\delta^{18}O$ anomalies and SST anomalies over a 15-year time period was, $r = 0.72$. A correlation with the Asian summer monsoon intensity was also seen. A long-term decrease in $\delta^{18}O$ was attributed to the increase in SST by about 0.8°C, which is slightly higher than the observed instrumental record. Cole et al. (2000) reconstructed a 194-year record of skeletal $\delta^{18}O$ from a coral at Malindi, Kenyan coast. The results accorded with Charles et al. (1997).

Studies, so far, have aimed to identify various modes of climate variability that influence the coral oxygen-isotope ratios. These indicate that on seasonal time scales the sea-surface temperature is the most significant factor that determines the oxygen isotope ratios, and on inter-annual time scales, the seawater oxygen isotopic ratio or salinity is the dominant forcing controlling coralline $\delta^{18}O$. This has a direct consequence to the SST-rainfall relationship, implying that SST takes a lead role in determining the precipitation in the northern Arabian Sea, but more work is still needed.

Speleothems

Potential of tropical speleothem has been investigated based on oxygen and carbon isotopes, and preliminary results indicate that these could be used as proxies for the past rainfall. Within the limitation of the dating of such deposits using U-series methods, the following events are seen: an arid phase before ~1200 yrs BP, a 14-year arid event around 2000 yrs BP and a high rainfall event around 600 yrs BP (Yadava and Ramesh, 2005). A speleothem mineralogical record from a dolomitic cave in Pokhara Valley (Central Nepal) documents the summer monsoon variability over the past 2300 yrs (Denniston et al., 2000). The aragonite annual layers formed between 2300 and 1500 yrs BP indicate reduced monsoon precipitation; whereas alternating calcite/aragonite laminae dated to 1500 yrs BP document increased monsoon conditions. Optically clear calcite layers deposited from 450 to 360 yrs BP indicate a less evaporative (i.e. cooler) environment possibly related to climatic change associated with the onset of LIA.

Future studies will require more calibration efforts to enable correlation of speleothem record with rainfall records on a regional basis, i.e. different meteorological stations in the same region, quantification of amount effect and multiple sample reconstruction from the same site. Recent studies raise some questions on the validity of speleothem reconstruction on account of kinetic isotope effects and suggests the need for a proper calibration of these records with the physical conditions such as temperature, pCO_2, drip rates, calcite precipitation rate, stalactite geometry and drip water chemistry (Mickler et al., 2005). Until then the rainfall reconstruction using speleothems like other proxies, will need caution. In addition, for annual reconstruction, chronologies with a sub-annual scale precision are needed, which is a non-trivial item to achieve. It is, therefore, important that a continuously-depositing speleothem is found if annual reconstruction is to be attempted. This is because, use of any dating inputs that introduces an error that is significantly higher than one year causes difficulties in correlation with instrumental records.

Sedimentary Records

Studies on young sediments from the dunes and lakes in the Thar Desert provided interesting insights into Holocene variability of the monsoon. The lacustrine record of the region indicates that the region experienced large scale hydrological changes and that the aeolian activity and the hydrological changes were out of phase by a few centuries. A synthesis of the Asian monsoon is provided by Morril et al. (2003), which indicates that abrupt climatic excursions towards arid phases occurred at 11.5 cal ka BP, 5-4.5 cal ka BP and at 1300 AD. This was based on multi-proxy analyses of data from 36 sites in the East Asian subcontinent and statistical analyses of the data. While useful, such integrated analyses of multiple data sets from a large region are to be treated with caution. This is because of differences in proxy response times, multi parameter dependences and spatial variability of the monsoon. It should therefore be somewhat inconceivable that any shift in climate over the entire Asian region was synchronous on human time scales, given that clear evidences on gradients and time lags between the forcing and the response times exist.

Aeolian Sediments

Studies on the aeolian sand sequences from the Thar Desert demonstrated the following:

1. The desert sands accretion is climate specific, and the "*window of opportunity*" for sand accretion occurred with a phase lag of over 8 ka

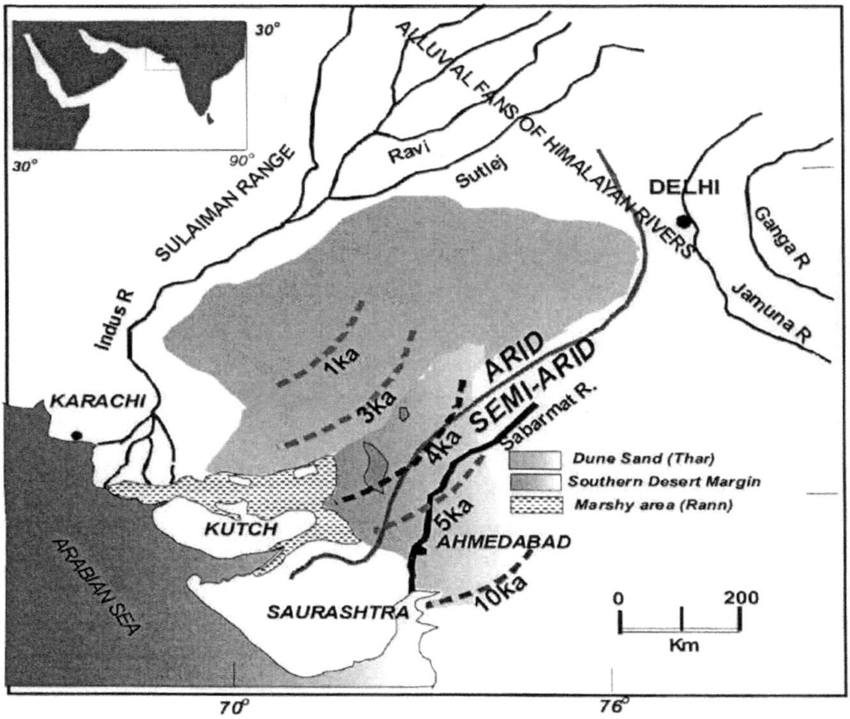

Fig. 10: Shift in dune stabilization through time. The contours indicate that aeolian activity south of the contours stopped at that time. Presently active dunes exist only in the north west of 1 ka boundary (Juyal et al., 2006).

from the glacial epoch and that also over small durations spanning perhaps only 10-20% of the time. Besides, sediment supply for sand accretion depends on wind and moisture. Both of these factors depend on the monsoon and associated pre-monsoon winds. Too much or too little of these two parameters goes against the formation of aeolian sedimentary record.

2. The dune accretion in the Holocene had a periodicity of 1500 years (Thomas et al., 1999).
3. The response time of landforms can be variable and this is a crucial aspect for global change and land use planning studies. Thus, it was seen that the stability of dunes occurred earlier in the eastern Thar and few hundred years later in the western Thar.
4. The dunes were stabilized in the southern limit at ~10 ka and the stabilization front progressively moved northward on millennial timescales, such that only the dunes in the core region of Thar are presently active (Fig. 10). This implies a progressive northward trajectory of the monsoon performance in terms of dune stabilization (Juyal et al., 2003; Singhvi and Kar, 2004), and shows a clearer control of vegetation and wind changes.

5. Vegetation also controls dune migration rates (Kar et al., 1996). In some regions, the dune migration rates during the recent centuries accelerated ten-fold as a result of human induced loss of vegetation cover.

Lacustrine Records

The lakes in the Thar Desert underwent considerable hydrological changes by nearly completing a full cycle of freshwater conditions, leading to desiccation and, finally, the present saline conditions. A synthesis of paleo-hydrological studies of the saline playas indicates that the difference of precipitation (P) and evaporation (E), i.e. $P - E$ balance resulted in a gradient in the lake hydrology from the Eastern to the Western Thar Desert. Thus, the lacustrine (perennial) phase of eastern playas existed for 8 ka. The north-western playa had a fresh water condition lasting for 2.5 ka only, whereas the playas in the extreme western desert remained perennially saline. Since the Holocene optimum, the western playas became ephemeral and saline at least 1 ka earlier than the north-western and 3 ka earlier than the eastern playas (Fig. 11). This implies that considerable spatial variability in terms of geochemistry and sedimentation style existed in the region (Roy et al., in preparation).

In the Ganga Plain, a 2.8 m deep sediment core from Lahuradewa Lake, in the Sant Kabir Nagar District (U.P.) provided palynological data that serves as a surrogate for rainfall (Fig. 12). This sequence shows that around

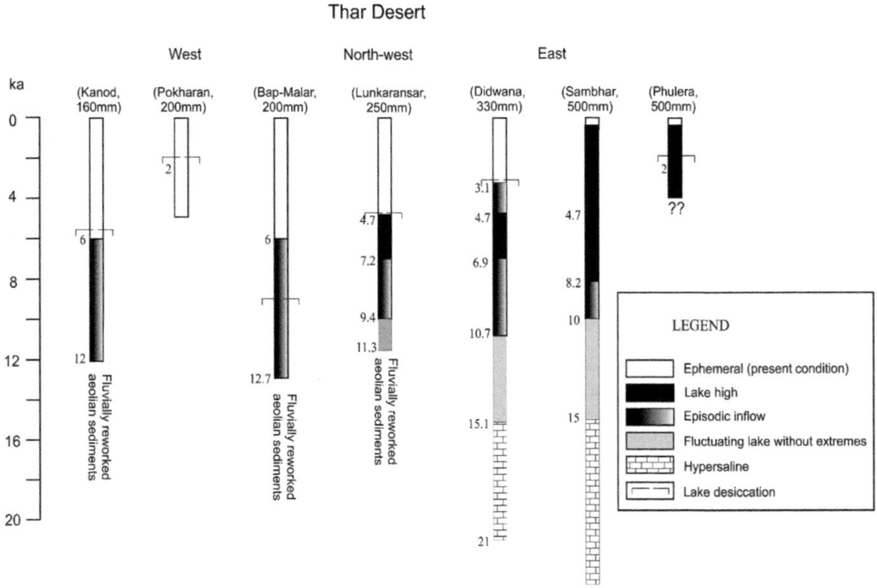

Fig. 11: Changes in the paleohydrology of Thar playas through time. The differences indicate changes in precipitation/evaporation through time. Prepared by PD Roy.

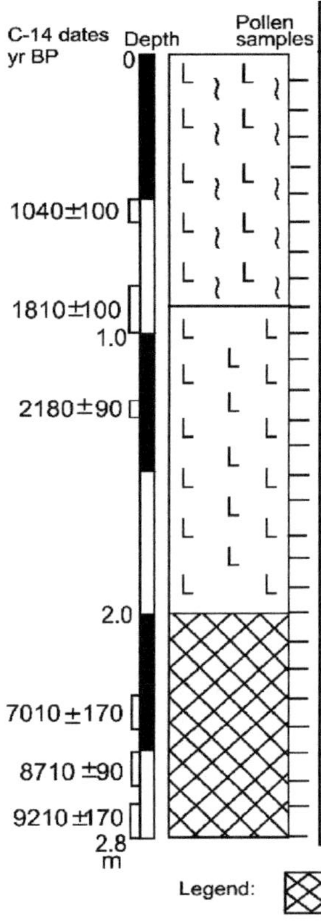

Period yr BP	Vegetation assemblage and salient features	Climate	Pollen zones
1300-Present	•Arboreals turned sparse as indicated by a sharp reduction in their number and frequency. •Agricultural practices continued with same pace as before since no singnificant change noticed in cerealia and other culture pollen taxa. •Lake gradually shrunk and assumed present ephimeral status as suggested by the meagre aquatic plants and ultimate disappearance of *Botryococcus*.	Reduction in monsoon rainfall (Dry climate)	LRD-V
4050-1300	•Groves of forest became dense and diversified as suggested by much increased frequency of arboreals (trees & shrubs) particularly *Madhuca indica*, Sapotaceae and *Barringtonia*. •Further increase in agricultural activities depicted by the steady occuurence of cerealia and other culture pollen taxa. •Transformation of lake into swamp commenced with the increase of sediment influx in the lake basin.	Further increase in monsoon rainfall	LRD-IV
6400-4050	•Establishment of forest groves with the considerable increase in *Bombax* and improvement in other tree taxa viz., *Madhuca indica, Holoptelea, Terminalia*, etc. •Expansion of agricultural practices as reflected by steady increase of cerealia pollen. •*Trapa* (Singhara) would have been a means of subsistence.	Increase in monsoon rainfall	LRD-III
9250-6400	•Invasion of open vegetation by few more trees viz., *Bombax*, *Emblica officinalis, Syzygium*, etc. •Appearance of cerealia suggestive of inception of agricultural activities. •Lake became wider as indicated by increase in aquatics and freshwater alga- *Botryococcus*.	Amelioration of climate	LRD-II
10600-9250	•Open vegetation dominated by grasses with scattered trees viz., *Holoptelea, Terminalia* , *Aegle marmelos* occupied the region. •Presence of pollen of *Potamogeton* and *Botryococcus* suggestive of existence of lake.	Cool and dry	LRD-I

Fig. 12: Lacustrine record from the Ganga Plains. Note the difference with the records from the Thar. The relationship is nearly out of phase.

10600 to 9250 cal yr. BP, open vegetation comprising mainly grasses, Chenopods/Amaranths, *Artemisia* together with occasional trees, occurred, implying a cool and dry climate in the region. Frequent occurrence of the fresh water alga-*Botryococcus* and other aquatic species suggest the existence of a lake. During 9250 to 6400 cal yrs BP, invasion of open vegetation by more trees depicts the amelioration of climate. Progressing preponderance of *Botryococcus* and other aquatic species indicates that the lake expanded due to increased monsoon rainfall. Interestingly, the first appearance of cerealia and other culture pollen at the level dated to ~7500 cal yrs BP suggests the inception of some sort of cereal-based agricultural practice in the region.

The lake expanded further, as indicated by improvement in aquatic flora. Between 6400 and 4050 cal yrs BP, establishment of groves of forests interspersed with open herbaceous vegetation reflects an increase in monsoon rainfall. Consistent presence of cerealia and culture pollen indicates an increase in agricultural practices under favourable climatic conditions. Between 4050 and 1300 cal yrs BP, the groves of forests became more diversified and dense on account of further enhancement in monsoon rainfall. At this time, increased sediment influx in the lake transformed it into a swamp. Maximum acceleration in agricultural practice took place during this period. Since 1300 cal yrs BP, there was a reduction in monsoon rainfall as the grove of forests turned sparse and less varied in composition than before. The lake shrunk and assumed its present ephemeral status, as seen by a sharp decline in aquatic plants and simultaneous expansion of marshy vegetation (Chauhan et al., 2006).

Sharma et al. (2004a) reconstructed climate using oxygen-isotope analysis of bovid teeth, and suggested that humid condition existed around 3600 cal years BP. Increased drier conditions then existed till 2800 cal yrs BP and a higher humidity period around 2500-1500 cal years BP occured. Finally a dry phase from 1300 cal yrs BP was seen. On the other hand, reconstruction from a lake in similar climatic region shows some disparities. Here the conclusion is 10000-5800 cal yrs BP reflects climatic optimum conditions, arid conditions from 5000-2000 cal yrs BP and climate amelioration from 1700 cal yrs BP (Sharma et al., 2004b, 2006). The contrasting records from the same region (arising possibly due to interpolations over longtime periods on account of paucity of dates), makes it difficult to examine the spatial variability of monsoon through time. In one scenario, the events in the Thar Desert and in Ganga Plain appear synchronous but in the other case an out-of-phase relation is seen. This calls for more, internally consistent datasets as also a comparison of present day meteorological records.

In two high altitude lakes of Nepal, viz., lake Piramide inferiore and LCN40 (86.48°E, 27.58°N' at 5000 masl), Lami et al. (1997) reconstructed climatic variation spanning the past 3000 years. In a comprehensive study involving algal pigments and fossil diatoms which indicated warm events at 2950, 2300-2000, 1700-1600, 700-400, 200, 60-cal year BP and cooler events at 1500-700, 400-300, 200-80 cal year BP. In our opinion at this altitude the lakes record suggests temperature changes via the glacial melt. Such archives are indeed important for pristine long-term paleo-temperature studies.

Peat Deposits

The Uttaranchal Himalaya lies within the northern limit of the SW monsoon. Hence, the multi-proxy and high-resolution Holocene climate records from pristine peat deposits (situated normally above 3500-m altitude in this part of the Himalayas) are most suitable for revealing the continental history and land-sea correlation of SW monsoon variability. The paleoclimate records, so far, are available only from three peat deposits. These are Dokriani and Dayara peats in Bhagirathi Valley of the Garhwal Himalaya and the Dhakuri peat deposit in Pinder Valley of the Kumaon Himalaya spanning the past 3500 years centennial-scale climate records. Analysis of pollen, diatoms, phytoliths, organic matter and magnetic susceptibility from the Dhakuri peat deposit (Phadtare and Pant, 2005, 2006) indicate rapid weakening of the SW monsoon at ca. 3200 and 2000 cal yrs BP (Fig. 13). The monsoon-influenced peat deposits of the Tibetan Plateau also indicate depleted monsoon around 3200 cal yrs BP (Hong et al., 2003). Except for a century of dry phase during 740-640 cal yrs BP, the climate progressively improved until present; with warmer temperatures during 1600-740 cal yrs BP, 640-460 cal yrs BP and 270-57 cal yrs BP. Intervening periods experienced relatively cool and dry conditions.

The rapid dry shift observed around 3200 cal yrs BP (Phadtare and Pant, 2006) reveals an out of phase relationship with the Arabian Sea monsoon record (Gupta et al., 2003). Subsequent dry events (viz. 2000-1600, 740-640, and 400-250 cal yrs BP), however, are comparable with weak monsoon phases over the Arabian Sea. The preliminary regional interpretation, however indicates significant dry shifts (i.e. weak SW monsoon) around 5900, 4600, 3300, 2000, 700 and 400 cal yrs BP (Phadtare, 2000, 2003; Rühland et al., 2005; Phadtare and Pant, 2005, 2006). In view of the complex and heterogeneous nature of Himalayan climate, high-density data particularly from the monsoon-influenced part of the Himalayas are indispensable.

Fluvial Archives

Studies on the hydrological changes in the Ganga Plain provide evidence of channel activity during the period 13-8 kyrs (stronger monsoon), formation of ponds during 8-6 kyrs indicating arid climate and finally aeolian aggradations (Srivastava et al., 2003a, b). Evidence of human impact on sedimentation rate is also documented. A first order channel in southern India, in River Pennar, which in the upper reaches responds only to the SW monsoon (Thomas et al., 2007) has provided interesting insights on land sea correlations. The paleo floodplain record of this river indicates that during the period 135 yrs to 85 yrs ago, the river experienced eight major flood

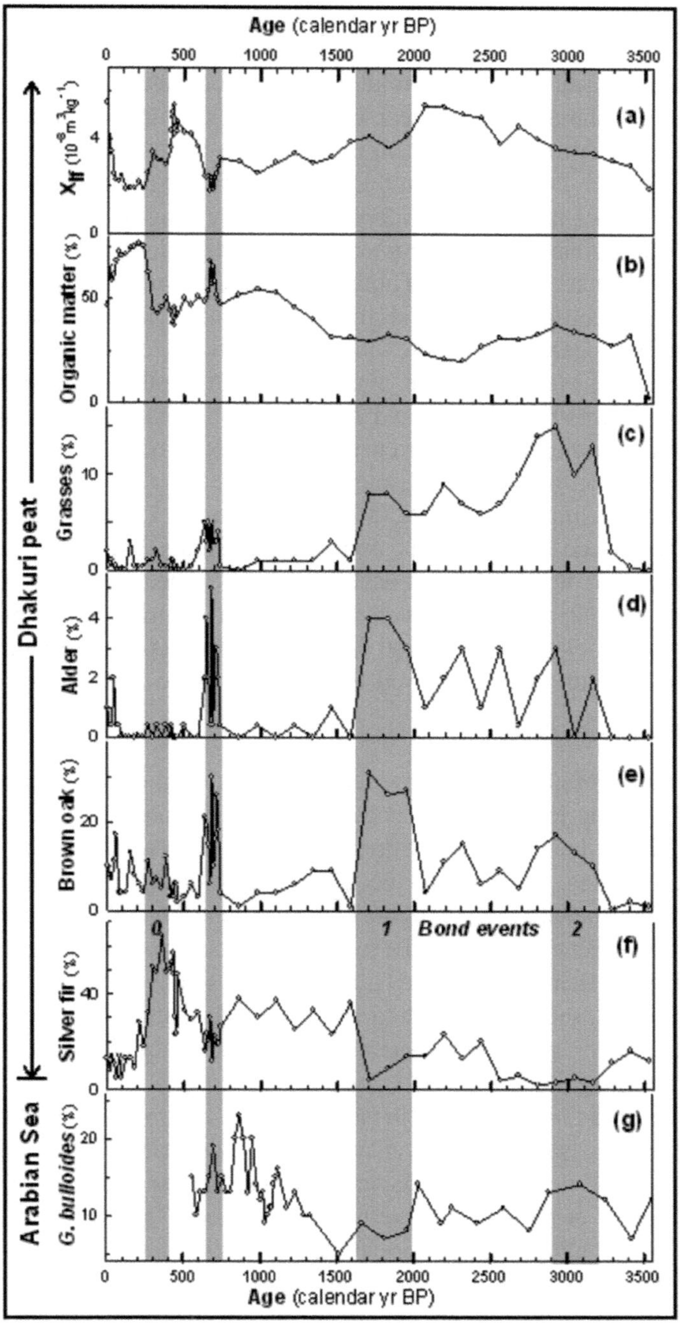

Fig. 13: Correlation of the climate proxy records from the Dhakuri peat sequence of Pinder Valley in Kumaon Higher Himalaya with *G. bulloides* (%) record from the Arabian Sea ODP Site 723A. Time series (a) – magnetic susceptibility of the detrital influx, (b) – organic matter content, (c) – grasses, (d) – Alder, (e) – brown oak, (f) – silver fir, and (g) – *G. bulloides.* The vertical gray bars indicate intervals of decreased summer monsoon in Himalaya. Kindly provided by Dr. N.D. Phadtare.

events and since the last flood 85 years ago, the river did not over top its banks to deposit sediment on its flood plains. The instrumental record of this region also indicates that the rainfall in this period exceeded 10% level about eight times. Should such a clear correlation exist, the implication of this record is that during the past century, monsoon intensification as suggested by oceanic records is not captured in the flood events of the region. The tree-ring record does not exhibit evidence of any warming in northern India and the rainfall reconstruction/instrumental record also does not show any such change. These are consistent. Though in the past marine records have been used to infer rainfall on the land, the present studies indicate that suggested correlation of oceanic monsoon wind proxy with the real rainfall on land, on the other hand are not tenable, *sensu stricto*. A synthesis of records from fluvial sediments by Chamyal and Juyal (2008) indicated that on account of their response time, fluvial record perhaps may not record century to millennial scale fluctuation of climate i.e. monsoon.

In the Pokhra basin (29°N, 84°E) in Nepal at 2500 masl, analysis of paleosols on colluvial debris and aeolian sands using stable isotope composition of its humus and accelerator mass spectrometric radiocarbon dating of organic matter indicated a humid phase during 6200-4500 cal yrs BP and a drier climate from 4500 cal yrs BP onwards. The vegetation cover in the humid period was inferred to be dense (Saijo and Tanaka, 1989).

Paleoflood Records

Floods and droughts commonly occur throughout the Indian subcontinent and at least one region experiences droughts or floods every year. The last two to three decades have witnessed increased variability in monsoon precipitation and this has enhanced the possibility of monsoon extremes. Just how unusual are these extreme climatic events? Are they rare events and should we expect events of similar magnitude to occur in the future? Terrestrial records of extreme climatic events include ‘slackwater flood deposits’. Stratigraphical, sedimentological and chronological studies of the slackwater deposits provide reasonably accurate information about paleofloods, their timing and magnitude.

Paleoflood investigations in some of the central and western Indian rivers, namely, Narmada (Ely et al., 1996; Kale et al., 1997; Kale et al., 2003a), Tapi (Kale, 1999), Godavari, Krishna, Pennar and Luni (Kale et al., 2000; Thomas et al., 2007), have been carried out. The slackwater records, augmented by gauge and historical records indicate significant changes in the frequency and magnitude of large floods during the last two thousand years (Fig. 14). In general, the records reveal the absence of large-magnitude floods during the late Medieval Warm Interval and the LIA. Historical and other proxy records indicate that this was also the period of increased frequency of droughts in South Asia (Fig. 14, of Kale, 1999). It is worth noticing that in the recent 50 years, most rivers have experienced at least one major, catastrophic event. In some cases perhaps, these represent the

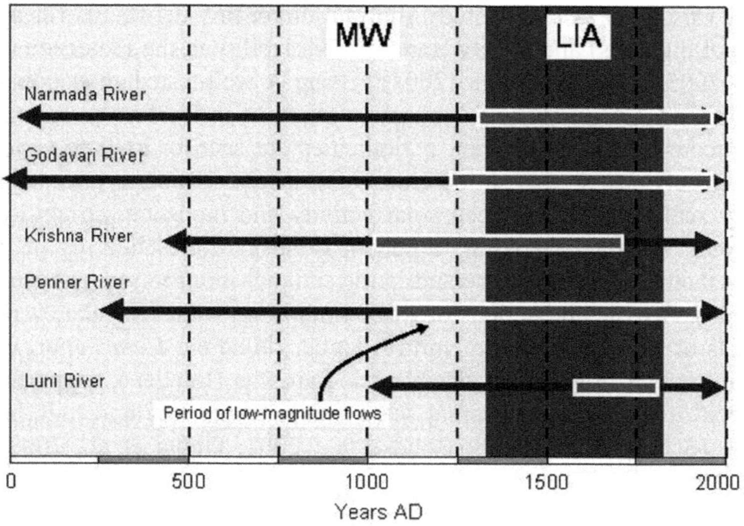

Fig. 14: Records of paleofloods in rivers Tapi, Luni, Godavari and Narmada along with identified mega Niño events and global temperature changes. Most rivers show weakened floods during the LIA, moderate to extreme floods during the period 2ka to ~600a and catastrophic floods during the past century. (From V.S. Kale, 2008).

millennium extremes. There is also evidence of clustering of low-frequency, extreme floods between 400 and 1000 AD, and into the most recent period (post-1950s). In addition, a catastrophic flood on the Tapi River at the beginning of LIA (ca. 300-400 yrs BP) and a highly erosive flood on the Narmada River at the commencement of the Christian era (ca. 2000 years ago) have also been inferred from the paleoflood records. On the basis of modern analogues, it appears that the century-scale variations in the flood magnitude and frequency in the two basins (Narmada and Tapi) are intimately linked to long-term fluctuations in the monsoon rainfall (Kale et al., 2003b; Kale and Baker, 2006).

Mangroves

An area that has not been fully exploited is the use of Mangroves for sea level change and climate reconstruction. Studies on a 50 m core enabled reconstruction of the depositional environment and climate, in terms of tidal, inter tidal-marine habitat changes during the past 10 ka (Hait and Behling, 2005). Kumaran et al. (2005) have examined mangroves in the western coast of India. In our opinion this archive has important possibilities of quantitative paleo reconstructions of sea level and vegetation changes and needs further exploration.

3.2 The Marine Record

Considerable efforts have been made towards high-resolution (high density sampling of the marine cores) reconstruction of proxy records of monsoon and its variability. A recent study shows a direct link between solar activity (Sunspot numbers) and the SW monsoon variability in the Holocene (Gupta et al., 2005; Thamban et al., 2007). Using a well-tested monsoon proxy foraminifer *Globigerina bulloides* past variations in the SW monsoon have been reconstructed. Significant periodicities are seen in its time series and these closely match the periodicities of Sunspot numbers, and suggest a century scale relation between solar activity and monsoon variability. The 1500-year periodicity, possibly a part of Daansgard-Oeschger cycles in the north Atlantic, is also documented in the summer monsoon record (Gupta et al., 2003). Grossly, the periods of weaker monsoon winds correlate well with intervals of cold spells in the North Atlantic, called the *Heinrich events* and the *Bond cycles*, for the last glacial period and the Holocene, supporting the evidence of a century-millennial scale link between low and high-latitude climate (Schulz et al., 1998; Bond et al., 2001; Gupta et al., 2003). The North Atlantic and monsoon link was explained through down-stream effect of North Atlantic cooling on the Eurasian and Tibetan region (Gupta et al., 2003).

Increased cooling means thicker glaciers in the Eurasian and Himalayan-Tibetan region and reduced land-sea thermal contrast that would weaken the summer monsoon. Thus forecast of abrupt changes in the North Atlantic region would translate to abrupt changes in the Indian monsoon. However a recent study by Gupta et al. (2005) suggests that the cold North Atlantic events (Bond cycles) and dry phases of the summer monsoon were synchronous indicating that the same mechanism simultaneously affected the two regions. High-resolution terrigenous proxy studies on a laminated sediment core from the Oxygen Minimum Zone of the eastern Arabian Sea margin, in combination with other high-quality cores from the Arabian Sea, suggests several abrupt events in monsoon precipitation throughout the Holocene (Thamban et al., 2007). The early Holocene monsoon intensification occurred in two abrupt steps at 9500 and 9100 years BP and weakened gradually thereafter, with significant weakening in precipitation recorded at ~7000 years BP, synchronous with similar conditions in India. Spectral analysis of the precipitation records reveals statistically significant periodicities

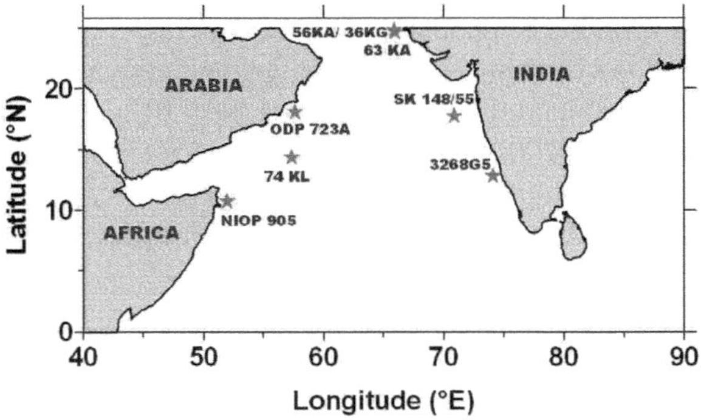

Fig. 15: Location of the marine cores summarized in Fig. 16.

at 2200, 1350, 950, 750, 470, 320, 220, 156, 126, 113, 104 and 92 years. Most of these millennial-to-centennial cycles exist in various monsoon records as well as the tree-ring $\Delta^{14}C$ data and/or other solar proxy records. Figure 15 provides the core locations for marine record reconstruction and Fig. 16 provides a summary of the more detailed published records from the seas around India. Only the Holocene records were considered and all the core data were transformed by the most recent estimate of the reservoir ages.

The Bay of Bengal is one of the larger fluvial influx sites in the world's oceans, with large seasonal sea-surface salinity variations. However, the large portion of the Bay is covered with the Bengal Fan, with altered sequences, and these have enabled continuous records for the Holocene. Figure 17 provides a reconstruction of inferred monsoon based on data from Bay of Bengal sediments (Chauhan, 2003). More recent carbon isotopic studies from the Bay of Bengal indicate that the Bay of Bengal received a mixture of C3/C4 plants which progressively became dominated by C3 plants since the mid Holocene (Galy et al., 2008). The paleoclimatic studies from the Bay suggest termination of Younger Dryas with a prominent fluvial pulse which altered the overall salinity regime in the Bay. The fluvial influx into the Bay was coeval with the meltwater pulse 1B in the Atlantic ocean (Fairbank, 1989; Chauhan et al., 2004), and further confirms that the glaciation in the Himalayas and monsoon are linked with global climate. The weathering intensity and sediment discharge into the Bay has also been found to be related to climate. Contrary to the general belief, the sediments input into the Bay was higher during the termination of Younger Dryas at a time when sea level was rising (Webber et al., 1997; Chauhan et al., 2002), with displays intensity of chemical weathering. The entrapment of sediments in the Bengal basin and in subaqueous prograding delta appears to have reduced the supply of the sediments into the Bay since 7300 cal yrs (Goodbred and Kuehl, 1999).

Some of the key inferences that have come out from these studies on marine/deltaic sediments are:

1. The monsoon strengthened in two steps at 9500 cal yrs and 9000 cal yrs BP.
2. Simultaneous effects of deglaciation and enhanced monsoon resulted in increased discharge into the Bay of Bengal during 9500-5000 cal yrs BP (Chauhan, 2003).
3. A decline in the monsoon occurred at 8200 cal. yrs BP. This coincides with the *Glacial Aftermath cool event* of Alley et al. (1997).
4. An arid phase occurred at 7300 cal yrs BP during which the paleo-circulation in the Bay of Bengal changed with an intensified cyclonic gyre (Chauhan and Vogelsang, 2006).
5. Another significant arid event can be seen at 6000 cal yrs BP, which is similar to cool pool dry tropics of Mayewaski et al. (2004).

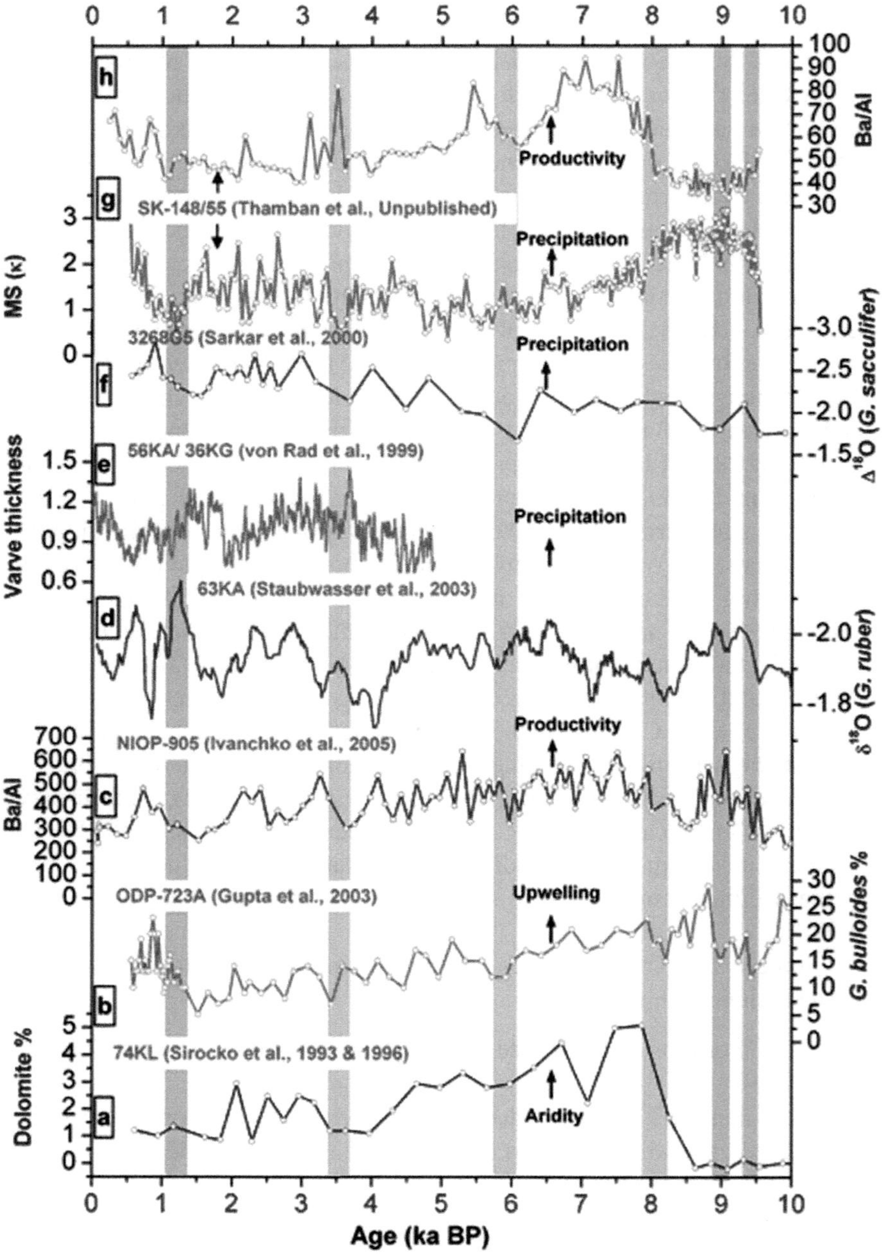

Fig. 16: Holocene monsoon synthesis from the Arabian Sea. Sedimentary records after Staubwasser et al. (2003), Gupta et al. (2003), Sarkar et al. (2000), Doose – Roselinki et al. (2001), Sirocko et al. (1993), Ivanochko et al. (2005), von Rad et al. (1999). The ages have been recalculated by MT using more recent estimates of reservoir ages (Thamban et al., 2007).

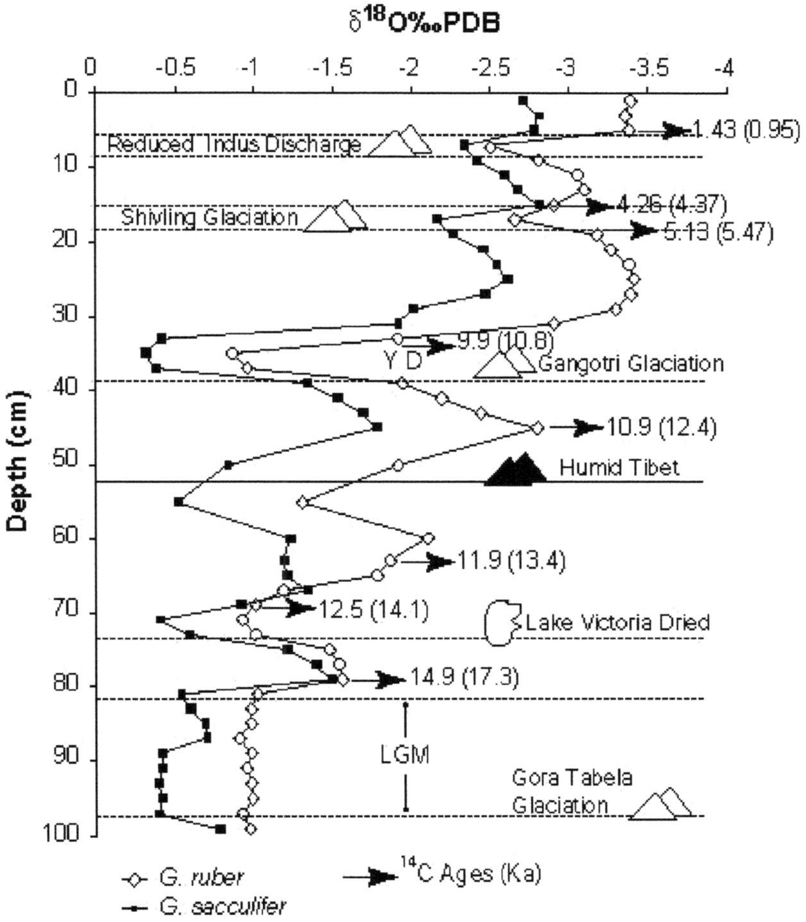

Fig. 17: $\delta^{18}O$ records of climate variability derived from a turbidity free core. For comparison other evidences are also cited (modified from Chauhan, 2003).

6. The arid episodes at 5000-4300 cal yrs BP engendered reduced fluvial input into the Bay of Bengal (Chauhan, 2003).
7. The 3500 cal yrs BP event of low productivity along the African and Arabian coast is comparable to the increasing rainfall along the Pakistan margin. Coastal signatures from India are however not well resolved.
8. An arid event of reduced fluvial influx occurred at 2000 cal yrs BP.
9. At 1500 cal yrs BP, intensification of monsoon along the western Arabian Sea is seen, but the Indian and Pakistan coasts show evidence of monsoon deterioration. The intensification signal here appears in these two places about 500 years later.
10. Regional differences are prominent for short events but overall the coast along India and Pakistan behaved differently.

It is considered that proxy responses and differences could be an artifact of different response time of proxies to the same forcing function and this aspect needs further elucidation.

4. LAND-SEA CORRELATIONS

As alluded to in the introduction, any comparison of marine and land records of the monsoon is a difficult proposition on account of the truncated nature of sediment response, difficulties in dating and differences in proxy responses. Questions such as, do intense summer monsoon winds as revealed by the marine record (e.g. Gupta et al., 2003), get transformed into increased monsoon precipitation on land and, if so, how does it account for the spatial variability of the monsoon? Like the land records, there are additional problems with marine records that arise from the oceanic reservoir ages, questions of their constancy through time and the large geographical region covered by the summer monsoon. Most of the radiocarbon age calibrations are reliable between 0 and 8000 years BP and the uncertainties become larger for older time intervals (Guilderson et al., 2005). For example, a comparison of marine record of the summer monsoon (Gupta et al., 2003) with the continental record (Fleitmann et al., 2003) shows a good parallelism from 0-8000 cal years BP, but marine ages get older by several hundreds to thousand years between 10000 and 8000 cal years BP. The past three millennia century scale record from the monsoon influenced peat deposits in the central Himalaya reveal four weak monsoon phases during 3200-2900, 2000-1600, 740-640 and 400-250 cal yrs BP. The 3900-2400 cal a BP weak monsoon event corresponds to a period of improved monsoon over Arabian Sea, whereas other weak monsoon periods are also seen in the Arabian Sea records.

Changes in sea levels, wind speeds, upwelling intensities and CO_2 production rates are some of the factors that cause variations in the marine reservoir age (Bard, 1988). A way to quantify the reservoir age variation through time would be to measure coexisting terrestrial organic matter (pieces of wood, charcoal, pollen, etc.) and planktonic foraminifera in deep-sea cores from the continental margins. Another way to measure the fossil reservoir age would be to compare the ^{14}C age of a short and well-defined event on land and in deep-sea sediments. The third possibility will be to date volcanic ash layers from within marine records and calibrate the radiocarbon age. Optical dating of sediments also offers a way out. These will perhaps be the best possibility given that these will then be directly correlatable to those on terrestrial sediment.

5. FORCING MECHANISMS

Presently, the India Meteorological Department uses 10 proxies for monsoon prediction (Table 1) (Rajeevan et al., 2004; Gadgil et al., 2005). These have met with reasonable to poor success in their annual monsoon predictions. Testing hypotheses regarding the monsoon variability seen in the paleo record is an active research area that is rapidly growing. One of the direct objectives of the paleo research has to be towards determination of the amplitude of various forcing functions. Reconstruction of spatial variability and determining the role of local/geographical factors in monsoon performance is another aspect that has not been investigated in detail. The Indian Ocean Dipole and ENSO have been implicated for interannual-decadal scale variability in the summer monsoon (Webster et al., 1999). However, the relation between ENSO and the monsoon has weakened over the last one and a half centuries (Kumar et al., 1999). Whether the weakening relation between ENSO and summer monsoon is linked to increased surface temperature over the past century remains debatable (Mann et al., 1999). It is also not established if this relationship is periodic or that it has a more complex time relationship.

Variations in solar output have been suggested as a key driver for century to millennial scale variations in Oman precipitation (Neff et al., 2001). Solar variations could also influence the monsoon regimes indirectly. Numerical simulations suggest that the effects of North Atlantic cooling can perhaps amplify this through teleconnections (Rind and Overpeck, 1993; Gupta et al., 2003). Alternately, solar variations could directly affect the upper tropospheric structure that drives the monsoon (Gupta et al., 2005). These observations indicate that the monsoon could be sensitive to relatively small changes in forcing (0.25% change in solar output, or a 2°C change in SST). A correlation between high-resolution record (~30 year interval) of the summer monsoon and sunspot numbers that serves as a direct proxy of solar activity (Solanki et al., 2004) during the Holocene, reveals that the solar variability directly controlled the multidecade to century scale abrupt changes in the summer monsoon (Gupta et al., 2005). The Arabian Sea record also indicates that over the past 11,100 years, almost every multi-decadal to centennial scale decrease in summer monsoon strength is tied to a distinct interval of reduced solar output (Gupta et al., 2005). Conversely, the increase in summer monsoon strength coincides with elevated solar output. The Arabian Sea record indicates that intense summer monsoon had high amplitude variability in the early Holocene. This variability had a statistically significant correlation with the solar proxies. The detrended records of monsoon from the cores and solar insolation (i.e. long-term trend removed), support the hypothesis that solar influence on the monsoon is direct (Gupta et al., 2005, Fig. 15). The sun-monsoon link can be explained by a direct solar influence on the Intertropical Convergence Zone (ITCZ) that controls the monsoonal precipitation.

Whatever are the causes of repeated abrupt changes in the summer monsoon, the increasing surface temperature could be a source of concern for climate modelers as well as policy planners because as the global temperatures rise, the summer monsoon may undergo surprising and dramatic changes.

6. SUMMARY

The key messages from the results presented above are:

1. The monsoon is a stable atmospheric phenomenon. It never failed at least during the Holocene. It however experiences: a high variability around a mean value, changes in spatial distribution and timing that can lead to agricultural droughts or floods.
2. Instrumental records indicate a warming trend during the past few decades but the amplitude is spatially variable. The maximum temperatures in the Himalayan regions have shown more increase and the minimum temperatures a lesser increase. The overall amplitude however is much smaller than the global estimates for the past few decades.
3. Precipitation has not shown much variability over the past century but the incidence of major flooding has increased during the past few decades. During this period some of the rivers in India have experienced floods of the highest magnitude for the millennium.
4. The frequency of cyclonic storms has decreased over the past century but the frequency of severe cyclonic storms has nearly doubled during the past few decades.
5. The paleo-reconstructions suggest lower temperatures during the Little Ice Age and higher temperature during the period 1500-500 yrs ago. The exact amplitude is not established but was < 0.5°C.
6. The flood events during the Little Ice Age (~500-200 years) were significantly lower compared to the preceding 1500 years in the western and peninsular India. A temperature-rainfall-flood relationship, however, is not established. This is desirable to obtain such information.
7. The paleohydrology of the Thar Desert shows indication of a differential evolution of precipitation and evaporation through space and time. The paleo record from the Ganga Plain shows an out of phase relationship with the Thar Desert data. This however needs to be reexamined due to internal discordance between records from the Ganga Plain. A spatial variability of monsoon is, however, clear.
8. The marine records from the Arabian Sea and Bay of Bengal show significant variability – some of which are seen on the land records and others are not found.
9. In the marine record, monsoon winds and solar variability are linked given that monsoon wind proxies show in-phase variation and similar cyclicity.
10. In terms of paleodata, an urgent need is to have few, spatially separated key sites dated with adequate chronologies and proxies. There is also an urgent quest for more paleo records from Nepal, Pakistan and Bangladesh.

7. FUTURE OUTLOOK

There are two aspects of looking at the monsoon/climate dynamics in a futuristic sense:

1. Societal, and
2. Scientific.

In respect of societal aspects, the critical need is of a realistic assessment of water and food security, disease and changes in them (with respect to the present), due to anticipated changes in rainfall and temperature due to global warming. The origin of agriculture and domestication of animals in South Asia have been linked to changes in the monsoon precipitation, which, in turn would have driven the beginning of agriculture and human civilizations in the region (Gupta, 2004; Gupta et al., 2006). As the arid phase (weak summer monsoon) began ~5,000 cal yrs BP, the societies in India migrated to more productive areas to the east and south and some may have developed mechanisms of adaptation to climate change (Pandey et al., 2003; Gupta et al., 2004, 2006). This can be viewed from the presence of ponds, tanks, and artificial reservoirs constructed during the late Holocene across India when the monsoon reached its Holocene minimum, and we find correlation between heightened historical human efforts for adaptation and the most recent minimum in the monsoon winds that occurred during the Maunder Minimum (1600 AD) (Anderson et al., 2002; Pandey et al., 2003). The monsoon record supports an emerging paradigm that at least in the tropics, the largest climatic changes and societal responses were driven by changes in precipitation rather than land temperature. A key factor for agriculture is the prediction of the onset date of monsoon and distribution of rainfall. These factors determine the agricultural yield and, with the exception of a few efforts (Gadgil and Sheshagiriao, 2000; Gadgil et al., 2002), these aspects have not been dealt with adequately by the scientific community. From the paleo perspective, it would be of interest to examine if some reliable proxies could be developed to help the monsoon modelers. Added to it will be an understanding of vegetation response to changes in temperature and precipitation. This is also not well constrained in the regional context.

In respect of an academic enquiry, the key issue will be to prepare a good paleo-record base to reconstruct spatial variability of rainfall through time in the past and then examine how best these could be modeled. The data base on paleo reconstruction is at best poor, on account of absence of adequate data density, rigour in interpretation and in high density dating. Then, there is an aspect of arriving at numerical estimates of annual rainfall reconstruction using some of the proxies. Increasing caution on their use has been advocated and need be exercised. There is often a tendency to correlate the local records with global events. Given the issues of sediment supply, transport, sedimentation and preservation, there is no *apriori* reason that all the land records will be created simultaneously. Thus, any such correlation is not tenable till such time when a causal connection of an event on land and in the marine realm is established, with a cause effect scenario. Infact a phase lag is most likely and there is a reasonable chance of high frequency events that are recorded in ocean and ice core are not even recorded on the land. Thus at the present level of understanding, it will be prudent to address to low frequency variations.

Paleo reconstructions serve two major roles. Firstly, they are good paleoanalogues for vegetational changes and landscape dynamics. These can be used for future climate projections under similar temperature conditions and without the human impact. Secondly, they provide a base line for the direction of changes in the present and thereby provide a mean for climate state parameters, to model the future. Some of the key questions that need to be addressed by the paleo-community are:

1. Can the temperature and rainfall record of the Indian sub-continent be obtained on an annual time scale for the past millennium or more? What tests should be prescribed to make these reconstructions robust. Can there be a data set for India similar to that created by Mann et al. (1999, 2008)? Can we combine existing tree-ring data to achieve this? How reliable are speleothem and coral records given the recent consideration on their validity in respect of kinetic fractionation effects?
2. What has been the spatial variability of monsoon rainfall during the past major global climatic epochs such as the medieval warming, the little and major ice ages? Can we establish a temperature-rainfall link? What was the change in the frequency of extreme rainfall and drought events during these epochs?
3. What role does the winter monsoon play? Can the signatures of winter and summer monsoons, be decoupled? Can we trace the southern extent of the influence of westerlies through time? What archives can be used?
4. What factors control the onset and duration of the monsoon? Are there any vegetation proxies that can define the duration and onset of monsoon season? Can other paleo-signatures help?
5. What signatures of climate do the trees in a region, record? Are they all similar or different? Is there a diagnostic parameter on the nature of climate signal they provide. Are there other archives that provide reliable paleo information on annual scale?
6. How monsoon wind has induced upwelling in the Arabian Sea, is related to the monsoon rain over the Indian subcontinent? How well can the ages on marine sediments be determined *vis a vis* a realistic estimate of reservoir ages? Does the Bay of Bengal respond to monsoon or it controls it as well?
7. How has vegetation responded to monsoon through time or vice versa? What are the interconnections between the vegetation and monsoon?

8. How have large and small river systems reacted to climatic change? What has been the sediment dynamics then and what are the controlling factors?
9. What kind of lacustrine archives across the landscape exist that can be used to reconstruct changes in rainfall gradients?
10. Can we reconstruct the manner in which the past societies were affected by and influence climatic changes?
11. How are the extreme events processed through the geomorphic records? What are the time scales, thresholds and amplitudes for such changes?
12. Could there be new proxies (such as molecular indices, biological proxies) that could be developed to solve the above mentioned questions?
13. What teleconnections affect monsoon system and how can these and their time dependence be elucidated? and
14. Can some regional climate models be built to explain paleodata and hence aid the predictive building of scenarios of landscape changes under expected climatic changes?

Much has been done BUT, more needs to be done.

ACKNOWLEDGEMENTS

A.K. Singhvi is thankful to the University of Heidelberg and the Av Humboldt foundation for support during the compilation of this article. His co-authors A.K. Gupta and N.R. Phadtare thank the Indian Department of Science and Technology, New Delhi for financial support. Singhvi also feels grateful for the excellent cooperation from all the authors. M. Thamban thanks NCAOR for the support. Comments by Professor Patrick Dedekker of Australian National University and Professor R.J. Wasson, Charles Darwin University helped improve the scientific context and presentation. The authors thank the START initiative that made it possible for them to come together and work synergistically.

REFERENCES

Abhayankar, H.G., 1987. Late Quaternary palaeoclimatic studies of western India. Unpublished Ph.D. Thesis, University of Pune, Pune.

Alam, M.M., Hossain, M.A. and Shafee, S., 2002. Statistical Analysis of cyclones and depressions in the Bay of Bengal during 1974-1999. *Journal of Bangladesh Academy of Sciences*, **26,** 207-218.

Alam, M.M., Hossain, M.A. and Shafee, S., 2003. Frequency of Bay of Bengal cyclonic storms and depressions crossing different coastal zones. *Int. J. Climatol.*, **23,** 1119-1125

Ali, A., 1979. Storm surges in the Bay of Bengal and some related problems. Ph.D. Thesis, University of Reading, England, pp 227.

Ali, A., 1996. Vulnerability of Bangladesh to climate change and sea level rise through tropical cyclones and storm surges. *Water, Air and Soil Pollution*, **92,** 171-179.

Alley, R.B., Maywaski, P.A., Sowers, T., Stuiver, M., Taylor, K.C. and Clark, P.U., 1997. Holocene climatic instability: a prominent widespread event 8200 years ago. *Geology*, **25,** 483-486.

Altabet, M.A., Higginson, M.J. and Murray, D.W., 2002. The effect of millennial-scale change in Arabian Sea denitrification on atmospheric CO_2. *Nature*, **415,** 159-162.

Anderson, D.M., Overpeck, J.T. and Gupta, A.K., 2002. Increase in the Asian Southwest Monsoon during the Past Four Centuries. *Science*, **297**, 596-599.

Ashahi, K. and Watanabe, T. 2000. Past and recent glacier fluctuations in Kanchenjunga Himal, Nepal. *J. of Nepal Geological Society*, **22,** 481-490.

Bard, E., 1988. Correction of accelerator mass spectrometry ^{14}C ages measured in planktonic foraminifera: Paleoceanographic implications. *Paleoceanography*, **3,** 635-645.

Bhalme, H.N. and Mooley, D.A., 1981. Cyclic fluctuations in the flood area and relationship with the double (Hale) sunspot cycle. *J. Appl. Meteorol.*, **20,** 1041-1048.

Bhattacharya, A., Lamarche, V.C. Jr. and Telewski, F.W., 1988. Dendrochronological reconnaissance of the conifers of North West India. *Tree ring Bulletin*, **48,** 21-30.

Bhattacharyya, A. and Chaudhary, V., 2003. Late-summer temperature reconstruction of the Eastern Himalayan Region based on tree-ring data of Abies densa. *Arctic, Antarctic, and Alpine Research*, **35(2),** 196-202.

Bhattacharyya, A., LaMarche, V. C. Jr and Hughes, M.K., 1992. Tree ring chronologies from Nepal. *Tree ring Bulletin*, **52,** 59-66.

Bond, G., Kromer, B., Beer, J., Muscheler, R., Evans, M.N., Showers, W., Hoffman, S., Lotti-Bond, R., Hajdas, I. and Bonani, G., 2001. Persistent solar influence on North Atlantic climate during the Holocene. *Science*, **294,** 2130-2136.

Borgaonkar, H.P., Pant, G.B. and Kumar, K.R., 1994. Dendroclimatic reconstruction of summer precipitation at Srinagar, Kashmir, India since the late-eighteenth century. *The Holocene*, **4,** 229-306.

Borgaonkar, H.P., Pant, G.B. and Kumar, K.R., 1996. Ring-width variations in *Cedrus deodara* and its climatic response over the western Himalaya. *International Journal of Climatology*, **16:** 1409-1422.

Briffa, K. R., Osborn, T.J., Schweingruber, F.H., Harris, I.C., Jones, P.D., Shiyatov, S.G. and Vaganov, E.A., 2001. Low frequency temperature variations from northern tree ring density network. *Journal of Geophysical Research*, **106,** 2929-2941.

Bryson, R.A. and Swain, A.M. 1981. Holocene variations of monsoon rainfall in Rajasthan. *Quaternary Research*, **16,** 135-145.

Buckley, B.M., Cook, B.I., Bhattachrya, A., Dukpa, D. and Chaudhary, V., 2005. Global surface temperature signals in pine ring width chronologies form southern monsoon Asia. *Geophys. Res. Lett.*, **32,** L20704. doi:10.1029/2005GL023745.

Chakraborty, S. and Ramesh, R., 1997. Environmental significance of carbon and oxygen isotope ratios of banded corals from Lakshadweep, India. *Quat. Int.*, **37,** 55-65.

Chakraborty, S. and Ramesh, R., 1998. Stable isotopes variations in a coral (*Favia speciosa*) from the Gulf of Kutch during 1948-1989 A.D.: Environmental implications. *Proc. Ind. Acad. Sci.* (Earth Planet Sci.), **107,** 331-341.

Chakraborty, S. and Ramesh, R., 1993. Monsoon induced sea surface temperature changes recorded in Indian coral. *Terra Nova*, **5,** 546-551.

Chalise, S.R., 1994. Mountain Environment and Climate Change in the Hindu-Kush Himalaya. *In:* M. Beniston (eds.), Mountain Environments in Changing Climates, London, Routeledge. 382-404.

Chamyal, L.S. and Juyal, N., 2008. Late Quaternary studies in parts of India: Implications for Monsoon variability. *J. Geol. Soc. India*, **71,** 611-629.

Charles, C.D., Hunter, D.E. and Fairbanks, R.G., 1997. Interaction between the ENSO and the Asian Monsoon in a coral record of tropical climate. *Science*, **277,** 925-928.

Chaudhary, V., Bhattacharya, A. and Yadav, R.R., 1999. Tree ring studies in the eastern Himlayan region: Prospects and problems. *IIWA Journal*, **20(3),** 317-324.

Chaudhary, V. and Bhattacharyya, A., 2002. Suitability of *Pinus kesiya* in Shillong, Meghalaya for tree-ring analyses. *Current Science*, **83(8),** 1010-1015.

Chaudhary, V. and Bhattacharyya, A., 2000. Tree ring analysis of *Larix griffithiana* from the Eastern Himalayas in the reconstruction of past temperature. *Current Science*, **79(12)**, 1712-1716.

Chauhan, M.S., Pokharia, A.K. and Singh, I.B., 2006. Holocene environmental scenario and human subsistence strategies in the Middle Ganga Plain: A palynological assessment from Lahuradewa Lake. Proceedings of International Seminar on First Farmers in Global Perspective. Uttar Pradesh State Archaeology Department, Lucknow, January 18-20, 2006, Abstract, 2-4.

Chauhan, O.S., 2003. Past 20,000 years history of Himalayan aridity: Evidence from the oxygen isotope record of Bay of Bengal. *Current Science.*, **84,** 90-93.

Chauhan, O.S., Patil, S.K. and Suneethi, J., 2004. Fluvial influx and weathering history of the Himalayas since last glacial maxima – Isotopic, sedimentological and magnetic records from the Bay of Bengal. *Current Science.*, **87(4),** 509-515.

Chauhan, O.S. and Vogelang. A., 2006. Climate induced changes in the circulation and dispersal patterns of fluvial sources during late Quaternary in the middle Bengal fan. *J. Earth System Science.*, **115(3),** 379-386.

Choudhury, A.M., Haque, M.A. and Quadir, D.A., 1997. Consequences of Global warming and sea level rise in Bangladesh. *Marine Geodesy*, **20,** 13-31.

Choudhury, A.M., Quadir, D.A., Neelormi, S. and Ahmed, A.U. 2003. Climatic changes and its impact on water resources of Bangladesh. *In*: Climatic Changes and its water resources in South Asia (A. Muhammad, ed.), pp 21-60. Asiatic AgroDev International, Islamabad.

Chowdhury, A., Dandekar, M.M. and Raut, P.S., 1989. Variability in drought indices in India - A statistical approach. *Mausam,* **37,** 471-482.

Clark, C.O., Cole, J.E. and Webster, P.J., 2000. Indian Ocean SST and Indian summer rainfall: predictive relationships and their decadal variability. *J. Climate*, **13,** 2503-2519.

Cole, J.E., Dunbar, R.B., McClanahan, T.R. and Muthiga, N.A., 2000. Tropical Pacific forcing of decadal SST variability in the western Indian Ocean over the past two centuries. *Science*, **287,** 617-619.

Cook, E.R., Krusic, P.J. and Jones, P.D., 2003. Dendroclimatic signals in long tree-ring chronologies from the Himalayas of Nepal. *International Journal of Climatology*, **23,** 707-732.

Denniston, R.F., Gonzalez, L.A., Asmerom, Y., Sharma, R.H. and Reagen, M.K., 2000. Speleothem evidences for changes in the Indian summer monsoon precipitation over the last 2300 years. *Quaternary Research*, **53,** 196-202.

Dhar, O.N. and Mandal, B.N., 1981. Greatest observed one-day point and aerial rainfall of India. *PAGEOPH*, **119,** 922-933.

Domroes, M., 1979. Temporal and spatial variations of rainfall, 1978. *J. Nepal Res. Cen.*, **2/3** (1978/79), 49-67.

Doose-Rolinski, H., Rogalla, U., Scheeder, G., Lückge, A. and von Rad, U., 2001. High-Resolution Temperature and Evaporation Changes during the Late Holocene in the Northeastern Arabian Sea. *Paleoceanography*, **16(4),** 358-367.

Dube, S.K., Chittibabu, P., Sinha, P.C. and Rao, A.D., 2004. Numerical modelling of storm surges in the Head Bay of Bengal using location specific model. *Natural Hazards*, **31,** 437-453.

Dube, S.K., 2005. Numerical modelling of storm surge in the Bay of Bengal and the Arabian Sea. Proceedings of the Seminar on 'Tropical Cyclones and Storm Surges in South Asian Region' held in Dhaka, 20-22 December 2003. Published by SMRC, Dhaka.

Dube, S.K., Rao, A.D., Sinha, P.C., Murty, T.S. and Bahulayan, N., 1997. Storm surge in the Bay of Bengal and Arabian Sea: The problem and its prediction. *Mausam*, **48,** 283-304.

Dubey, B., Yadav, R.R., Singh, J. and Chaturvedi, R., 2003. Upward shift of Himalayan pine in western Himalaya, India. *Current Science*, **85,** 1135-1136.

Dvorak, V.F., 1984. Tropical Cyclone Intensity Analysis using satellite data. NOAA Tech. Rep NESDIS 11, US Dept. Commerce, Washington DC, 20233, 47 pp.

Easterling, D.R., Horton, B., Jones, P.D., Peterson, T.C., Karl, T.R., Parker, D.E., Salinger, M.J., Razuvayev, V., Plummer, N., Jamason, P. and Folland, C.K., 1997. Maximum and minimum temperature trends for the globe. *Science*, **277,** 364-366.

Ely, L.L., Enzel, Y., Baker, V.R., Kale, V.S. and Mishra, S., 1996. Changes in the magnitude and frequency of Holocene monsoon floods on the Narmada River, Central India. *Bulletin of the Geological Society of America*, **108,** 1134-1148.

Emanuel, K.A., 1987. The dependence of hurricane intensity on climate. *Nature*, **326,** 483-485.

Emanuel, K.A., 2005. Increasing destructiveness of tropical cyclones over the past 30 years. *Nature*, **436,** 686-688.

Esper, J., 2000. Long term tree ring variation in Juniperus at the upper timer line in the Karakoram (Pakistan). *The Holocene*, **10,** 253-260.

Esper, J., Schweingruber, F.H. and Winiger, M., 2002. 1300 years of climatic history for western Central Asia inferred from tree rings. *The Holocene*, **12,** 267-277.

Fairbanks, R., 1989. A 17,000-year glacio-eustatic sea level record: Influence of glacial melting rates on the Younger Dryas event and deep ocean circulation. *Nature*, **342,** 637-642.

Fleitmann, D., Burns, S.J., Mudelsee, M., Neff, U., Kramers, J., Mangini, A. and Matter, A., 2003. Holocene forcing of the Indian monsoon recorded in a stalagmite from southern Oman. *Science*, **300,** 1737-1739.

Gadgil, S. and Seshagiriarao, P.R., 2000. Farming strategies for a variable climate – A challenge. *Current Science*, **78(10),** 1203-1215.

Gadgil, S., Seshagirirao, P.R. and Narharirao, K., 2002. Use of climate information for farm level decision making: rainfed groundnut in southern India. *Agricultural Systems*, **74,** 431-457.

Gadgil, S., Rajeevan, M. and Nanjundiah, R., 2005. Monsoon prediction – Why another failure? *Current Science*, **88,** 1389-1400.

Gadgil, S., 2003. The Indian Monsoon and its variability, *Ann. Rev. Earth and Planet. Sci.*, **31,** 429-467.

Galy, V., Fracois, L., France-Lanord, C., Faure, P., Kudrass, H., Palhol, F. and Singh, S.K., 2008. *Quat. Sci. Revs*., **27,** 1396-1409.

Ghosh, S.K. (1977). Prediction of Storm surges on the coast of India. *Ind. J. Meteo. Geophys*., **28,** 157-168.

Goodbred, S. L., Jr. and Kuehl, S.A., 1999. Holocene and modern sediment Budget for the Ganges-Brahmaputra river system: Evidence of high stand dispersal to flood-plain, shelf, and deep-sea depocenters. *Geology*, **27,** 559-562.

Goodbred, S. L., Jr. and Kuehl, S.A., 2000. Enormous Ganges-Brahmaputra sediment discharge during strengthened early Holocene monsoon. *Geology*, **28,** 1083-1086.

Goswami, B.N. and Xavier, P.K., 2005. ENSO control on the south Asian Monsoon through the length of the rainy season. *Geophys. Res. Lett*., **32,** L18717, doi:10.1029/2005GL023216..

Goswami, B.N., Venugopal, V., Sengupta, D., Madhusoodanan, M.S. and Xavier, P.K., 2006. Increasing trend of extreme rain events over India in a warming environment. *Science*, **114,** 1442-1445.

Guilderson, T.P., Reimer, P.J. and Brown, T.A., 2005. The boon and bane of radiocarbon dating. *Science*, **307,** 362-364.

Gupta, A.K., 2004. Origin of agriculture and domestication of plants and animals linked to early Holocene climate amelioration. *Current Science*, **87,** 54-59.

Gupta, A.K., Anderson, D.M. and Overpeck, J.T., 2003. Abrupt changes in the Asian southwest monsoon during the Holocene and their links to the North Atlantic Ocean. *Nature*, **421,** 354-357.

Gupta, A.K., Anderson, D.M., Pandey, D.N. and Singhvi, A.K., 2006. Adaptation and human migration, and evidence of agriculture coincident with changes in the India summer monsoon during the Holocene. *Current Science*, **90,** 1082-1090.

Gupta, A.K., Das, M. and Anderson, D.M., 2005. Solar Influence on the Indian summer monsoon during the Holocene. *Geophysical Research Letters*, **32,** L17703.

Gupta, A.K., Singh, R.K., Thomas, E. and Joseph, S., 2004. Indian Ocean high-productivity Event (10-8 Ma): Linked to Global Cooling or to the Initiation of the Indian Monsoons? *Geology*, **32,** 753-756.

Hait, A. and Behling, H., 2005. Responses of Mangroves to Holocene climatic changes, western Ganga-Brahmaputra delta, India. Poster presented at the open science meeting of IGBP-PAGES core project, Beijing Aug 9-11. www.pages.unibe.ch.

Hingane, L.S., Patil, S.D. and Rupa Kumar, K., 1985. Long term trends of surface air temperatures in India. *J. Clim.*, **5,** 521-528.

Hong, Y.T., Hong, B., Lin, Q.H., Zhu, Y.X., Shibata, Y., Hitota, M., Uchida, M., Leng, X.T., Jiang, H.B., Xu, H., Wang, H. and Yi, L., 2003. Correlation between Indian Ocean summer monsoon and North Atlantic climate during the Holocene. *Earth and Planetary Science Letters*, **211,** 371-380.

Hormann, K., 1994. Computer-based Climatological Maps for High Mountain Areas, New methods and their application with examples from the Himalayas. MEM series 12, ICIMOD, Katmandu Nepal, 1-33.

Hughes, M.K. and Davies, A.C., 1987. Dendroclimatology in Kashmir using tree ring widths densities in subalpine conifers. *In:* Methods of Dendrochronology: East-West Approaches, L.A. Kairiukstis, Z. Bednarz and E. Feliksik (eds). IIASA/ Polish Academy of Sciences, pp. 163-176.

Hughes, M.K., 1992. Dendroclimatic evidence from western Himalaya. *In:* Climate since AD 1500, R.S. Bradley and P.D. Jones (eds). Routledge, London, pp. 415-531.

Ivanochko, T.S., Ganeshram, R.S., Brummer, G.A., Ganssen, S.J.A., Jung, S.G. and Moreton, D. Kroon, 2005. Variations in tropical convection as an amplifier of global climate change at the millennial scale. *Earth Planet. Sci. Lett.*, **235,** 302-314.

Jha, P.K., 1992. Environment and Man in Nepal. Craftsman Press, Bangkok.

Jones, R.N., Hennessy, K.J., Page, C.M., Pittock, A.B., Suppiah, R., Walsh, K.J.E. and Whetton, P.H., CSIRO Atmospheric Research Report to the South Pacific.

Juyal, N., Kar, A., Rajaguru, S.N. and Singhvi, A.K., 2003. Luminescence chronology of aeolian deposition during the late Quaternary on the southern margin of Thar Desert. *Quat. Int.*, **104,** 87-98.

Juyal, N., Chamyal, L.S., Bhandari, S., Bhushan, R. and Singhvi, A.K., 2006. Continental record of the south west monsoon during the last 130ka: evidence for the southern margin of the Thar Desert, India. *Quat. Sci. Revs.*, **25,** 2632-2650.

Kale, V.S., 1999. Late Holocene temporal patterns of palaeofloods in central and western India. *Man and Environment*, **24,** 109-115.

Kale, V.S., 2008. Palaeoflood Hydrology in the Indian context. *J. Geol. Soc. India*, **71,** 56-66.

Kale, V.S., Mishra, S. and Baker, V.R., 1997. A 2000-year palaeoflood record from Sakarghat on Narmada, central India. *Journal of the Geological Society of India*, **50,** 285-288.

Kale, V.S., Singhvi, A.K., Mishra, P.K. and Banerjee, D., 2000. Sedimentary records and luminescence chronology of Late Holocene palaeofloods in the Luni River, Thar Desert, northwest India. *Catena*, **40,** 337-358.

Kale, V.S., Mishra, S. and Baker, V.R., 2003a. Sedimentary records of palaeofloods in the bedrock gorges of the Tapi and Narmada Rivers, central India. *Current Science*, **84,** 1072-1079.

Kale, V.S., Gupta, A. and Singhvi, A.K., 2003b. Late Pleistocene-Holocene palaeohydrology of Monsoon Asia. *In:* K.J. Gregory and G. Benito (Eds.)

Palaeohydrology: Understanding Global Change. John Wiley and Sons Ltd, Chichester, 213-232.

Kale, V.S. and Baker, V.R., 2006. An Extraordinary Period of Low-magnitude Floods coinciding with the Little Ice Age: Palaeoflood Evidence from Central and Western India. *The Journal Geol. Soc. Ind.*, **68,** 477-483.

Kang, S., Mayewski, P.A., Qin, D., Yan, Y., Hou, S., Zhang, D., Ren, J. and Kreutz, K., 2002. Glaciochemical records from a Mt. Everest ice core: Relationship to atmospheric circulation over Asia. *Atmospheric Environment*, **36,** 3351-3361.

Kar, A., Felix, C. Rajaguru, S.N. and Singhvi, A.K., 1996. Late Holocene growth and mobility of a transverse dune in Thar Desert. *J. Arid. Env.*, **38(2),** 175-185.

Karl, T.R., Jones, P.D., Knight, R.W., Kukla, G., Plummer, N., Razuvayev, V., Gallo, K.P., Lindseay, J., Charlton, R.J. and Peterson, T.C., 1993. Asymmetric trends of daily maximum and minimum temperature. *Bull. Amer. Met Soc.*, **74,** 1007-1023.

Karmaker, S. (Shrestha M.L. Ed.), 1998. The impact of tropical cyclones on the coastal regions of the SAARC countries and their influence in the region. Report No.1, SAARC Meteorological Research Centre (SMRC), 329 pp.

Karmaker, S. and Shrestha, M.L., 2000. Recent climate change in Bangladesh. Report No. 4, SAARC Meteorological Research Centre (SMRC), Dhaka, Bangladesh.

Khan, T.M.A., Quadir, D.A., Akhtar, M. and Alam, M., 1999. Sea level changes and coastal erosion problems along Bangladesh coast. *Journal of Remote Sensing and Environment*, **3,** 87-103.

Klein, R., Tudhope, A.W., Chilcott, C.P., Patzold, J., Abdul Karim, Z., Fine, M., Fallick, A.E. and Loya, Y. 1997. Evaluating southern Red Sea corals as a proxy record for the Asian Monsoon. *Earth Planet. Sci. Lett.* **148**, 381-394.

Kothawale, D.R. and Rupa Kumar, K., 2002. Tropospheric temperature variation over India and links with the Indian summer monsoon: 1971-2000. *Mausam*, **53,** 289-308.

Kothawale, D.R. and Rupa Kumar, K., 2005. On the recent changes in surface temperature trends over India. *Geophys. Res. Lett.*, **32,** L18714, doi. 10.1029/ 2005GL023528.

Kripalani, R.H., Inamdar, S. and Sontakke, N.A., 1996. Rainfall variability over Bangladesh and Nepal: Comparison and connection with features over India. *Int. J. Climatol.*, **16,** 689-703.

Krishna Kumar, K., Deshpande, N.R. and Rupa Kumar, K., 1997. Long-term changes in the heavy rainfall events over India. Workshop on Indices and Indicators for Climate Extremes, NCDC, Ashville, USA, June 3-6, 1997.

Krishna Kumar, K., Rajagopalan, B. and Cane, M.A., 1999. On the weakening relationship between the Indian monsoon and ENSO. *Science*, **284,** 2156-2159.

Krishna Kumar, K., Soman, M.K. and Rupa Kumar, K., 1995. Seasonal forecasting of Indian summer monsoon rainfall. *Weather*, **50(12),** 449-467.

Krishnan, R. and Ramanathan, V., 2002. Evidence of Surface Cooling from Absorbing Aerosols. *Geophy. Res. Lett.*, **29(9),** 10.1029/2002GL014687.

Kroon, D., Steens, T. and Troelstra, S.R., 1991. Onset of monsoonal related upwelling in the western Arabian Sea as revealed by planktonic foraminifers. *In:* Prell, W.L., Niitsuma, N., et al. Proceedings of the Ocean Drilling Program, Scientific results, 117: College Station, Texas, Ocean Drilling Program, 257–263.

Kudrass, H.R., Hofmann, A., Doose, H., Emeis, K., and Erlenkeuser, H., 2001. Modulation and amplification of climatic changes in the Northern Hemisphere by the Indian summer monsoon during the past 80 k.y. *Geology*, **29,** 63-66.

Kumar, K.K., Rajagopalan, B. and Cane, M.A., 1999. On the weakening relationship between the Indian monsoon and ENSO. *Science*, **284,** 2156-2159.

Kumaran, K.P.N., Nair, K.M., Shindikar, M., Limaye, R.B. and Padmalal, D., 2005. Stratigraphical and palynological appraisal of the late Quaternary mangrove deposits of the west coast of India. *Quaternary Research*, **64,** 418-431.

Lal, M., 2001. Tropical cyclone in a warmer world. *Current Science*, **80**(9), 1103-1104.

Lami, A., Guilizzoni, P., Marchetti, A., Bettinetti, R. and Smith, D.J., 1997. Paleolimnological evidence of environmental changes in some high altitude Himalayan lakes (Nepal). *Mem. Ist. Ital. Idrobiol.*, **57,** 103-130.

Liu, X. and Chen, B., 2000. Climate Warming in the Tibetan Plateau during recent decades. *International Journal of Climatlogy,* **20,** 1729-1792.

Malla, U.M., 1968. Climatic elements and seasons in Katmandu valley. *The Himalayan Review*, Special Issue, Nepal Geographical Society, 53-77.

Mandal, G.S., 1989. Low frequency oscillation and seasonal variability of tropical cyclones. Topics chairman and rapporteur's Report of the second WMO workshop on Tropical Cyclones (IWTC-II), WMO/TD No. 319.

Mandal, G.S., 1990. Low frequency oscillation and seasonal variability of tropical cyclones of north Indian Ocean. Ph. D. Thesis, Jadavpur University, pp. 406-444.

Mandal, M., Mohanty, U.C. and Rahman, S., 2004. A study on the Impact of Parameterization of physical processes on prediction of tropical cyclones over the Bay of Bengal with NCAR/PSU Mesoscale Model. *Natural Hazards*, **31,** 391-414.

Mani, A., 1981. The Climate of the Himalaya. *In:* J. Lall and S. Moddie (Eds.), The Himalaya: Aspects of Change. Oxford University Press, Delhi, 3-15.

Mann, M.E., Bradley, R.S. and Hughes, M.K., 1999. Northern hemisphere temperatures during the past millennium: Inferences, uncertainties and limitations. *Geophys. Res. Lett.*, **26,** 759-762.

Mayewski, P.A., Rohling, E., Stager, C., Karlén, K., Maasch, K., Meeker, L.D., Meyerson, E., Gasse, F., van Kreveld, S., Holmgren, K., Lee-Thorp, J., Rosqvist, G., Rack, F., Staubwasser, M. and Schneider, R., 2004. Holocene climate variability. *Quaternary Research*, **62,** 243-255.

Mickler, P.J., Stern, L.A. and Banner, J.L., 2006. Large kinetic isotope fractionation in modern speleothems. *Bull Geol. Soc. Amer.*, **118,** 65-81.

Mishra, D.K. and Gupta, G.R., 1976. Estimation of maximum wind speed in tropical cyclones occurring in India Sea. *Indian J. Met. Hydr. and Geophys.*, **27(3),** 285-290.

Mitra, A.P., 2003. Climate in south Asia and water resources – Introductory note. *In:* Climate changes and Water resources in south Asia (A. Muhammad, Ed.). Asiatic AgroDev International, Islamabad.

Mooley, D.A. and Parthasarathy, B., 1984. Fluctuations of All-India summer monsoon rainfall during 1871-1978. *Climatic Change*, **6,** 287-301.

Morril, C., Overpack, J.T. and Cole, J.E., 2003. A synthesis of abrupt changes in Asian summer monsoon since the last deglaciation. *The Holocene*, **13(4),** 465-473.

Mukherjee, B.K., 1990. Middle Atmospheric Dynamics and Monsoon variability. *Mausam*, **41**, 203-208.

Murty, T.S., 1984. Storm surges: Meteorological Ocean tides. Department of Fisheries and Oceans. Ottawa, Canada.

Murty, T.S., Flather, R.A. and Henri, R.F., 1986. The storm surge problem in the Bay of Bengal. *Prog. Oceanog.*, **16,** 195-233.

Nayava, J.L., 1974. Heavy monsoon rainfall in Nepal. *Weather Roy. Met. Soc.*, **29,** 443-450.

Nayava, J.L. 1980. Rainfall in Nepal. *The Himalayan Review*, Nepal Geographical Society, **12,** 1-18.

Neff, U., Burns, S.J., Mangini, A., Mudelsee, M., Fleitmann, D. and Matter, A., 2001. Strong coherence between solar variability and the monsoon in Oman between 9 and 6 kyr ago. *Nature*, **411,** 290-293.

Neumann, C.J., 1993. Global Guide to Tropical Cyclone Forecasting. WMO, Geneva.

OCED, 2003. Development and Climate Change: Focus on water resources and hydropower. Organization for Cooperation and Development, Paris. 66 pp.

Pandey, D.N., Gupta, A.K. and Anderson, D.M., 2003. Rainwater harvesting as an adaptation to climate change. *Current Science*, **85,** 46-59.

Pant, G.B. and Rupa Kumar, K., 1997. Climates of South Asia. John Wiley & Sons. Chichester, UK, 320 pp.

Pant, G.B., 1979. Role of tree ring analysis and related studies in palaeoclimatology: Preliminary survey and scope for the Indian region. *Mausam*, **30,** 439-448.

Pant, G.B. and Hingane, L.S., 1988. Climatic Change in and around the Rajasthan desert during the 20th Century. *J. Climatol.*, **8,** 391-401.

Pant, G.B. and Parthasarathy, B., 1981. Some aspects of an association between the southern oscillation and Indian summer monsoon. *Arch. Meteorol. Geophys. Biokl.*, **Sr. B., 29,** 245-251.

Pant, G.B., Rupa Kumar, K., Parthasarathy, B. and Borgaonkar, H.P., 1988. Long term variability of the Indian summer monsoon and related parameters. *Adv. Atmos. Sci.*, **5,** 469-481.

Parnetta, J.C., 1991. Cities and Oceanic Islands: A case study of Male, capital of the Republic of Maldives. *In:* Impact of Sea level rise on on cities and regions: Cities on Water (ed. R. Frassetto), Venice, Italy.

Parthasarathy, B., Rupa Kumar, K. and Munot, A.A., 1991. Evidence of secular variations in Indian summer monsoon rainfall-circulation relationships. *J. Climate*, **4,** 927-938.

Parthasarathy, B., 1984. Interannual and long-term variability of Indian summer monsoon rainfall. *Proc. Indian Acad. Sci.* (Earth Planet. Sci.), **93,** 371-385.

Parthasarathy, B., 1958. Some aspects of the rainfall in India during southwest monsoon season. *In:* Monsoons of the World, New Delhi, Met Dept., 185-194.

Patwardhan, S.K. and Bhalme, H.N., 2001. A study of cyclonic disturbances over India and the adjacent ocean. *Int. J. Climatol.*, **21,** 527-534.

Pfeiffer, M., W.-C. Dullo and Eisenhauer, A., 2004. Variability of intertropical convergence zone recorded in coral isotopic record from the central Indian Ocean (Chagos Archipelago), *Quaternary Research*, **61**, 245-255.

Phadtare, N.R., 2003. Pollen-inferred full Holocene climate history of the Dokriani Glacier area in Garhwal Himalaya, India. *In:* Late Quaternary Environment Change: Emerging Issues. French Institute, Pondicherry, pp. 169-172.

Phadtare, N.R., 2000. Sharp decrease in summer monsoon strength 4000-3500 cal yr BP in the central Higher Himalaya of India based on pollen evidence from alpine peat. *Quat. Res.* **53,** 122-129.

Phadtare, N.R. and Pant, R.K., 2005. High-resolution studies on the Holocene climate changes and monsoon variability in the Kumaon-Garhwal Himalaya. Unpublished project report (ESS/23/VES/089/2001), Department of Science & Technology, Government of India, pp. 42.

Phadtare, N.R. and Pant R.K., 2006. A century-scale pollen record of vegetation and climate history during the past 3500 years in the Pinder Valley, Kumaon Higher Himalaya, India. *J. Geol. Soc. India,* **68(3),** 495-506.

Pielke, R.A., Jr. and Landsea C.W., 1999. La Niña, El Niño, and Atlantic Hurricane Damages in the United States. *Bull. Amer. Meteor. Soc.*, **80,** 2027-2033.

Prasad, O., 2005. Storm Surge Prediction in North Indian Ocean. Proceedings of the Seminar on 'Tropical Cyclones and Storm Surges in South Asian Region' held in Dhaka, 20-22 December 2003, published by SMRC, Dhaka.

Quade, J., Cerling, T., Bowman, J.R., 1989. Development of Asian monsoon revealed by marked ecological shifts during the latest Miocene in the northern Pakistan. *Nature*, **342,** 163-166.

Quadir, D.A., 2006a. Tropical Storms and surges over South Asia. Paper for contribution to the START MAIRS, SAARC Meteorological Research Centre (SMRC), Dhaka, Bangladesh.

Quadir, D.A., 2006b. Climate variability of Bangladesh. Paper for contribution to the START MAIRS, SAARC Meteorological Research Centre (SMRC), Dhaka, Bangladesh.

Quadir, D.A., Khan, T.M.A., Hossain, M.A. and Iqbal, A., 2003. Study of climate variability and its impacts on rice yield in Bangladesh. *SAARC Journal of Agriculture*, **1,** 69-83.

Quadir, D.A., Shrestha, M.L., Khan, T.M.A., Ferdousi, N., Rahman, M. and Mannan, A.M., 2004. Variation of surface air temperature over the land areas in and around the Bay of Bengal. *Natural Hazards*, **32,** 561-584.

Quadir, D.A., Rahman, A., Osman, S., Saha, G.C. and Bhatacharjee, S.R., 2002. Final Report of the project on 'Initial National Communication in Response to the UN Framework Conventions on Climate Change (UNFCCC)', Ministry of Environment, Government of Peoples Republic of Bangladesh.

Rajeevan, M., Pai, D.S., Dikshit, S.K. and Kelkar, R., 2004. IMDs new operational models for long range forecasting of south west monsoon over India and their verification for 2003. *Currrent Science*, **86(3),** 422-431.

Ramage, C., 1971. Monsoon meteorology. International Geophysical Series, Academic Press, San Diego, Calif., Vol. 15, 296 pp.

Ramanathan, V., Chung, C., Kim, D., Bettge, T., Buja, J.T., Kielh, W.M., Washington, W.M., Fu, Q., Sikka, D.R. and Wild, M., 2005. Impact of Atmospheric brown cloud: Impact on south Asian climate and hydrological cycle. *Proc. National Acad. Sci. (USA)*, **102(15),** 5326-5333.

Ramaswamy, C., 1962. Breaks in the Indian summer monsoon as a phenomenon of interaction between the easterly and the sub-tropical westerly jet streams. *Tellus*, **14,** 337-349.

Rao, Y.P. and Desai, B.N., 1973. The Indian summer monsoon. *Meteor. Geophys. Rev.*, **4,** 1-18.

Rawson, R.R., 1963. The monsoon lands of Asia, Hutchinson Educational Ltd., London, 256p. Regional Environment Programme, 1999, p. 69.

Rind, D. and Overpeck, J.T., 1993. Hypothesized causes of decade- to century-scale climatic variability: Climate model results. *Quaternary Science Reviews*, **12,** 357-374.

Robinson, H., 1976. Monsoon Asia. Macdonald and Evans Ltd., Estover, Plymonth (Third Edition), 510 pp.

Roy, P.D., Juyal, N., Smykazkloss, W. and Singhvi, A.K. (in prep). Paleo-hydrology of the saline playas of the Thar Desert during the late pleistocene: A synthesis based on Geochemical attributes.

Roy, S., Harris, R.N., Rao, R.U.M. and Chapman, D.S., 2002. Climate change in India inferred from geothermal observations. *J. Geophys. Res.*, **107(B7),** 10.1029/2001JB000536.

Rühland, K., Smol, J.P., Phadtare, N.R., Pant, R.K. and Sangode, S.J., 2005. Changes in climate over the last 3000 years triggered marked changes in a Western Himalayan peatland, India. PAGES Open Science Meeting II, Beijing, China (poster).

Rupa Kumar, K., Hingane, L.S. and Ramana Murthy, Bh.V., 1987. Variations of tropospheric temperatures over India during 1944-85. *J. Climate Appl. Meteorol.*, **26,** 304-314.

Rupa Kumar, K., Krishna Kumar, K. and Pant, G.B., 1994. Diurnal asymmetry of surface temperature trends over India. *Geophys. Res. Lett.*, **21,** 677-680.

Rupa Kumar, K., Pant, G.B., Parthasarathy, B. and Sontakke, N.A., 1992. Spatial and subseasonal patterns of the long-term trends of Indian summer monsoon rainfall. *Int. J. Climatol.*, **12,** 257-268.

Rupa Kumar, K., Krishna Kumar, K., Ashrit, R.G., Patwardhan, S.K. and Pant, G.B., 2002. Climate Change in India: Observations and Model Projections. *In:* Climate Change and India (Eds: P.R. Shukla, S.K. Sharma and P.V. Ramana). Tata McGraw-Hill Ltd., New Delhi, pp. 24-75.

Sarkar, A., Ramesh, R., Somayajulu, B.L.K., Agnihotri, R., Jull, A.J.T. and Burr, G.S., 2000. High resolution Holocence monsoon record from the Eastern Arabian Sea. *Earth Planet Sci. Lett.*, **177,** 209-218.

Schulz, H., von Rad, U. and Erlenkeusser, H., 1998. Correlations between Arabian Sea and Greenland climate oscillations of the past 110,000 years. *Nature*, **393,** 54-57.

Sharma, S., Joachimski, M.M., Sharma, M., Tobschall, H.J., Singh, I.B., Sharma, C., Chauhan, M.S. and Morgenroth, G., 2004a. Lateglacial and Holocene environmental changes in Ganga plain, Northern India. *Quaternary Science Reviews*, **23,** 145-159.

Sharma, S., Joachimski, M.M., Tobschall, H.J., Singh, I.B., Tiwari, D.P. and Tewari, R., 2004b. Oxygen isotopes of bovid teeth as archives of plaoeclmatic variation in archeological deposits of Ganga plain. *Quat. Res.*, **62,** 19-28.

Sharma, S., Joachimski, M.M., Tobschall, H.J., Singh, I.B, Sharma, C. and Chauhan, M.S. 2006. Correlative evidence of monsoon variability, vegetation change and human habitation in Senai lake deposit, Ganga plain. *Curr. Sci.*, **90,** 973-978.

Sheikh, M.M. and Manzoor, N., 2005. Region-wise Climate Changes in Pakistan (1951-2000). Proceedings of the APN National Workshop: "Global Change

Perspective in Pakistan: Challenges, Impacts, Opportunities and Prospects", 28-30 April, 2005 Islamabad, Pakistan.

Sheikh, M.M., Manzoor, N., and Khan, A.M., 2009. Climate Profile and Past Climate Changes in Pakistan. Global Change Impact Studies Centre (GCISC), Islamabad Research Report: GCISC-RR-01, June, 2009.

Sheikh, M.M., Manzoor, N., Ashraf, J., Adnan, M., Khan, A.M. and Hameed, S., 2009. Trends in Extreme Daily Rainfall and Temperature Indices over South Asia. Global Change Impact Studies Centre (GCISC) Research Report (under publication).

Shrestha, A.B., Cameron, P.W., Paul, A.M. and Jack, E.D., 1999. Maximum temperature trends in the Himalaya and its vicinity: An analysis based on temperature records from Nepal for the period 1971-94. *J. Climate*, **12,** 2775-2786.

Shrestha, A.B., Cameron, P.W., Paul, A.M. and Jack, P.D., 2000. Precipitation fluctuations in the Nepal Himalaya and its vicinity and relationship with some large scale climatological parameters. *Int. J. Climatol.*, **20,** 317-327.

Shrestha, K.L., Shrestha, M.L., Shakya, N.M., Ghimire, M.L. and Sapkota, B.K., 2003. Climate change and water resources of Nepal. *In:* Climate Change and Water Resources in South Asia. (A. Muhammed, Ed.), pp. 259. Asianics Agro Dev International, Katmandu.

Shrestha, M.L., 2000. Interannual variation of summer monsoon rainfall over Nepal and its relation to Southern Oscillation Index. *Meteorl. Atoms. Phys.*, **75,** 21-28.

Shukla, J., 1975. Effect of Arabian Sea surface temperature anomaly on Indian summer monsoon: A numerical experiment with the GFDL model. *J. Atmos. Sci.* **32,** 503-511.

Singh, J. and Yadav, R.R., 2005. Spring precipitation variations over the western Himalaya, India, since A.D. 1731 as deduced from tree rings. *Journal of Geophysical Research*, **110,** D01110, doi: 10.1029/2004JD004855.

Singh, J., Yadav, R.R., Dubey, B. and Chaturvedi, R., 2004. Millennium-long ring width chronology of Himalayan cedar from Garhwal Himalaya and its potential in climate change studies. *Current Science*, **86,** 590-593.

Singh, N. and Sontakke, N.A., 2002. On climatic fluctuations and environmental changes of the Indo-Gangetic plains, India. *Climatic Change*, **52,** 287-313.

Singh, N., Pant, G.B. and Muley, S.S., 1992. Spatial variability of aridity over northern India. *Proc. Indian Acad. Sci.* (Earth Planet Sci.), **101,** 201-213.

Singh, S.V., Mohile, C.M. and Inamdar, S.R., 1987. Relationship of southern oscillation and other large scale features with the Bay of Bengal cyclones during the post-monsoon seasons. *Advance in Atmospheric Science*, 169-174.

Singh, O.P. and Khan, T.M.A. 1999. Changes in the frequencies of cyclonic storms and depressions over the Bay of Bengal and the Arabian Sea. SMRC Report-2 SAARC Meteorological Research Centre, Agargaon, Dhaka, Bangladesh.

Singh, O. P., Khan, T.M.A. and Rahman, M.S., 2000. The vulnerability assessment of the SAARC coastal region due to sea level rise; Bangladesh case. SMRC publication No. 3, 111 pp.

Singh, O.P., Khan, T.M.A. and Rahman, M.S., 2001. Has the frequency of intense tropical cyclones increased in the north Indian Ocean? *Current Science*, **80,** 575-580

Singhvi, A.K. and Kar, A., 2004. The Aeolian sedimentation record of the Thar Desert. *In:* Quaternary History and Paleoenvironmental record of the Thar Desert, India (A.K. Singhvi, Ed.). *Proc. Ind. Acad .Sci.* (Earth and Planetary Science), **113,** 371-401.

Sirocko, F., Sarnthein, M., Erlenkeuser, H., Lange, M., Arnold, M. and Duplessey, J-C, 1993. Century scale events in monsoon climate over the past 24000 years. *Nature*, **364,** 322-324.

Solanki, S.K., Usoskin, I.G., Kromer, B., Schüssler, M. and Beer, J., 2004. Unusual activity of the sun during recent decades compared to the previous 11,000 years. *Nature*, **431,** 1084-1087.

Soman, M.K. and Krishna Kumar, K., 1990. Some aspects of daily rainfall distribution over India during southwest monsoon season. *Int. J. Climatol.*, **10**, 299-311.

Srivastava, P., Singh, I.B., Sharma, M. and Singhvi, A.K., 2003a. Luminescence chronometry and late Quaternary Geomorphic history of the Ganga Plain. *Paleogeography, Paleoclimatology, Paleoecology,* **197,** 15-41.

Srivastava, P., Singh, I.B., Sharma, S., Shukla, U.K. and Singhvi, A.K., 2003b. Late Pleistocence-Holocence hydrologic changes in the interfluve areas of central Ganga plains. *Geomorphology*, **54,** 279-292.

Staubwasser, M., Sirocko, F., Grootes, P.M. and Segl, M., 2003. Climate change at the 4.2 ka BP termination of the Indus valley civilization and Holocene south Asian monsoon. *Geophys. Res. Lett.* **30(8),** 1425.

Thamban, M., Rao, V.P. and Kawahata, H., 2007. Indian summer monsoon variability during the Holocene as recorded in sediments of the Arabian Sea: Timing and implications. *Jour. Oceanography*, **63,** 1009-1020.

Thomas, J.V., Kar, A., Kailath, A.J., Juyal, N., Rajaguru, S.N. and Singhvi, A.K., 1999. Late Pleistocene-Holocene Aeolian accumulation in Thar Desert. *Z.Geomorph. suppl.* **116,** 181-194.

Thomas, P.J., Juyal, N., Kale, V.S. and Singhvi, A.K., 2007. Luminescence chronology of late Holocene extreme climatic events in the Upper Pennar basin. *J. Quat. Sci.,* DOI: 10.1002/jqs1097.

Thompson, L.G., Yao, T., Moseley-Thompson, E., Davis, M.E., Henderson, K.A. and Lin, P. N., 2000. A high-resolution millennial record of the south Asian Monsoon from Himalayan ice core. *Science*, **289,** 1916-1919.

Tudhope, A.W., Lea, D.W., Shimmield, C.B., Chilcott, C.P. and Head, S., 1996 Monsoon climate and Arabian Sea coastal upwelling recorded in massive corals from Southern Oman. *Palaios*, **11**. 347-361.

Torrence, C. and Webster, P.J., 1999. Interdecadal changes in ENSO – Monsoon system. *J. Climate*, **12,** 2679-2690.

Tyson, P., Steffen, W., Mitra, A.P., Fu, C. and Lebel, L., 2001. The Earth system: regional-global linkages. *Reg. Environ. Change*, **2,** 128-140.

Von Rad, U., Schaaf, M., Michels, K.H., Schulz, H., Berger, W.H. and Sirocko, F., 1999. A 5000 year record of climate change in varved sediments of oxygen minimum zone off Pakistan, Northeastern Arabian Sea. *Quat. Res.*, **51,** 39-53.

Webster, P.J., Moore, A.M., Loschnigg, J.P. and Leben, R.R., 1999. Coupled ocean-atmosphere dynamics in the Indian Ocean during 1997-98. *Nature*, **401,** 356-360.

Williams, M.A.J., Pal, J.N., Jaiswal, M. and Singhvi, A.K., 2006. River response to Quaternary fluctuations: evidence from the Son and Belan valleys, south central India. *Quat. Sci., Revs.*, **25,** 2619-2631.

Yadav, R.R. and Park, W.-K., 2000. Precipitation reconstruction using ring-width chronology of Himalayan cedar from western Himalaya: *Preliminary results. Proceeding of Indian Academy of Sciences* (Earth and Planetary Sciences), **109(3),** 339-345.

Yadav, R.R. and Singh, J., 2002. Tree-ring-based spring temperature patterns over the past four centuries in western Himalaya. *Quat. Res.*, **57,** 299-305.

Yadav, R.R., Park, W.-K. and Bhattacharyya, A., 1997. Dendroclimatic reconstruction of April-May temperature fluctuations in the western Himalayan region of India since A.D. 1698. *Quat. Res.*, **48,** 187-191.

Yadav, R.R., Park, W.-K. and Bhattacharyya, A., 1999. Spring temperature fluctuations in the western Himalayan region as reconstructed from tree-rings: A.D. 1390-1988. *The Holocene*, **9(1),** 185-90.

Yadav, R.R., Park, W.-K., Singh, J. and Dubey, B., 2004. Do the western Himalayas defy global Warming? *Geophysical Research Letters*, **31(17),** L17201101029/2004GLO202.

Yadava, M.G. and Ramesh, R., 2005. Monsoon reconstruction from radiocarbon dated tropical Indian speleothems. *The Holocene*, **15,** 48-59.

Yoshino, M.M., 1984. Climate and agricultural land use in monsoon Asia. Univ. of Tokyo Press, Japan, 398p.

4 Land Transformation and Its Consequences in South Asia

V.K. Dadhwal and A. Velmurugan

1. LAND TRANSFORMATION

1.1 Introduction

Concerns about global and regional land use/cover change arose from the realization that land transformation influences climate change and reduces biotic diversity; hence the interest in deforestation, desertification, and other changes in natural vegetation. The more recent focus on issues related to ecosystem goods and services, sustainability, and vulnerability has led to a greater emphasis on the dynamic coupling between human societies and their ecosystems at a local scale.

The pace, magnitude and spatial reach of human alterations of the earth's land surface is ever changing and it is driven by the interplay of various causes and, hence, the land surface modification or conversion varies globally. These changes are complex and disjunctive process. Changes in land cover and in the way people use the land have become increasingly recognized over the last 15 years as important global environmental changes in their own right (Turner, 2002). The concept of land and land use/land cover change need to be described in clear terms so as to synthesize and compare various data sets on land use/land cover at various scales.

Land is the most important natural resource, which embodies soil, water and associated flora and fauna involving the total ecosystem. According to FAO (1999), land refers to a delineable area of the earth's terrestrial surface, encompassing all attributes of the biosphere immediately above or below this surface, including those of the near-surface climate, the soil and terrain forms, the surface of hydrology (including shallow lakes, rivers, marshes and swamps), the near-surface sedimentary layers and associated groundwater and geo-hydrological reserve, the plant and animal populations, the human settlements pattern and physical results of past and present human activities (terracing, water storage or drainage structure, roads, buildings, etc.).

The term "land use" (LU) is often used improperly to describe some regional to global datasets which contain a mixture of both "land use" and "land cover" information. "Land use" is in reality quite distinct from "land cover". de Bie (2000) defines LU as "A series of operations on land, carried out by humans, with the intention to obtain products and/or benefits through using land resources". In contrast, land cover is defined as "the observed bio-physical cover on the Earth's surface" (FAO, 2000). According to various sources quoted in Briassoulis (2000), land cover may be described as the physical, chemical, ecological or biological categorization of the terrestrial surface, while land use refers to the human purposes that are associated with that cover.

Comprehensive information on the spatial distribution of land use/land cover categories and the pattern of their change/transformation is a prerequisite for planning, utilization and management of the land resources. Land transformation takes two forms: *conversion* from one category of land use/ land cover (LU/LC) to another and *modification* of condition within a category. Conversion is the better documented and more readily monitored of the two, but too great an emphasis on it obscures important forms of land modification.

There is a functional complexity within types of land-cover change, and a structural complexity between types of land-cover change, both in terms of spatial arrangements and temporal patterns of change. The relationship between land cover and various LU occurring within a given land cover unit is complex. Therefore, land-cover change needs to be measured in its complexity in order to fully understand it. In the present discussion we describe land transformation as land use/land cover conversion and/or modification.

1.2 Scope and Importance

One of the clearest manifestations of human activity within the biosphere has been the conversion of natural landscapes to highly managed ecosystem, such as croplands, pastures, forest plantations, and urban area (Ramankutty et al., 2002). Land use and land cover change (land transformation) are significant to a range of themes and issues central to the study of global environmental change. Alterations in the earth's surface hold major implications for the global radiation balance and energy fluxes, contribute to changes in biogeochemical cycles, alter hydrological cycles, and influence ecological balances and complexity. Through these environmental impacts at local, regional and global levels, land-use and land-cover changes driven by human activity have the potential to significantly affect food security and the sustainability of the world agricultural and forest product supply systems.

The South Asian region as a whole is experiencing expansion and intensification of crop land, shrinking forest and grass lands, rapid urban growth, land degradation of different kind and intensity coupled with higher population growth and poverty apart from changes in socio-political framework. Hence, the understanding of land cover/land use change, historical as well as current, is vital to formulate appropriate policy option towards sustainable development of the region. But, land should be recognised as valuable in its natural state and is not simply seen as a raw material for developmental activities alone.

1.3 World Trends

Over the last few decades, numerous researchers have improved the measurements of land cover change at regional to global scale, the understanding of the causes and predictive models of land use/cover change. Historical changes in permanent cropland is estimated at a global scale during the last 300 years by spatializing historical cropland inventory data based on a global land cover classification derived by remote sensing, which used a hind casting approach (Ramankutty and Foley, 1999), or based on historical population density data (Goldewijk, 2001). The area of cropland has increased globally from an estimated 300–400 million ha in 1700 to 1500–1800 million ha in 1990, a 4.5- to five-fold increase in three centuries and a 50% net increase just in the twentieth century. Europe, the Indo-Gangetic Plain, and eastern China experienced first the most rapid cropland expansion during the eighteenth century. Meanwhile the intensification of use of lands already cultivated (modification) was accelerated (Meyer and Turner, 1992). The 1.97-fold increase in world food production from 1961 to 1996 was associated with only a 10% increase of land under cultivation but also with a 1.68-fold increase in the amount of irrigated cropland and a 6.87- and 3.48-fold increase in the global annual rate of nitrogen and phosphorus fertilization (Tilman, 1999).

The area under pasture, for which more uncertainties remain, increased from around 500 million ha in 1700 to around 3100 million ha in 1990 (Goldewijk and Ramankutty, 2003). These increases led to the clearing of forests and the transformation of natural grasslands, steppes, and savannas. Forest area decreased from 5000-6200 million ha in 1700 to 4300-5300 million ha in 1990. The net global decrease in forest area was 9.4 million hectares per year from 1990 to 2000 (FAO, 2001).

Steppes, savannas, and grasslands also experienced a rapid decline, from around 3200 million ha in 1700 to 1800-2700 million ha in 1990 (Ramankutty and Foley, 1999; Goldewijk, 2001) and the details are given in Table 1.

During the 1990s, forest-cover changes were much more frequent in the tropics than in the other parts of the world. In particular, the Amazon basin and Southeast Asia contain a concentration of deforestation hotspots. The Asian continent includes most of the main areas of degraded dryland (Lepers et al., 2005).

Table 1: Historical changes in land use/cover at a global scale over the last 300 years

Year	*Forest/woodland (10^6 ha)*	*Steppe/savanna/grassland (10^6 ha)*	*Cropland (10^6 ha)*	*Pasture (10^6 ha)*
1700	5000 to 6200	3200	300 to 400	400 to 500
1990	4300 to 5300	1800 to 2700	1500 to 1800	3100 to 3300

Apart from crop land and natural cover, built-up or paved-over areas are roughly estimated to occupy from 2% to 3% of the earth's land surface (Young, 1999; UN, 2002). In 2000, globally, towns and cities sheltered more than 2.9 billion people (Grubler, 1994). Urbanization affects land in rural areas through the ecological footprint of cities which includes, but is not restricted to, the consumption of prime agricultural land in peri-urban areas for residential, infrastructure, and amenity uses, which blurs the distinction between cities and countryside, especially in western developed countries.

South Asia is one of the most densely populated and developing regions of the world and due to various driving forces the land use/land cover has been undergoing rapid transformations which have implications on the production base and sustainability of the entire region and environmental consequences ranging from regional to global scale.

2. BIO-PHYSICAL SETTINGS OF SOUTH ASIA

The location of South Asia is one of the unique in the world. The region is bounded in the south by the Indian Ocean, in the south-east by the Bay of Bengal and in the south-west by the Arabian Sea. The South Asian region covering Bangladesh, Bhutan, India, Maldives, Nepal, Pakistan and Sri Lanka, spans an area of about 4.5 million km^2 and accounts for a population of about 1.3 billion (FAO, 2006). They represent almost 22% of the world population from only 3.31% of the world's total land mass and have GDP of 565 billion (US $) in 1998 which is only 1.97% of world GNP (WB, 2000).

2.1 Bio-geophysical Zoning

South Asia occupies a major portion of the Indo-Malayan realm and a smaller portion of the Palaearctic realm. This region is representative of five of the fourteen major ecological regions called 'biomes', which demonstrate the biodiversity and vegetation patterns of this region as determined by climate, water, geology, soil and diverse topography. Hence, a land transformation has a major impact on biomes of this region.

2.2 Physiography and Relief

South Asia's topography consists of an amazing variety of mountains, plateaus, dry regions, intervening structural basins, etc. The South Asian region can be divided into two main land units: the ancient land mass of Peninsular India, and the geologically young Himalayas and associated ranges. Peninsular India, including Sri Lanka, consists of a single tectonic structure which originally formed a part of Gondwanaland. The diversity in the region's physiography is exhibited in some of the most spectacular natural sites like Mount Everest, the world's highest mountain, in Nepal; Sunderbans, the largest mangrove swamp in the world shared by India and Bangladesh; the temperate forests in Bhutan; the deserts and arid regions of north-west India and south Pakistan; high altitude cold deserts; major plateau regions in south India and the Arabian Peninsula; great structural basins and river plains; and beautiful coral reef lagoons and atolls in Maldives and the Lakshadweep islands (India), which dot the Indian Ocean, and the Andaman islands (India) in the Bay of Bengal.

The South Asian region, in general, slopes from the north to the east and the west. While the north-western part is home to the flat Indus plains and the Balochistan plateau, partly covered by deserts. The kingdom of Bhutan in the north is mostly mountainous (with a few fertile valleys and savannas), while southern India rests on the relatively flatter Deccan plateau. Sri Lanka, on the other hand, is a tropical island with three-quarters of plains in the south, and massive mountains covering the rest of its land area. Maldives, at the southern end of the region, is a country of lowlying islands rich in corals. The region has a long stretch of coastline extending to about 10,000 km from Pakistan to Bangladesh.

2.3 Climate

A monsoon climate, characterised by wet summers and dry winters, generally prevails over South Asia. The south-west monsoons (late May to October) bring the maximum rainfall to the region, followed by the northeast monsoons. Due to the variations in land forms, precipitation and climate vary significantly from place to place in different countries within the region. Cyclones, brought forth by the south-west monsoons, are a common phenomenon in coastal areas of India, Bangladesh and Sri Lanka. The climate also varies from the semi-arid in Pakistan to the tropical monsoon and hot-dry, humid-dry in the rest of the region. However, localised climatic conditions prevail in hilly regions. The region also witnesses marked variations in temperature as well, ranging from as low as –20°C in the cold deserts to a scorching 48°C in desert areas and some plains.

2.4 Rivers

Several important river systems in South Asia originate in upstream countries and then flow to the other countries. The Indus River originates in China and the basin spans north-western India and Pakistan; the Ganga-Brahmaputra river systems originate partly in China, Nepal and Bhutan, and flow to India and Bangladesh; some minor rivers drain into Bangladesh and Nepal. In Sri Lanka, there are 103 rivers draining water in a radial pattern from the high watersheds. The longest river, the Mahaweli, provides water to 16 per cent of the island. Maldives does not have any rivers, but small brackish ponds can be found on some islands.

2.5 Vegetation and Forests

The diversity in the latitude, altitude, climate and topography has resulted in a variety of vegetation in the region, ranging from the temperate and the tropical to the dry (desert) vegetation. A monsoon rainfall pattern, which is characteristic and typical of the region, plays a crucial role in determining the vegetation type. The topography and soil type of the region do not have much role to play in determining the vegetation type except in the Himalayan region and the deserts. The vegetation of the South Asian region is principally distinguished into four types based on rainfall i.e., Evergreen forests, Deciduous monsoon forests, Dry forests and scrubland and Desert and semi-desert vegetation.

Forests occupy 18.6 per cent of the total land area of the region and account for 2.73 per cent of the total forest area in the world. Protected areas cover about five per cent of the region's land area, offering shelter to some of the most endangered and threatened species of plants and animals. Owing to the diversity of climate, soil and vegetation types, South Asia houses approximately 15.6 per cent of the world's flora and 12 per cent of its fauna (State of the Environment – South Asia, 2001).

3. LAND TRANSFORMATION/CHANGES IN SOUTH ASIA

3.1 Geographical Distribution

South Asia is one of the largest regions of Asia and as stated in the earlier section it has mountains, river valleys, plateau etc., with wide variety of land use and land cover categories. The geographical area distribution (Fig. 1) of the South Asian nations is as wide as its physiography and India is the largest nation (73.3%) and Maldives the smallest (0.01%) one. Pakistan is the second largest country (17.7%) followed by Nepal (3.3%) and Bangladesh (3.2%). The land use/land cover and its transformation to some extent are also decided by the size of the country (in terms of per capita land availability for various purposes), population pressure and geographical location among other parameters. The smallest island nation like Maldives has neither big

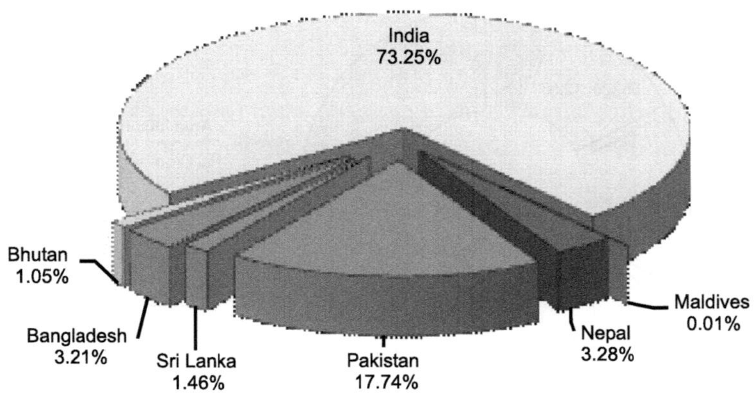

Fig. 1: Land area distribution in South Asia.
(Total geographical area 4.5 million km^2)
(*Source:* FAO, 2006)

mountain ranges nor larger forest area cover. Sri Lanka is larger island than Maldives and Nepal is dominated by mountains. India is the largest country with nearly one billion population and has diverse landscape and climate, which offers much scope for land transformation.

3.2 Land Use/Land Cover Change in South Asia

In order to assess the land use/land cover changes over time reliable and comparable data sources are essential. Remote sensing data can be used to derive land use/land cover for the last three decades at varying scales; however, the historical data sets are derived/synthesized from different sources. When compared to the other regions of the world as well as global level studies, synthesis of such data sets with reliability is lacking in South Asia. In the present discussion, we made use of some of the global data sets and country level statistics especially for India.

The major land use/cover identified for assessment are crop land, forest, shrub, savanna/grass, urban and other land use. Figures 2 and 3 are derived from FAO and SAGE data sources on South Asia and indicate that crop land is the dominant land use followed by forest cover. Crop cover occupied nearly 50% of the land area in 2003 as against 38.6% in 1900 indicating a 11.4% increase in crop land area (Fig. 4). A close observation of the trend indicated that the percentage of agricultural area has enormously increased in the region since the past one-and-half decades (SACEP, 2001). Other than crop land and built up area all other land use/land cover classes have decreased during 1900 to 2003. A small increase in other land use is mostly due to change in the definition of various classes and its interpretation along with grouping of new classes found as a result of increase in the accuracy of assessment.

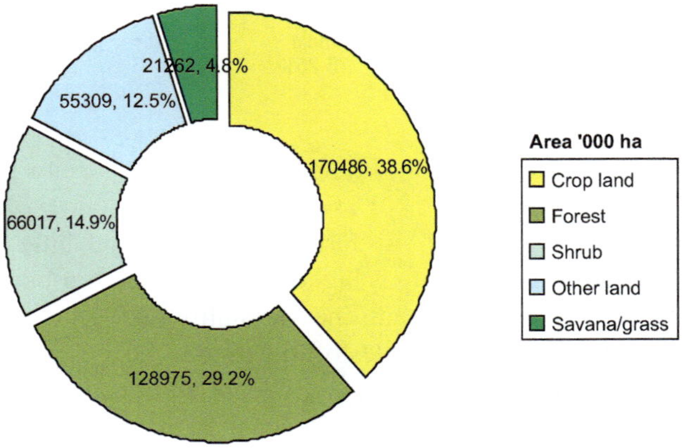

Fig. 2: Land use/land cover distribution in South Asia (1900).
Data source: www.sage.wisc.edu

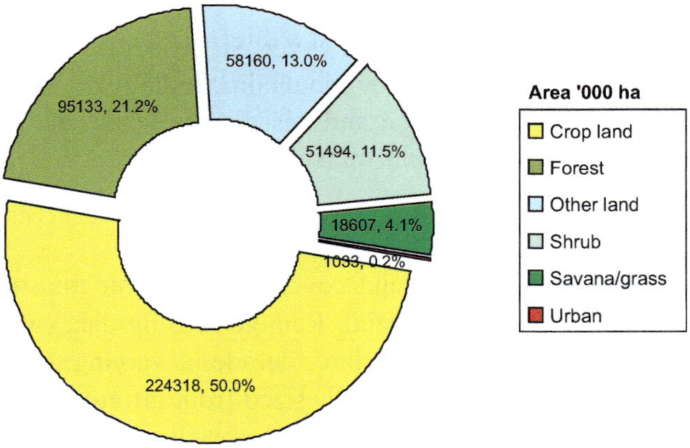

Fig. 3: Land use/land cover distribution in South Asia (2003).
Data source: FAO, 2006 and www.sage.wisc.edu

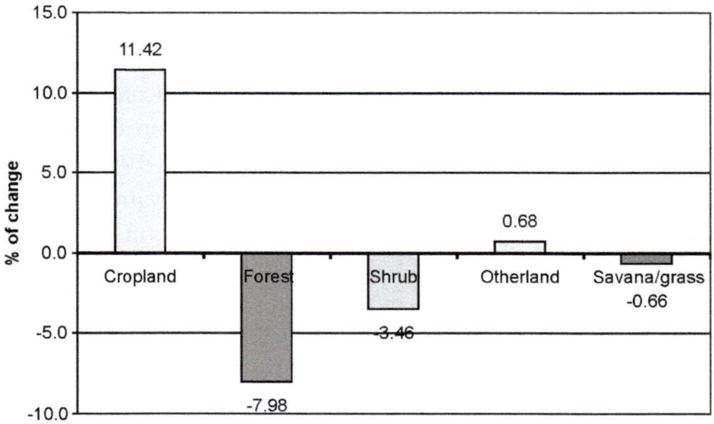

Fig. 4: Land use/land cover change in South Asia (1900 to 2003).

Deforestation occurs when forest is converted to another land cover or when the tree canopy cover falls below a minimum percentage threshold of 10% (FAO, 2001). Deforestation or decrease in forest density occurs in South Asian region as the need for crop and urban land use is increasing especially in the last century and at a much faster rate after their independence from foreign rulers. The forest cover was found to be 29% and 21% of total geographical area in 1900 and 2003 respectively, which recorded an 8% decrease in area under forest (Houghton et al., 2001). The rate of clearing of forest for crop land in South and Southeast Asia in 1952, 1980 and 1991 were 1, 4 and 3 million ha/annum, respectively. Due to national and international efforts the rate of change has been decreasing from 1980 onwards. However, FAO (1999) data indicated that 18.7% of South Asian land area is covered with forest which is more than Asian average (16.4%) and less than the world average (25%).

In South and Southeast Asia the land in shifting cultivation cycle was 4.46, 6.11 and 6.69 Mha/year during 1952, 1980 and 1991 respectively, indicating an increasing trend (Houghton et al., 2001). This showed that the policy on forest protection and rehabilitation of shifting cultivators did not bring desired results; as a consequence, new areas are being brought under shifting cultivation cycle especially in tribal belts. To change this trend national and international efforts have to be mobilized to a greater degree and local population has to be provided with alternate source of livelihood, involving them in forest conservation activities, among other measures. However, Upadhyay (1995) indicated that shifting cultivation is not at all the unmitigated disaster it is usually made out to be. Shifting cultivation, when practiced using traditional methods, is significantly less destructive than current, more intensive agricultural practices. There are both advantages and disadvantages that need to be appreciated before other land use practices can be considered more suitable.

Historically South Asian region had wide variety of grazing animals both wild and domesticated which was supported by savanna and grasslands. But, owing to land conversion either to crop land or urban land uses the land under savanna/grasses got reduced by 0.7%. Other than the above mentioned categories of land cover/land use, wetlands accounts for 13.4 Mha in South Asia. According to the estimates of Asian Wetland Bureau, 15 per cent of all wetland habitats in South Asia are afforded some legal protection, but only 10 per cent is totally protected. Hence, the land got transformed into other uses as influenced by the drivers of such transformation which are described in the subsequent sections.

Key regional trends in South Asia indicated that the area cultivated is expected to show only a marginal increase while irrigated land area will continue to grow slowly and area under shrub and culturable wasteland will decrease. Use of inorganic fertilizers has expanded rapidly in recent decades, from 3 kg of plant nutrients per ha in 1970 to 79 kg/ha in the mid-1990s, and is expected to continue to increase, albeit more slowly. With higher incomes, meat consumption and demand for dairy products are expected to continue their significant expansion which will demand more land under feed and fodder crops. However, the large ruminant population is likely to stabilize, or even decline, as tractors replace both draught buffalo and oxen. The region's 1999 population of about 1344 m is expected to reach 1920 m by 2030, and the proportion living in cities will rise to 53 per cent with its consequent impact on the urban land use (FAO, 2001).

3.3 Country Level Changes

South Asian countries widely differ in their geographical extent and natural settings apart from cultural factors. As a consequence, this change in land use/cover and its complexities also varies with countries. Figure 5 depicts the land use/land cover of South Asian countries for three time periods (1900, 1950 and 1992). However, it is to be noted that there are inconsistencies among various national and international estimates of land use/land cover changes. The consistency and reliability of the data depends largely on the definition, classification scheme and method of data collection. Because of this, figures and reports published by international organisation like FAO, sometimes do not match with the data given by the respective countries and other research organisations. In some cases, international change studies focus or make available only grouped data sets ignoring region level geographical and political changes. Such a situation occurs while assessing land use and land cover transformation for South Asia and India.

Crop land is the major land use in India, Bangladesh and Pakistan followed by forest, except in Pakistan where shrub land is the second largest land cover during these periods. In Bhutan and Nepal forest is the major land cover. This indicates that the primary sector is the major source of livelihood and economic activity in South Asia. This trend in land use over the years coupled with population growth led to over-exploitation or dependence on land resulting in various kinds of land degradation with implications for regional as well as global level.

3.3.1 Agricultural Land Use Patterns

Agricultural land-use data are important for many of the regional to global activities e.g. the validation of agricultural land evaluation; the preparation of perspective studies on agricultural production and food security; early warning for food security; natural disaster relief operations; farming systems studies; and policy formulation (http://www.fao.org/ag/agl/agll/landuse/landusedef.stm). Crop land area has increased almost in all the countries of South Asia (Fig. 6). Sri Lanka recorded a maximum increase in crop land area (120%) followed by Bhutan (92%) and Nepal (77%). This indicates that at the expense of forest cover area, crop land has expanded while these countries undergo economic transformations. Interestingly, majority of these

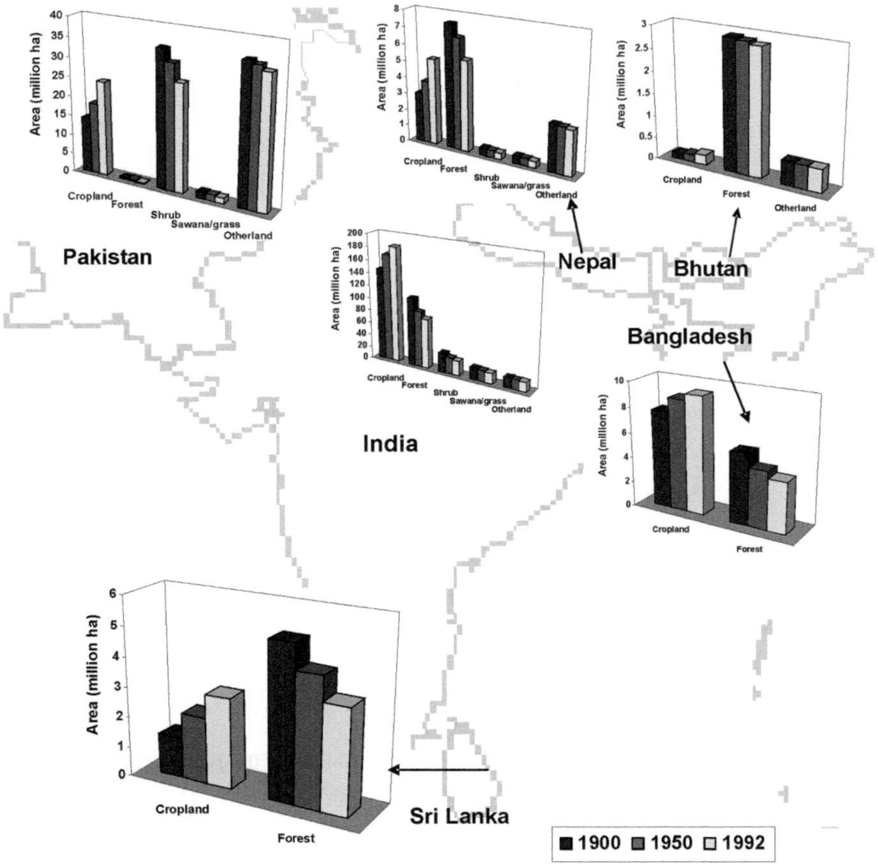

Fig. 5: Land use/land cover changes in South Asian countries. (1900, 1950 and 1992)

area expansions occurred in mountainous or sloppy lands giving ample evidence for possible increase in sediment load in water bodies originating from these countries. However, in terms of absolute area expansion India (38 Mha) and Pakistan (10 Mha) account for 90% of crop land expansion and rest of the countries only 10% of area. Even though cropland area in South Asia has increased but in Bangladesh and India, it has shown a decreasing rate of expansion (CDIAC, 2001; SAGE, 1999). In South Asia, agriculture is characterised by small landholdings in alluvial lowlands, too many people on too little land (population density of Maldives, Bangladesh and India are 1540, 1008 and 314 persons/km^2 respectively), production largely for

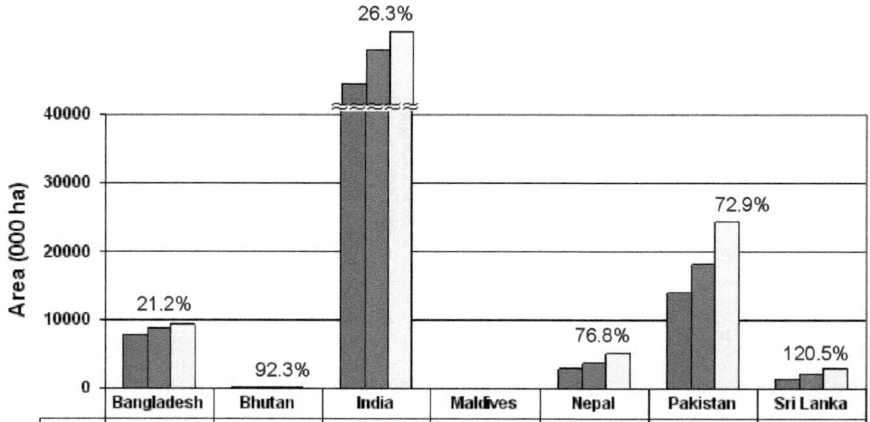

	Bangladesh	Bhutan	India	Maldives	Nepal	Pakistan	Sri Lanka
1900	7709	117	144329	0	2951	14028	1352
1950	8820	158	169108	0	3793	18159	2186
1992	9340	225	182298	0	5216	24258	2981

Values in percentage indicates change in area in1992 over 1900

Fig. 6: Changes in cropland area in South Asia.
Compiled from Ramankutty, SAGE

subsistence, high rates of tenancy, heavy dependence on cereals and other food staples, and pre-modern technologies. Rice followed by wheat is the staple food crops (FAO, 2006).

Farming Systems

In South Asia, during the last three decades, majority of the land transformations are driven by food security and industrial development, the two major goals of any developing region. Hence, it is essential to understand the people and their food supplying system to predict the trend in future land use changes. In South Asia there are eleven farming systems which are depicted in Fig. 7 and most important of these systems from the perspective of population, extent of poverty and potential for growth and poverty reduction i.e., rice, rice-wheat and mixed farming system are briefly described below.

(i) *Rice System* is dominated by intensive wetland rice cultivation in fragmented plots, with or without irrigation. The system is concentrated in Bangladesh and West Bengal, India. Poor farmers operate extremely small areas, and often rely on off-farm income for survival.

(ii) *Rice-Wheat System* is characterized by a summer paddy crop followed by an irrigated winter wheat crop (sometimes a short spring vegetable) which forms a broad swathe from Northern Pakistan through the Indo-Gangetic plain to Northwest Bangladesh. Total area is 97 Mha with an estimated 62 Mha under cultivation, of which around 78 per cent is irrigated. This is the major system in terms of its contribution to the regional food basket.

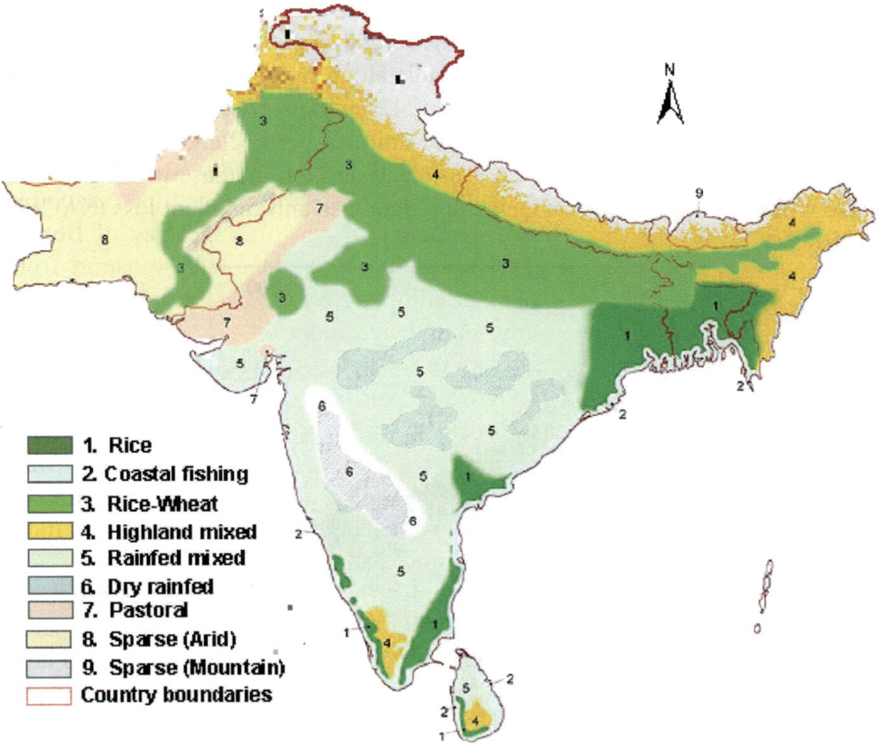

Fig. 7: Major farming systems of South Asia.

(iii) *Mixed System* combines both crop and livestock farming system and is practiced in highland and rainfed land. Highlands are generally intermediate between the rice-wheat plains of the lowlands and the sparsely populated high mountain areas above, and extends along the length of the Himalayan range, as well as in pockets in Southern India and Sri Lanka. Rainfed Mixed System is crop and livestock farming system, occupies the largest area within the sub-continent and is confined almost entirely to India. Vulnerability of the system stems from the substantial climatic and economic variability.

According to FAO estimates, over 60% of the population of the region depends on agricultural sector for their livelihoods and shifting cultivation is practiced in some parts of the region. Bhutan and Nepal have more than 90% of their population involved in agriculture, while in Sri Lanka it is around 37% only. In Bangladesh nearly 15 persons depends on one ha of cultivated land and in India it is 7.1 persons. As a result of this, land is put into intensive use sometimes in unsustainable way to meet the food demand leading to agricultural land degradation. Mono cropping, over-exploitation and heavy use of fertilizers have led to increased soil quality depletion in the region. The fractional change in crop land in 1900 and 1992 are given in Fig. 8 which shows the above mentioned trend. As a result of mixed farming system and current white revolution, India alone supports 20% of the world's livestock population, and an average of 42 animals graze in one hectare of land against the threshold level of five animals per hectare. In Bhutan and Nepal animals graze on the open grass land and the increase in animal herd size will have implication on the state of land use and sediment loss (SACEP, 2001).

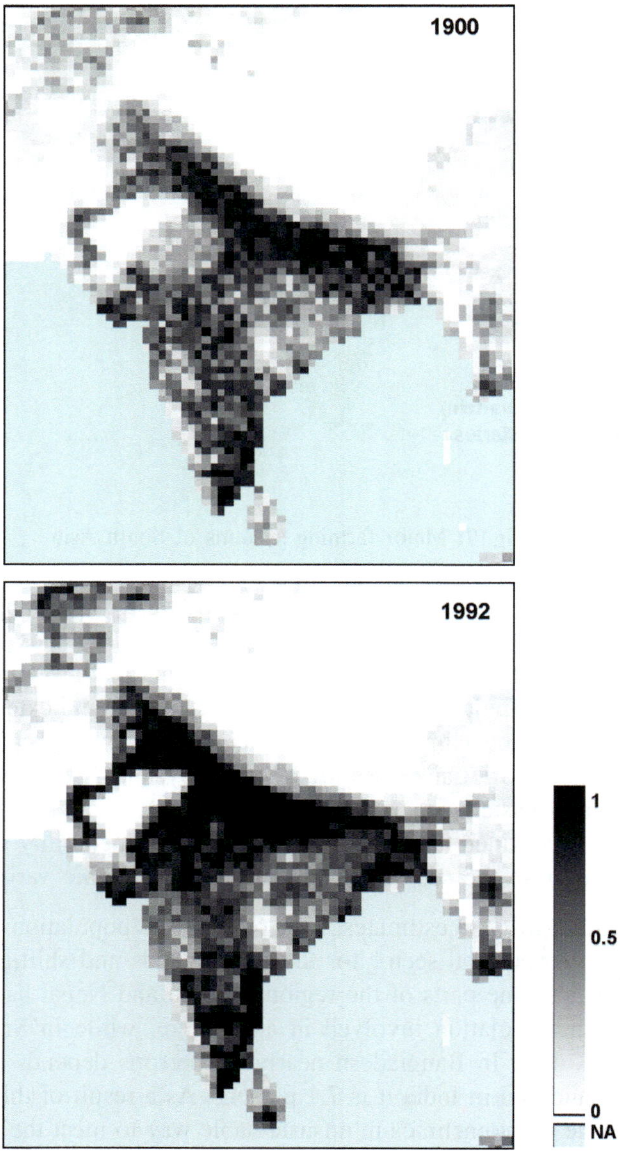

Fig. 8: Distribution of cropland in South Asia.
Source: SAGE, 1999

3.3.2 Forest Cover Change

South Asian countries account for only two per cent of the world's forest area but, not without uncertainty, due to scale and various sub-classes within the forest class at regional and global level. When compared to the world average forest cover (25%) the region's forest cover estimate (Table 2) is lower (18.7%). But "trees outside the forest", particularly on agricultural land, are an important source of wood and non-wood forest products, notably in Bangladesh, Sri Lanka, Pakistan and parts of India. In terms of percentage of land area under forest cover Bhutan (58.6%) followed by Nepal (33.7%) have more area under forest. But area extent under forest cover is more in India (65 Mha) followed by Nepal (4.8 Mha).

Changes in forest cover of South Asian countries is shown in Fig. 9 and indicated that during 1900-1992 the forest cover had decreased in all the South Asian countries (SAGE, 1999). The highest rate of decrease was observed in Pakistan (69%) followed by Bangladesh (29%). In absolute values India (28 million ha) followed by Nepal (two million ha) accounted for nearly 89% of loss of forest land in South Asia. Annual deforestation rate in India is 0.34 Mha, while in Bangladesh and Nepal 3.3 and 1.7 per cent annual forest decrease takes place. Bangladesh records the highest annual deforestation rate from Asia. However, State of Forestry in Asia and the Pacific-2003 report of FAO (2003) indicated that in the last decade India and Bangladesh has shown an increase in their total forest cover due to area expansion under plantation and wood lots. The respective country level statistics also indicated an increase in area under forest plantation mainly to cater to the increasing demand for forest products and to compensate the forest loss. Forestry has been accepted as a farming practice, but at slower

Table 2: State of forest cover distribution in South Asia

Country name	*Forest area (000 ha)*			*#Annual rate of change (%) (1990-2000)*	*@Percentage of total area*	*@Per capita forest (ha)*
	#1990	*@1995*	*#2000*			
Bangladesh	1169	1010	1334	1.3	7.8	0.02
Bhutan	3016	2756	3016	n.s.	58.6	0.75
India	63732	65005	64113	0.1	21.9	0.06
Maldives	1	1	1	n.s.	n.a.	n.a
Nepal	4683	4822	3900	-1.8	33.7	0.2
Pakistan	2755	1748	2361	-1.5	2.3	0.01
Sri Lanka	2288	1796	1940	-1.6	27.8	0.1
South Asia	**77644**	**77137**	**76665**	**-1.2**	**18.7**	**0.06**
Asia		503001			16.4	0.1
World		3454000			25	0.64

@ State of the world's forests, FAO (1999)

State of forestry in Asia and the Pacific – 2003, FAO (2003)

pace and has become limited to the bigger farmers. But, the introduction of fast-growing exotic tree species has changed the composition of the local vegetation to some extent and has resulted in large-scale monocultures of teak, sal, eucalyptus, Mexican pine, etc. Hence, it doesn't contribute to the biodiversity conservation. Gautam et al. (2003) have reported an increase in broadleaf forest, conifer forest and winter-cropped lowland agricultural area and decrease in upland agriculture in Central Nepal between 1976 and 2000, indicating a change in composition within land uses (modification).

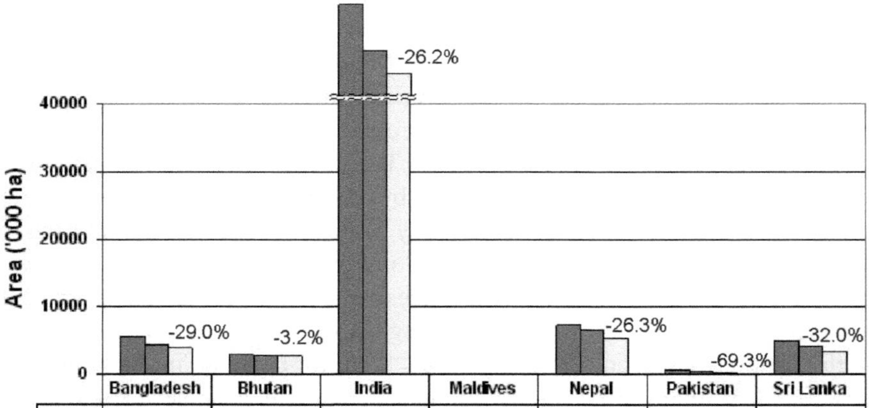

	Bangladesh	Bhutan	India	Maldives	Nepal	Pakistan	Sri Lanka
1900	5624	2917	107339	0	7461	537	5097
1950	4513	2881	88385	0	6732	397	4263
1992	3993	2823	79187	0	5496	165	3469

Values in percentage indicates change in area in1992 over 1900

Fig. 9: Changes in forest area in South Asia.
Compiled from Ramankutty, SAGE

Percapita availability of forest land is affected by the growth rate of population, afforestation and level of forest protection which vary among the South Asian countries. As a result of this Pakistan has only 0.01 ha and Bhutan has 0.75 ha compared with world average of 0.64 ha and Asian average of 0.1 ha. The legally protected land area in Bhutan is 26%, while Sri Lanka, Nepal and Bangladesh has over 10% of land protected. The recorded protected area from India, Pakistan and Maldives is less than 5% of the total land mass.

3.3.3 Wetlands

Wetlands cover approximately 13.4 Mha in South Asia, and include floodplains, marshes, estuaries, lagoons, tidal mudflats, reservoirs, rice paddies, saline expanses, freshwater marshes and swamps. The loss of wetlands worldwide has been estimated at 50% of those that existed since 1900. Since the 1950s, tropical and sub-tropical wetlands have been increasingly degraded or lost through conversion to other use. Worldwide agriculture is the principal cause for wetland loss. By 1985 it was estimated that 57% of available wetland had been drained for intensive agriculture in Europe and North America, 27% in Asia, 6% in South America and 2% in Africa, a total of 26% loss to agriculture (OECD, 1996). In Bangladesh, wetlands cover almost 50 per cent of the total land surface, and are an important source of income and livelihood for several thousands of its people. In Sri Lanka, wetlands account for 15 per cent of the land area (Scott and Poole, 1989). The biological diversity of South Asia's wetlands is high and the region is the global centre of diversity for a number of ecosystems or species groups. Transformation of wetlands have triggered a sequential disaster for large variety of flora and fauna of this region.

3.3.4 Urban Land Use

The South Asian region as a whole is experiencing rapid urban growth, but is still predominantly rural. The urban growth rate remained high throughout the 1970-90 periods. During the 1980s, urban population of the region grew at the rate of 3.0 to 6.5 per cent per annum, which was the second fastest urban growth rate in the world after Africa. Currently, 28.33 per cent of the South Asian population lives in urban areas within which 42% live in slums (World Development Indicators, 2000). It is to be noted that world wide the cities experiencing the most rapid change in urban population between 1990 and 2000 are mostly located in developing countries and are mainly located along the coastal zones and major waterways (Deichmann et al., 2001). India, the major developing country, accounts for majority of the urban area of South Asia followed by Pakistan and Bangladesh (Fig. 10). However, in terms of rate of urbanization Bhutan (5.9%) followed by Nepal (5.2%) recorded the highest growth rate (Fig. 11) which are situated in mountainous areas and urbanization is relatively a recent phenomena. This has implication for other land use.

It is estimated that 1 to 2 million ha of cropland are being taken out of production every year in developing countries to meet the land demand for

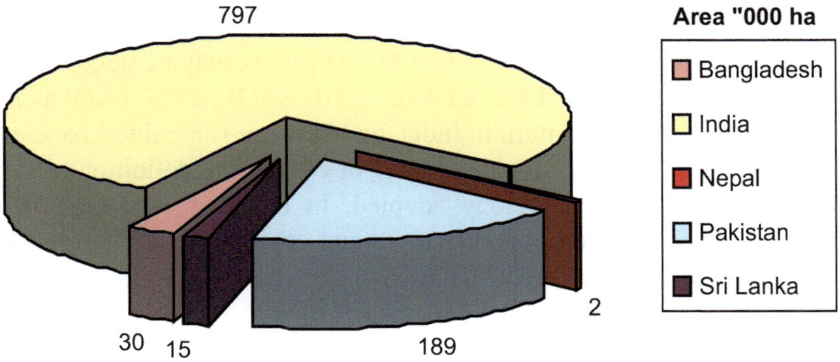

Fig. 10: Distribution of urban area in South Asia, 2003.
Source: FAO, 1999

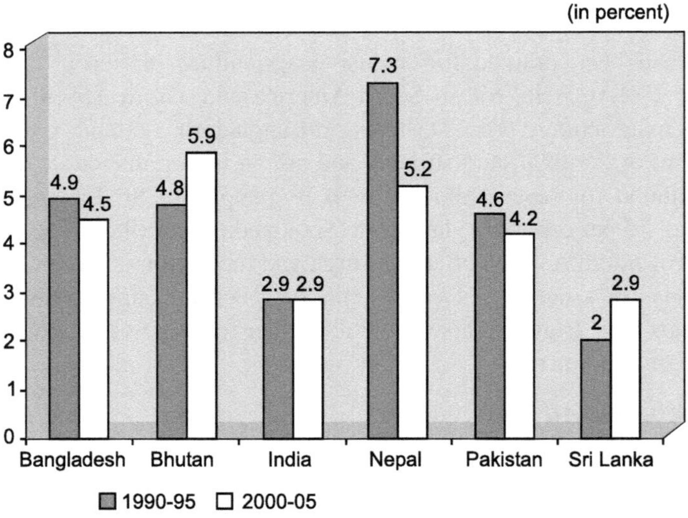

Fig. 11: Rate of urbanization.
Source: State of Envt in South Asia, 2001

housing, industry, infrastructure, and recreation (Doos, 2002). This is likely to take place mostly on prime agricultural land located in coastal plains and in river valleys.

4. INDIAN SCENARIO

The pattern of land use of a country at any particular time is determined by the physical, economic and institutional framework taken together. Land use and land cover pattern of India has undergone tremendous changes due to the impact of green revolution, urbanization and industrialization. Changes in the land use pattern bring associated ecological changes; therefore, it is important to give an overview of land use pattern. In India, out of the total geographical area of 329 Mha, only 305 Mha is reporting area (DES, 2006). Table 3 brings out the changes in land use pattern and Fig. 12 gives distribution of land use pattern from 1950-51 to 1992-93 for six major categories. There are different data sources (DES; CDIAC, 2001; SAGE, 1999; FAO) available to analyse the land use pattern in India, but there are some discrepancies and inconsistency of data mainly due to scale of estimation, definition of various categories and the methodology adopted. In this, we used DES data to analyse the land use pattern and changes and other global and regional data sources to compare the trend.

4.1 Agriculture Intensification

Area under agriculture emerges as the dominant category with an increase in its share from about 42% in 1950-51 to 47% in 1992-93 of reported area

Table 3: Change in land use/land cover in India from 1950 to 1992

Source	*Land use/land cover*	*% of change*
DES (2001)	[1]Cultivated area	13
	Forests	57
	Built-up/non agricultural	118
	[2]Waste land/barren	-49
	Pasture and grazing land	57
	[3]Shrub land	-66
	Geographical Area	329
*CDIAC (2001); Houghton et al. (2001)	Net cultivated area	16
	Forested area	-22
	Settled/build up area	69
	Grass/shrub complex	-4
	Barren/sparsely vegetated	-8
	Wetlands (Non-forested)	-11
	Geographical Area	311
SAGE (1999); Ramankutty and Foley (1999)	Crop land	8
	Forest	-10
	Shrub	-11
	Savanna/grass	-5
	Other land	-2
	Geographical Area	318

* Based on the trend (from 1920 to 1980) estimates for 1992 are extrapolated
[1] includes Land distribution class 5+4b of DES, 2001
[2] includes Land distribution class 2b+3c of DES, 2001
[3] includes Land distribution class 3b+4a of DES, 2001

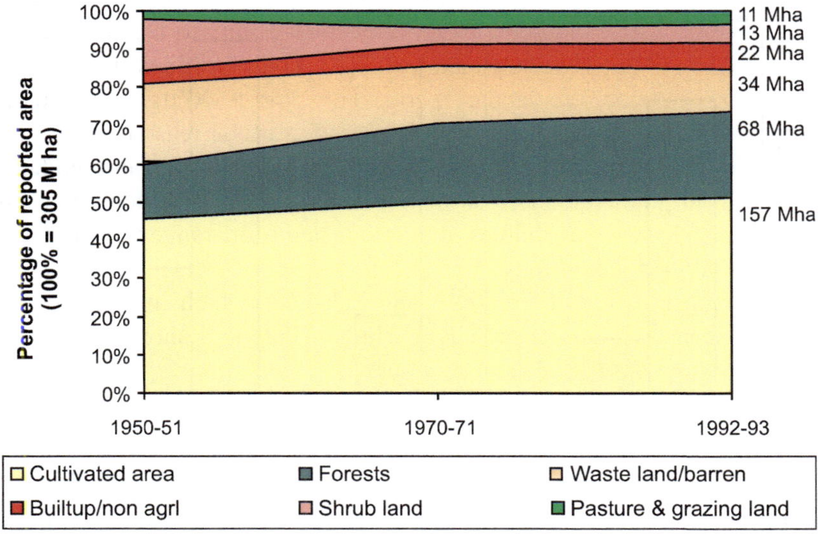

Fig. 12: Distribution of land use in India (1950-51 to 1992-93).

(305 Mha) – an increase of 13% in actual area. However, the trend indicated that the net area sown has increased till 1970-71; afterwards it remained more or less constant. The increase in population and substantial income growth demand an extra about 2.5 million tonnes of food grains annually apart from significant increase needed in the supply of livestock, fish and horticultural products. As a result of these pressures and industrialization, there has been a marked change in supply and use of land, water, fertilizer, seeds and livestock. The important changes which have occurred in agricultural scenario of the country during the period from 1950-51 to 2000-01 were an increase of 23 Mha in net sown area (118.75 to 141.1 Mha) and 18 per cent increase in cropping intensity, resulting in an increased gross cropped area by 55.58 Mha (from 131.89 Mha to 187.94 Mha). This led to a tremendous increase in the food grain production from 50 Mt to 212 Mt, and productivity from 522 kg/ha to 1707 kg/ha during 1950-51 to 2003-04. At the same period net irrigated area has increased from 21 Mha to 51 Mha (Fig. 13). However, these conversions and modifications of land use, generally as a result of green revolution, vary across states.

4.2 Forest Cover Change

Forest is the second major land use category of India which accounted for 22% of reported area in 1992-93. In area extent it increased from 41 Mha in 1951-52 to 68 Mha in 1992-93, an increase of 57% although at decreasing rate, mainly due to afforestation efforts (32 Mha, FAO, 2003) in wasteland/ barren land apart from conservation measures. According to FSI in 1987 the forest cover of India was 19.49% (1:1000,000 scale) and it increased to 20.5% (1:50,000 scale) in 2001. Notwithstanding the positive change in

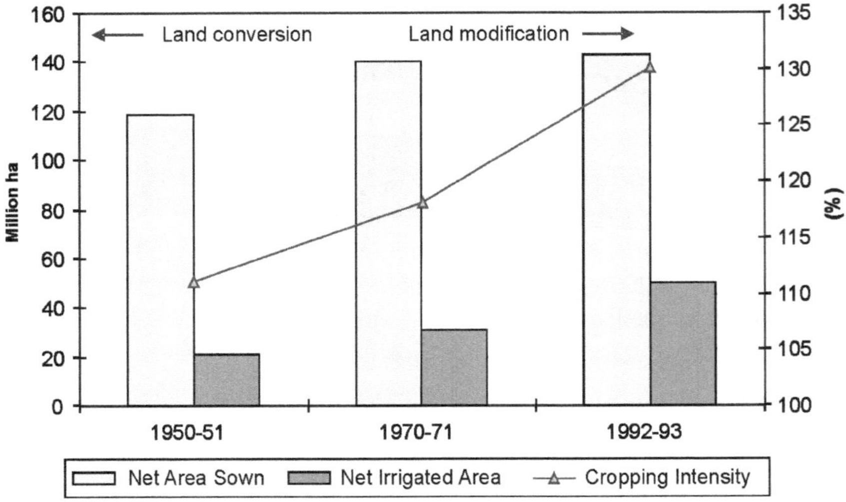

Fig. 13: Intensive and extensive transformation of agricultural land use in India. Compiled from DES, 2001

forest area the fragmentation and disturbance status have increased in various parts of India. But, other data sources have indicated a decreasing trend in forest cover and FAO has shown an increasing trend from 1992 onwards. In line with the decreasing trend Bose et al. (1997) have reported that the annual average rate of forest land diversions was of the order of 0.155 Mha between 1952 and 1980. The official figures on non-forestry diversions relate only to planned and authorized alienation of forest land, illegal encroachments are estimated to be 7 Mha of forest land in various states and union territories (SFR, 1987). In spite of the land diversion there are potential areas in culturable wastelands and land other than current fallows for expansion of forest cover (FSI, 1999).

Along with agriculture and industrial activity, shifting cultivation is a major cause for deforestation and fragmentation. Shifting cultivation (Jhum) is practiced in many states, particularly in the North-Eastern region of the country. There are varying estimates of areas affected by shifting cultivation by different agencies. The area affected by shifting cultivation in the North-Eastern region was reported to be 3.81 Mha (Task Force on shifting cultivation, 1983), 2.80 Mha (North-Eastern Council, 1975) and 7.40 Mha (FAO, 1975). Using remote sensing data FSI (2006) has estimated cumulative area of 1.73 Mha under shifting cultivation. The extent of area under shifting cultivation is maximum (0.39 Mha) in Nagaland followed by Mizoram (0.38 Mha) and Manipur (0.36 Mha).

4.3 Urbanization and Non-agricultural Uses

This category includes all lands occupied by buildings, roads and railways, or under water which accounted for 22 Mha in 1992-93. During recent years, there has been an increase in the area put to non-agricultural uses (118% increase from 1950-51 to 1991-92). Predictably as a result of increased developmental activities, more and more land is being used for industrial sites, housing both in rural and urban areas, transport systems, recreational purposes, irrigation systems etc. The states which account for more than two-thirds of the land under non-agricultural uses are Andhra Pradesh, Madhya Pradesh, Uttar Pradesh, Bihar, Tamil Nadu, Rajasthan, Orissa and Karnataka.

Contrary to popular concepts of a predominantly rural India, an increasingly larger percentage of Indian population lives in the urban areas. Today, India's urban population is second largest in the world after China. According to the 2001 census 307 million people are living in urban areas, which was 30.5% of total population, with the decadal growth rate of 41% resulting in the faster growth rate of million cities in India. Coupled with this, growth of industrial, commercial and institutional activities resulted in large scale land transformations presently witnessed in and around major cities of India. Some of these are very dynamic and rapid particularly in the urban fringe. This helps in spatial extension of urban areas engulfing particularly urban-rural fringe.

4.4 Coastal Transformation

Coastal resources are very critical to the socio-economic development of countries in South Asia as vast majority of the population live and depend on coastal areas. They mostly depend on local resources for their livelihoods and increasingly, coastal areas have become attractive for tourism and recreational purposes in addition to their other uses. But, the balance between the environmental, economic and social values of the region derived from different land uses has often been sub-optimal due to underlying processes that consider short-term economic gains but fail to account for long-term costs and benefits. As a result of this mismanagement or over-exploitation of natural resources, the region is facing degradation of coastal and marine ecosystems at various degrees.

Aquaculture in coastal areas play a significant role in development of poor communities and the area is increasing (Edwards, 2000). But, over-exploitation of coastal areas has accelerated the degradation of mangrove forests which provide ecological benefits such as protection from erosion, flooding, cyclones, typhoons and tidal waves (Primavera, 2000). Over-exploitation by other types of local industry has taken place on land in the many coastal areas, especially the Sundarbans, of India and Bangladesh, the largest mangrove areas in the world. Here, excessive wood harvesting and conversion of coastal areas to fish ponds have caused degradation, and increased flooding (Yeung, 2001; Hinrichsen, 1990). The destruction of the Chakaria Sundarbans in Bangladesh after the spread of extensive shrimp farming activities is well documented by Hossain (2001).

The destruction of mangroves also alters the regional water system and coastal habitats, affecting fisheries (Baran and Hambrey, 1998), coastal ecosystem balances, and climate stability. All of these effects reduce the fertility of land located in nearby areas (Biksham et al., 1996). The leaf litter detritus from mangroves is important to fisheries because it provides an essential source of nutrients for the trophic food web and juvenile fish. It is estimated that 90% of all marine organisms spend some portion of their life cycle within mangrove systems (Adeel and Pomeroy, 2002).

The coral reefs of South Asian region are also at high risk to disturbance status mainly due to a combination of pressures, including destructive fishing, mining, pollution and sedimentation, tourism, trade in coral and global warming (WRI, 2000).

Major industrial cities and towns are situated on or near the coastline and after the liberalized economic policy of India, setting up of special economic zones and export processing zones in coastal areas are increasing. They discharge large amounts of untreated effluents daily apart from over-exploitation of water resources in the coastal areas. Marine-based tourism which are on rise in South Asia especially in Maldives, India and Sri Lanka, also leads to environment degradation through the construction of hotels, beach clubs and marinas involving infilling, dredging and resuspension of contaminated silts. All these have influenced the changes in coastal habitats.

4.5 Other Land Use Changes

About 11.8 Mha of land in the country is recorded as permanent pastures and grazing land which are under community use. The states which have considerable proportions of areas under permanent pastures and grazing lands are Himachal Pradesh, Karnataka, Madhya Pradesh, Gujarat and Rajasthan. Madhya Pradesh accounts for a large percentage of land under pastures and grazing lands.

The other types of areas, which are covered under barren and unculturable lands, are generally unsuitable for agricultural use either because of the topography or because of their inaccessibility. Instances are the desert areas in Rajasthan, the saline lands in parts of the Rann of Kutch in Gujarat, the weed infected and ravine lands in Madhya Pradesh and alkaline lands in Uttar Pradesh. The states of Rajasthan, Gujarat, Uttar Pradesh, Madhya Pradesh, Meghalaya, Assam and Maharashtra together account for more than 67 per cent of the land under this category in the country. Barren and unculturable land, fallow and other uncultivated land have decreased from 1950-51 to 1992-93.

The wet land area (non-forested) in India has also decreased by 11% from 1950-51 to 1992-93 (CDIAC, 2001). The total area under wetland was estimated to be 5.4 Mha (Scott and Poole, 1989) which supports large variety of flora and fauna. Wetlands are continuously under pressure from agricultural land use, erosion induced siltation and eutrophication which spell ecological disaster for the wetlands.

4.6 Degraded Lands

The steady growths of human as well as livestock population, the widespread incidence of poverty, and the current phase of economic and trade liberalization, are exerting heavy pressures on India's limited land resources for competing uses in forestry, agriculture, pastures, human settlements and industries. This has led to very significant land degradation. Land in India suffers from varying degrees and types of degradation stemming mainly from unstable use and inappropriate management practices. According to Sehgal and Abrol (1994) about 187.8 Mha (57% approximately) out of 328.73 Mha of land area has been degraded in one way or the other. It appears, therefore, that most of our land is degraded, is undergoing degradation or is at the risk of getting degraded. There is no land use as degraded land but it occurs across all land uses. Among the different categories, lands under cultivation face the biggest problem followed by grazing land and pastures, forests, barren lands, and unculturable lands in decreasing order. As degraded lands are result of land cover/land use change dominated by anthropogenic activity, details are given in the 'consequences of LU/LC change' section.

5. DRIVERS OF LAND TRANSFORMATION

Land transformation is driven by a combination of the following fundamental and high-level causes (Lambin et al., 2003):

- Resource scarcity causing pressure of production on resources
- Changing opportunities created by markets
- Outside policy intervention
- Loss of adaptive capacity and increased vulnerability
- Changes in social organization, in resource access, and in attitudes

Some of the fundamental causes leading to land-use change are mostly endogenous, such as resource scarcity, increased vulnerability and changes in social organization, even though they may be influenced by exogenous factors as well. Some of them are experienced as constraints. They force local land managers into degradation, innovation, or displacement pathways.

High-level causes, such as changing market opportunities and policy intervention, are mostly exogenous. Each of these high-level causes can apply as slow evolutionary processes that change incrementally at the time-scale of decades or more, or as fast changes that are abrupt and occur as perturbations that affect human-environment systems suddenly. But, only a combination of several causes, with synergetic interactions, is likely to drive a region into a critical trajectory (Puigdef´abregas, 1998). The various typology and drivers of land use/land cover change in the South Asia are summarized in Table 4 and two most important of them are discussed below.

5.1 Demography

Available case studies highlight that there is a complex relationship between population and land-use/cover change (Lambin et al., 2001). On the one hand, case study evidence at the time scale of a few decades supports the conclusion that the answers found in population growth, poverty, and infrastructure rarely provide an adequate understanding of land change. On the other hand, insights from studies at longer time scale show that both increases and decreases of a given population always had and still have tremendous impacts upon land-cover changes. Therefore, a population analysis of more nuances is required, by considering specific demographic variables and life cycle features. In many of the South Asian countries in-migration into forested, low-population density areas is the main demographic driver leading to initial or frontier conversion of forest cover.

In India, the increase in population and substantial income growth, demand an extra about 2.5 million tonnes of food grains annually besides significant increase needed in the supply of livestock, fish and horticultural

Table 4: Typology of the causes of land use change

S.No.	*Typology*
1.	*Resource scarcity causing pressure of production on resources* • Increasing population growth • Loss of land productivity due to misuse • The dependency on fuel wood for meeting basic energy need place enormous stress on the forest resources and is a common problem within the region. • Coral mining • Encroachment from other land uses • Migration
2.	*Changing opportunities created by markets* • Intensive agriculture development – technological innovation, commercialization and globalization • Changes in market prices • Infrastructural development – Road and other infrastructure development in mountain areas • Capital investment • Increased pressure from tourism
3.	*Outside policy intervention* • Economic development • Subsidies and incentives • Insecurity of land tenure • Poor governance, unstable government and war. A well-defined land use policy is either absent or if at all such policy exist, its poor implementation in all the countries and most of existing laws are overlapping with other jurisdictions
4.	*Loss of adaptive capacity and increased vulnerability* • Climate change and associated natural disasters • Impoverishment • Social discrimination • Natural hazards
5.	*Changes in social organization, in resource access and in attitudes* • Changes in governing institutions managing land resources • Urban aspiration • Loss of entitlement to environmental resources • Growth of individualism and breakdown of extended family

products. Under the assumption of 3.5% growth in per capita GDP demand for food grains (including feed, seed, wastage and export) is projected to be 256 million tonnes in 2020 (Paroda and Kumar, 2000). As a result of these pressures there has been a marked change in supply and use of land, water, fertilizer, seeds and livestock with a consequence upon increased GHG's emission.

5.2 Market Forces

Land scarcity-driven agricultural intensification occurs in economies which are not yet fully integrated in the market, and is usually linked to growth in population and its density (whether caused by natural increase, migration, incursion of non-agricultural land uses or institutional factors such as land tenure regime). Land use intensification is a common response not only to pressures but also to opportunities. Opportunities and constraints for new land uses are created by markets and policies, increasingly influenced by global factors. In many cases individual and social responses follow from changing economic conditions, mediated by institutional factors. Extreme biophysical events occasionally trigger further changes. With increasing interconnected market forces in South Asia (WTO, emergence of SAFTA), it operates at sub-regional level, and the impact of institutional drivers extends to the global level. However, the impact of globalisation on land use/cover change are still speculative and yet to be recorded in uncertain terms.

5.3 Individual and Institutional Forces

Urban land uses are underlain by circulatory migration and demographic change, with inner city and peri-urban developments having rather distinct underlying patterns. The growth of urban aspirations, the urban-rural population distribution, and the impact of rapidly growing cities on ecosystem goods and services are likely to become dominant factors in land-use change in the decades to come, be it in major urban or peri-urban areas (INSA et al., 2001) or in remote hinterland or watershed areas (Humphries, 1998).

While many land-use changes are due to ill-defined, weak institutional enforcement, one should not assume that indigenous institutions are always inefficient. Land degradation is more prominent when macro policies – be it capitalist or socialist – undermine local (adaptation) policies. There is often a mismatch between environmental signals reaching local populations and the macro-level institutions.

The mix of driving forces of land-use change varies in time and space, across scales. Biophysical drivers may be as important as are human drivers. Trigger events, whether these are biophysical (a drought or hurricane) or socio-economic (a war or economic crisis), play also a major role in driving land-use changes. From the growth of systematic case study comparisons and descriptive models important generalisations on the driving forces of land-use change start to emerge. This in turn will lead to improved modelling of the changes in land cover (Veldkamp and Lambin, 2001).

6. METHOD/TECHNIQUES OF LAND TRANSFORMATION ANALYSIS

Our proper understanding of land-cover change is limited for two reasons: (i) lack of accurate measurements of its rate, geographic extent, and spatial pattern, and (ii) very poor capability of modelling from empirical observations. Hence, an interdisciplinary approach is required for analysing land-cover change by coupling empirical observations and diagnostic models using modern inputs such as remote sensing and the integrating capabilities of GIS. This is discussed in this section under perspectives, modelling and remote sensing in land transformation analysis.

6.1 Perspectives of Land Transformation

To develop models of land-use change, and to generate projections, requires a good understanding of the major human causes of land-cover changes in different geographical and historical contexts. Because, various human-environment conditions react to and reshape the impacts of drivers differently, leading to specific pathways of land-use change. It also requires an understanding of how changes in climate and global biogeochemistry affect both land use and land cover, and vice versa, to integrate feedback loops. Such understanding is gained through a collection of case studies of land-use dynamics, which highlight how people make land use decisions. In modelling there are different perspectives of understanding the land-use/land-cover changes and in general, they are agent based, system based, narrative and integrated approaches (Lambin et al., 1999). Each perspective approaches the impact on land, of the interactions between macrostructure and micro agency, from a different vantage point.

Agent-Based Perspective

The agent-based perspective is centered on the general nature and rules of land-use decision making by individuals. It represents the motivations behind decisions and the external factors that influence decisions about land use (Leemans et al., 2003; Ahmed, 2001). Economic models of land-use change, for example, assume that land managers attempt to fulfill their needs and meet their expectations by accommodating economic, social, and environmental constraints (utility optimization). Microeconomic approaches to land-use changes explain spatial configurations of changes and usually assume that the agents have the ability to make informed predictions and plans and those are risk minimizers (Chomitz and Gray, 1996).

Systems Perspective

A systems perspective tends to focus on gradual and progressive processes of change at the scale of large entities through the organization and institutions of society (Ostrom, 1990). Institutions operate interactively at different spatial and temporal scales; the institutions link local conditions to global processes and vice versa. It has to cope with issues that include technological innovations, policy and institutional changes, collective ownership of land resources, rural-urban dynamics, and macroeconomic transformations.

Narrative Perspective

The narrative perspective adopts a much longer time horizon and focusses on critical events and abrupt transitions through historical detail and interpretation (Richards, 1990). It includes changing political economies, environmental feedback on land use, and external shocks and it overcomes the error of interpretation based on the present and immediate past. Scenarios generated to project future land-use changes are based on drawing analogies between historical and current situations.

Integrated Approach

This approach combines other approaches based on the scenarios, location and future need in an integrated way. Most of the present-day models of climate and land use change follow integrated approach as the drivers of land use/cover varies across time and space.

6.2 Modelling Land Transformation

To analyse causes and impacts of these land-use changes, modelling is becoming increasingly more important as these factors of change interact over time and space in complex ways and evidenced by the evolution of a plethora of models since the 1990s. Models can be of various types that are potentially useful for the study of land use/cover changes. This is the result of the diversity of the data used, objectives, methods and expected outputs, as well as that of land use/cover changes themselves and related environmental conditions.

Economic Models

Models which are put under this category are called single disciplinary (Barbier, 2001; Kaimowitz and Angelsen, 1998; Brown and Pearce, 1994). Kaimowitz and Angelsen (1998) reviewed about 150 economic models for deforestation used in different parts of the globe. The reviewed models were basically related to individual choices on decision making with simplification of complex multidimensional processes, and highlight only a few of many variables and causal relations involved in land-use change giving answers to why, where, when and how much land has been converted from one use to other.

Multidisciplinary Models

These models are primarily developed for taking space and/or ecological interactions into account (Agarwal et al., 2002; Lambin, 1994). Spatially explicit land-use change models with integration of socio-economic and bio-physical variables are considered important for the projection of alternative future pathways and for conducting experiments to enhance understanding of key processes in land-use changes (Veldkamp and Lambin, 2001). Under this category various types such as descriptive models designed to project future land-use changes, empirical models to explain land-cover changes, spatial statistical models for projecting the future spatial patterns of changes and dynamic ecosystem models to test the scenarios on future changes in land-cover under different scales of time and space can be placed.

Land Use/Land Cover Changes and Land Degradation Models

Various types of models are available to deal with the effect of land use/land cover change on land degradation. Some of the model studies with some relevance/application in South Asian region are Conceptual models (Lambin, 1994; Thapa and Weber, 1995), Analytical models (Kaimowitz and Angelsen, 1998; Bluffstone, 1995; Amacher et al., 1996), Empirical regression models (Kohlin and Parks, 2001; Sah, 1996 and Amacher et al., 1996) and Linear/ non-linear programming and simulation models (Sankhayan et al., 2003; Banskota and Sharma, 1995). Recently, a detailed review of these models is given by Upadhyay et al. (2006).

Apart from the above mentioned categories of models, based on the scale, there are models which deal at global level and regional or local level. ***Global models*** focus mostly on "global" changes and their effects. Such models, which are collectively called global X models [IMAGE 2 (Alcamo, 1994); Global land use/land cover model (GLM) such as MOIRA and IIASA/ FAP (Robinson, 1994), provided us with useful conceptual tools and frameworks of the study of land use/cover changes and demonstrated the limitations and other problems involved in global modelling, apart from improving our understanding in the land use/land cover changes. ***Regional models*** are certainly needed for the study of land use/cover changes, but it will not be achieved without the modelling effort at finer scale, at which human aspects play more important roles. CLUE-CR model (Wageningen University) is a multi-scale model and LUGEC model (Kagatsume and Kitamura, 1999) works at local level. If fact, the growing concern over the human aspects of land use/cover changes demands models that are operational and useful at regional, national and even more local levels.

These studies have applied different kind of models, give diverse findings, and also are to a varying degree characterised by disciplinary bias. Although they undoubtedly have added to the understanding of the factors and processes responsible for land-use change behaviour and their consequences, it is beyond the scope of this topic if we discuss each one of them; rather, some of the important models are mentioned here so as to understand possible future improvements and its application in South Asia.

6.3 Role of RS and GIS

In the past, research methods applied in land-use/cover change research were largely influenced by advances in remote sensing. This technology has led to an emphasis on short timescales, because earth observation data have been available only for a few decades. Recently, a wide range of other methods have been used to reconstruct long-term changes in landscapes. This change in temporal frame has led to a greater consideration of the long-term processes of ecological restoration and land-use transition.

Direct observations of land-cover change can be made using remote sensing. It is a promising tool for objectively making these measurements at different spatial and temporal scales, from large-scale assessments of regional trends to local-scale analysis of complex dynamics. Additional attributes information can be linked to spatial data in GIS. By directly measuring land-cover change, it is possible to explicitly quantify its rate and spatial pattern. This information can be used for specific analysis of land-cover fragmentation as well as for analysis of spatial trends in, and geometric patterns of, land-cover change. Direct observations provide a quantitative assessment of rates of change, which can be used as forcing functions for a variety of biophysical and socio/demographic/economic models. These observations of spatial trends and rates can be used to develop empirical diagnostic models and short-term prognostic models (Lambin, 1994).

Remote sensing has an important contribution to make in documenting the actual change in land cover on regional and global spatial scales from the mid-1970s (Achard et al., 2002; DeFries et al., 2002; Lambin et al., 2003). It also has a role to play in evaluating indices of change in ecological processes, such as net primary production and rainfall use efficiency (Prince et al., 1998). Remote sensing information is found in a widely scattered literature, some of it refereed, some in the gray literature, and some unpublished as yet. There is also an obvious need for good inventory data and statistics about land cover and land-cover change at subnational, national, and international scales, augmented by a need for subnational and national indicators of condition, status, and trends of the global environment. There is a need to determine the interrelationships of remotely sensed and statistical inventory data, to integrate heterogenous data sources. Linking household-level information, particularly how people make land use decisions, to remote sensing data is becoming a major tool to increase our understanding of land-use dynamics.

The tremendous investment in scientific analysis of remote sensing data over the last decade, and the profusion of studies based on other data sources, provides a basis for a synthesis. Although information is not complete globally, several products are now available that depict the land cover of Earth globally in the 1990s and in 2000–2001. The same is true for snapshots of many important regions with substantial land-cover change and some of the data sources are already discussed.

There are multiple examples of studies and resultant databases of rapid land-cover change and ecosystem disturbances in South Asia. Gautam et al. (2003) have used satellite images from 1976, 1989 and 2000 to study the land use dynamics and landscape change pattern in Nepal. Thenkabail et al. (2005) have identified net irrigated area by source and season in the Ganges and Indus river basin using MODIS data. Rai and Sharma (2004) have studied the impact of land transformation on carbon emissions to the atmosphere by field collected carbon density and flux data to each land use/ cover class derived from remote sensing data. Ramankutty and Foley (1999) have studied the changes in land use patterns over the last three centuries from 1700 to 1992 using a combination of historical land use data and satellite imagery to monitor the changes in the extent of croplands over different time periods.

7. CONSEQUENCES OF LAND TRANSFORMATION

Sustaining an increasing population requires a continuous and increasing rate of flow of agricultural and forest products, such as food, fodder, timber and fuel wood which essentially requires land. Productivity can increase through intensification of agricultural practices, improved cropping systems and increase in agricultural areas. Such changes influence the biogeochemical properties of land, emission of GHG's and climate change, natural disasters, land degradation and desertification. As discussed earlier South Asian region has been experiencing land use/land cover changes at various degrees and type, which draws its own consequences which are described below.

7.1 GHG's Emission and Regional Climate Changes

Carbon is exchanged naturally between terrestrial ecosystems and the atmosphere through photosynthesis, respiration, decomposition, and combustion. Human activities change carbon stocks in these pools and exchanges between them and the atmosphere through land use, land-use change and forestry, among other activities. Emissions of CO_2, methane (CH_4), nitrous oxide (N_2O) and other GHG's are influenced by land use, land-use change, and forestry activities (e.g., restoration of wetlands, biomass burning, and fertilization of forests).

Biomass Burning

In many countries, vegetation, forests, savannahs and agricultural crops are burnt down to clear land and change its use. The extent of biomass burning has increased significantly over the past 100 years. It is now recognised as a significant global source of green house gases (GHG's). Further, substantial amounts of carbon have been released from forest clearing at high and middle latitudes over the last several centuries, and in the tropics during the latter part of the 20th century contributing more than half of all the carbon released into the atmosphere (IPCC, 2000).

C and N Pools and Fluxes

The total global C budget indicated that a significant sink is needed to balance the global C cycle called missing sink. This missing sink is assumed to be located in the terrestrial biosphere (Kauppi et al., 1992; Fisher et al., 1994), but the actual location and the processes involved are difficult to localize and observe. The missing sink emphasizes the importance of land use, land cover and land management in assessing sources, sinks and fluxes of GHGs (Leemans, 1996).

During recent decades deforestation rates have been accelerating and the resulting flux of CO_2 accounted for 25% of the total increase of atmospheric CO_2 (Watson et al., 1992). Based on watershed scale measurement of land-use/cover change, Rai and Sharma (2004) have estimated the annual carbon release from the entire Indian Himalayan forests area (6.7 Mha) as 52 Tg C. Recently more attention has been paid to the balance between deforestation and forestation in the tropics (Skole and Tucker, 1993). This balance is important with respect to the total global fluxes. The resulting heterogenic spatial and temporal patterns make a precise assessment of the total flux from deforestation difficult.

The above discussion has mainly focused on C, but similar conclusions can be drawn for other greenhouse gases, such as CH_4 and N_2O (Bouwman, 1995). For example, changes in land cover alter the uptake of CH_4 by soils; different agricultural practices lead to changed CH_4 emissions; and N_2O emissions are influenced by the timing and amount of fertilizer applications. Such examples indicate that the spatial pattern of GHG emissions from the terrestrial biosphere is very heterogeneous and influenced by physical, biogeochemical, socio-economic and technical factors. Actual land use and its resulting land cover are important controls especially when mitigation policies are evaluated. State-of-the-art assessments should be dynamic, geographical and regionally explicit and include the most important aspects of the physical subsystem; the biogeochemical subsystem and land use and changes therein (FAO, 1996).

To feed the burgeoning population and limited agricultural land expansion, the South Asian countries have to rely on increasing the productivity (intensification) and diversification. This increase in crop productivity is coupled to increases in the use of agricultural inputs, mainly N fertilizers and pesticides use apart from management factor. In general, Indian soils are low in total N content as well as nitrogen use efficiency (NUE) of major crops especially rice and wheat. As a consequence of this Nr enters into various components of the ecosystem causing 'N cascade effect' which is defined as the sequential transfer of Nr through environmental system, which results in environmental changes as Nr moves through or is temporarily stored within each system (Galloway et al., 2004). The total loss from about 10 Mha under rice-wheat cropping system alone is likely to be 0.2 Tg/yr. Pathak et al. (2004) predicted on an average ammonia volatilization loss of 12-15%, denitrification loss of 25-30%, and leaching loss of 15-16% of applied N from rice fields with an application of 120 kg N/ha as urea from rice fields in north-west India.

Table 5: Effect of Green revolution in Nr creation and GHG emission from India (Tg of N)

S.No.	*N pools*	*1950-51 level*	*1995-96 level*	*Magnitude of Nr added/Difference*
1.	N fertilizer	0.06	10.8	10.74
2.	BNF	0.55	1.14-1.18	0.59-0.63
3.	Crop production	2.94	12.47	9.53
4.	Livestock	0.97	1.62	0.65
5.	Land use change (Mha)[@]			
	Net sown Area	118	141	23
	Gross cropped area	132	188	56
6.	GHG production			
	N_2O	-	0.26	-
	NO_x	-	3.46	-
	NH_3	-	7.4	-

[@] This accelerates N mineralization and increased N fertilizer use

In India, through N fertilizer and BNF nearly 11.4 Tg of Nr is additionally added in 1995-96 when compared to 1950-51 level of Nr use (Table 5). The excessive N fertilizer use, increased irrigation facilities and intensification of agriculture have led to more Nr losses especially leaching losses of NO_3^- and increased the nitrate content in ground water in many parts of the country (Uttar Pradesh, 300-694 ppm; Punjab, 362-567 ppm and Haryana, 300-1310 ppm) (Singh and Singh, 2004).

7.2 Natural Disasters

In South Asia the incidence of natural disasters is increasing in frequency and magnitude and many of the disasters occur or are exacerberated by human activities. Countries that face severe deforestation, erosion, over-cultivation and over-grazing of marginal lands are hit hardest by catastrophes. For example, deforestation results in an increased concentration of surface runoff and hence flooding, and destabilised slopes can result in devastating landslides.

Flood

The effects of environmental degradation can also transcend national boundaries. For example, increasing soil erosion in the hills of Nepal is resulting in heavy siltation of the river beds in India and Bangladesh, which is raising river bed levels and causing more frequent flooding. Tiwari (2000) has reported reduced groundwater recharge, increased run-off and soil erosion in the Himalayan mountain areas coupled with recurrent floods and decreased irrigation potential in the Indo-Gangetic plains. In the coastal areas, human vulnerability has increased due to degradation of the natural buffers, such as coral reefs and mangroves. This is particularly true in the region's island nations and the coastal areas, such as in Bangladesh. The region also witnesses the tremors of global ecological changes. For instance, the loss of natural vegetation – particularly forests – around the world is also a major cause of preventable natural disasters that afflict South Asia (State of the Environment – South Asia 2001).

Drought

A drought triggers a crisis, but does not cause it. Over-cultivation and over-grazing weaken the land, allowing no margins when a drought arrives. Many countries in South Asia are prone to drought, including Pakistan, Nepal, India, Sri Lanka and parts of Bangladesh. In India, about 33 per cent of the arable area is considered to be drought-prone. The 1981-82 in Nepal and 1996 drought in Sri Lanka were very severe causing serious economic loss.

Landslides

Incidence of landslides is very common in the hills and mountainous regions of South Asia. Topography alone could be the primary cause; however, in most cases, landslides are aggravated by human activities such as deforestation, cultivation and construction which destabilise the already fragile slopes. For instance, as a result of combined actions of natural and man-made factors, as many as 12,000 landslides occur in Nepal every year. In 1998, floods and landslides in Nepal killed about 273 people and the country suffered a loss of about US $28,854,000 (SOE Report 2000, Nepal).

7.3 Urban Problems

Ever increasing urban populations and their growing amounts of wastes have over-taxed the natural recycling capabilities of local rivers and lakes. Of the many problems associated with urban effluents, nutrient loading or eutrophication of local waters is one of the most serious. Nutrients are essential plant foods, but excessive amounts can cause radical plant growth – such as massive algal blooms, for example – that block the sunlight that other organisms need. As the plants die and decompose, the dissolved oxygen in the bottom waters is depleted – a condition that is deadly for fish and other aquatic life.

The vulnerability to natural disasters is also growing due to increasing population and inadequately planned urban growth. The number and density of people living in cities within earthquake and tropical cyclone zones have risen dramatically in the past two decades. This growth has mostly been haphazard and uncontrolled. Physical infrastructure has expanded rapidly as well, and has generally hindered sustainable construction practices and safe building standards, particularly from the standpoint of mitigating events such as floods and earthquakes. Poor planning decisions have led to the establishment of potentially hazardous facilities, such as nuclear power plants, chemical factories and major dams, in earthquake prone zones and densely populated areas.

7.4 Land Degradation and Desertification

As a result of land use/land cover changes, intensification of agriculture and unscientific management practices at varying degrees, South Asian region is suffering from land degradation and desertification. Data from GLASOD survey carried out during the 1980's by UNEP and ISRIC were used to prepare the status of land degradation in South Asia and presented in Fig. 14. As per GLASOD, in South Asia nearly 90 Mha of land (20% of total geographical area of South Asia) is affected by degradation and 72 Mha is under severe to very severe degradation. The principal causes of land

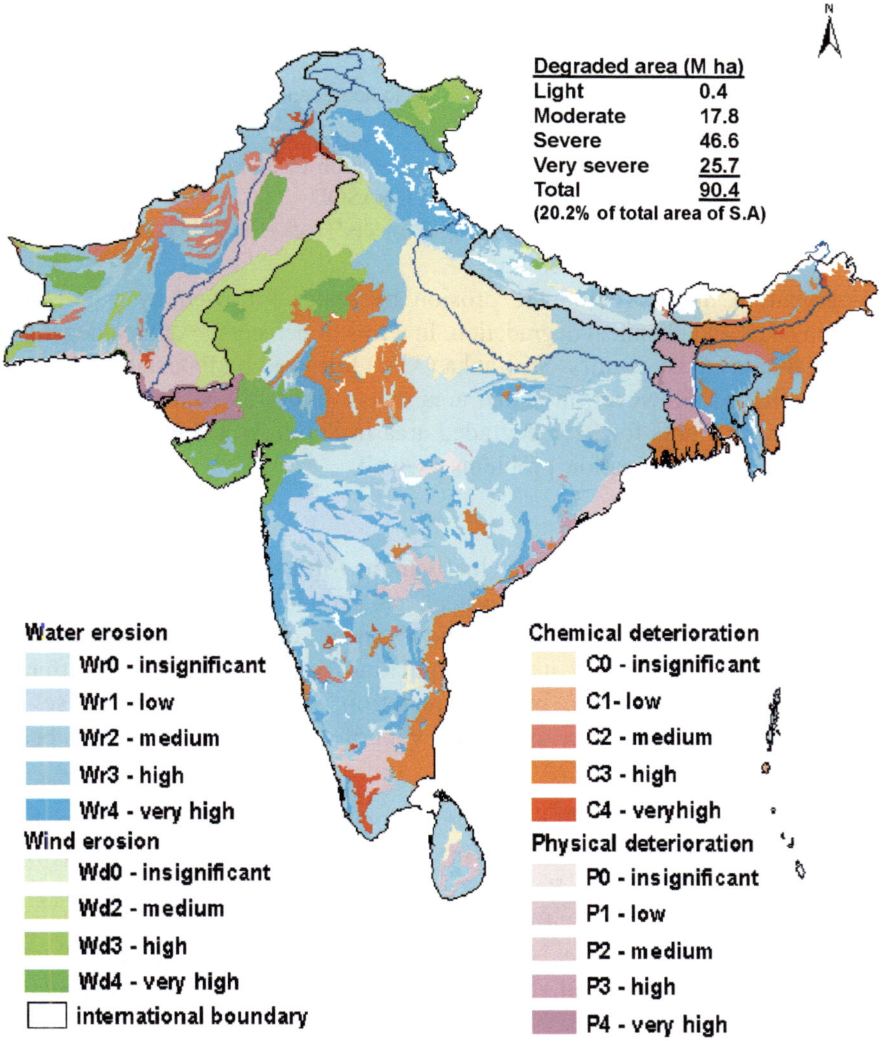

Fig. 14: Status of land degradation in South Asia (Dominant type).

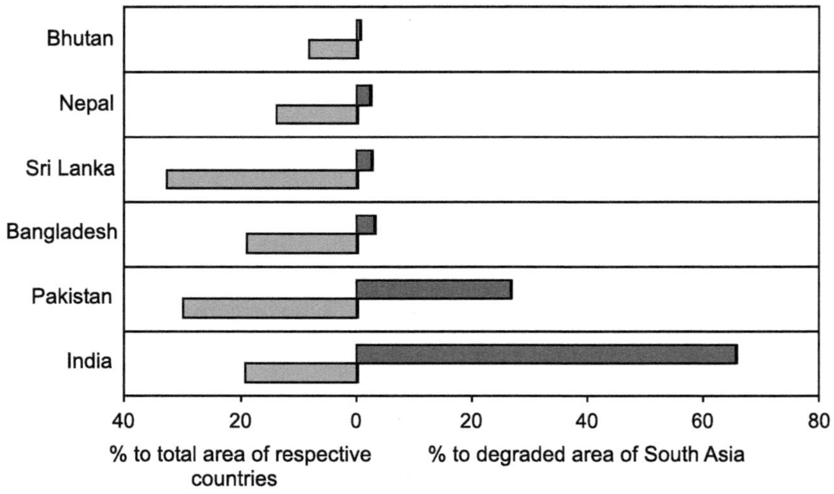

Fig. 15: Area under land degradation in South Asia.
Data source: GLASOD
http://www.isric.org/UK/About+ISRIC/Projects/Track+Record/GLASOD.htm

degradation in the region are erosion by water followed by wind; and biophysical and chemical degradation. In terms of percentage of area degraded to total area of the country, Sri Lanka (33%) followed by Pakistan (30%) are most affected. However, in actual area extent India (59 Mha) and Pakistan (24 Mha) accounts for 92% degraded area in South Asia (Fig. 15) and they suffer from desertification problems as well.

Erosion

Land degradation manifest itself chiefly in the form of **erosion** (water erosion followed by wind erosion), biophysical, and chemical deterioration. In India, approximately 57 per cent of the land is under some form of degradation, while Bhutan, because of its low population density, has not yet suffered severe land degradation, but deforestation, often the initial cause of degradation, is taking place and 10 per cent of the agricultural land has been affected by soil erosion (SACEP).

- Water erosion across India is the major cause of top-soil loss (in 132 Mha) and terrain deformation (in 16.4 Mha). In general, countries with humid climatic zones – Bangladesh, Nepal, Sri Lanka and the greater part of India – are severely affected by water erosion on their rain-fed lands, by soil fertility decline, and by deforestation. In parts of the hill and mountain areas of Nepal, deforestation and water erosion have reached an extreme degree.
- Wind erosion is extensive in India and Pakistan, affecting about 25 million ha of land. Wind erosion is dominant in the western part of India causing a loss of top soil and terrain deformation in 13 Mha.

- The most devastating form of water-borne land degradation in Bangladesh is riverbank erosion, and the active floodplains of the Ganges, the Brahmaputra-Jamuna, the Tista and the Meghna rivers are most susceptible to this problem.

Deforestation and Erosion by Shifting Cultivation

Shifting cultivation is traditionally practiced in 13 states of the country and more extensively in the northeastern hill states, Orissa and the Eastern Ghats on an estimated forest area of about 4.35 Mha. This contributes significantly towards forest land degradation. With the progressive reduction in the land to population ratio, the fallow period between cultivations has fallen from 30 years to about 2 to 3 years. This in turn does not permit the natural processes of recuperation to repair the disturbed ecosystem resulting in erosion and a decline of soil fertility. An estimated 0.7 Mha of forest lands are encroached upon for agriculture by the people who live in their vicinity, such lands are mostly of a marginal nature, susceptible to degradation.

Overgrazing Induced Erosion and Land Degradation

The increasing pressures of livestock population and migratory graziers are a potent force responsible for the continual degradation of forests, grasslands, village common lands and alpine pastures causing widespread devastation. A livestock population of 467 million grazes on 11 Mha of pastures. This implies an average of 42 animals grazing in a hectare of land against the threshold level of five animals per hectare (Sahay, 2000). In the absence of adequate grazing land, nearly a third of the fodder requirement is met from forests in the form of grazing and cut fodder for stall-feeding (MoEF, 1999). An estimated 100 million cow units graze in forests against a sustainable level of 31 million per annum. A sample survey by the FSI estimates that the impact of grazing affects approximately 78% of India's forests. Over-grazing and over-extraction of green fodder, both lead to forest and land degradation through a loss of vegetation and physical deterioration in the form of compaction and reduced infiltration, and increase in soil erodibility.

Chemical Degradation

Chemical soil degradation in the region is mainly caused by agricultural mismanagement especially increased use of fertilizers, pesticides and mismanagement of water resources. Over-exploitation of water resources had led to increased soil salinity in many parts of the region. Waterlogging and salinization affects between 2 and 3 million ha in India and Pakistan respectively. In Pakistan, salt build-up in the soil is known to reduce crop yields by 30 per cent. In Bangladesh, over 30% of the land available for cultivation is situated in the coastal belt, and most of the land is not utilized for crop production due to increased soil salinity.

Mining

Land degradation is the inevitable result of any form of mining, particularly opencast mining, which thoroughly disturbs the physical, chemical, and biological features of the soil and alters the socioeconomic features of the area. Although there are no reliable data available for the area actually affected by mining and quarrying, mining lease area is approximately 0.8 Mha, which may be taken as degraded directly due to mining activities in addition to the areas affected indirectly. Gem mining and inland coral mining in Sri Lanka have led to land degradation (http://www.sacep.org/html/environment).

Forest Fire

Apart from the destruction of vegetation, high intensity forest fires alter the physico-chemical and biological properties of the surface soil and leave the land prone to erosion and with a lowering of soil quality. The occurrence of frequent forest fires has been a major cause of degradation of forest land in many parts of India.

Desertification

Desertification is a land degradation process that involves a continuum of change, from slight to very severe degradation of the plant and soil resource, and is due to man's activities. Some analysts believe desertification is only a phase in a natural climatic process that does not receive attention because it occurs slowly and over the long term. Others believe that "Drought triggers a crisis, but does not cause it". It can be defined as (Dregne, 1986):

"Desertification is the impoverishment of terrestrial ecosystems under the impact of man. It is the process of deterioration in these ecosystems that can be measured by reduced productivity of desirable plants, undesirable alterations in the biomass and the diversity of the micro and macro fauna and flora, accelerated soil deterioration, and increased hazards for human occupancy."

In South Asia the countries suffering most from desertification are Bangladesh and India. About 12 per cent of India suffers from the threat of desertification in the arid northwest and in a broad semi-arid zone from the Punjab in the northwest to Tamil Nadu in the south. Because desertification is a long-term process – long in terms of its development and impact – it is difficult to pin-point its cause and effect relationships.

India's Rajasthan region is just one of a number of examples of the impact of population growth in the desertification process. Data from FAO show that while only 20 per cent of the arid land in Rajasthan could be cultivated in the 1970s, 30 per cent was being cultivated in 1951 and 60 per cent was being cultivated in 1971, mainly at the expense of grazing lands and traditional long fallow periods. By 1972, sand dunes had increased in height by as much as five metres and the water productivity of wells was declining (Table 6).

7.5 Siltation

Higher erosion rates have resulted in the sedimentation of river beds, siltation of drainage channels, irrigation canals and reservoirs. Siltation has changed

Table 6: Land use change in the Thar desert (Mha)

Land use	*1980*	*1990*	*1993*
Desert area of no or minimal value			0.4
Area in danger of desertification			13.4
Cultivated area	12.8	13.2	12.3
Pasture land	5.9	5.4	5.2
Others	0.3	0.4	0.4

Report of MEA, submitted to CSD in 1995

the hydrology of several watersheds of the country, resulting in a greater frequency and severity of floods, and reducing water availability in dry season. The storage capacity of many reservoirs has been reduced drastically due to accelerated erosion and deposition. Siltation of major river courses and spillover sections due to excessive deposition of silt is observed extensively in Bihar and Uttar Pradesh since many flood-prone rivers flow through them. The total area affected by this problem is estimated to be 2.73 Mha (Das, 1977; Mukherjee et al., 1985). The Ganga and Brahmaputra carry the maximum sediment load, about 586 and 470 million tonnes, respectively, every year. Approximately 6000 to 12,000 million tonnes of fertile soil are eroded annually and a significant proportion of it is deposited in the reservoirs resulting in a reduction of their storage capacity by 1%-2%.

7.6 Degradation of Coastal and Marine Resources

The rich marine environment in the South Asian region is subjected to great pressure leading to coastal hazards and other impacts which varies from place to place because of differences in exposure to monsoons and storms, differences in local tectonics and subsidence, and variations in air and sea climates.

- India and Bangladesh, with their vast coastline, is often struck by cyclones and the resultant coastal storm surges. The recent tsunami in Indian Ocean has forcefully added a new dimension to the natural calamities affecting India and it is argued that majority of losses are due to improper planning and implementation of coastal land use.
- Transformation of rice fields into shrimp farms has changed the land use/ land coverage of the densely populated coastal areas of Bangladesh. During 1975–2000, the country's shrimp pond area has increased from less than 20,000 to 141,000 ha causing salinity, fertility degradation and ecosystem damage (Shajaat Ali, 2006).
- Mangroves have been exploited for timber, fuel wood and other purposes, while large areas have been cleared for agricultural activities and for shrimp farming. Further, freshwater interceptions for agricultural schemes have severely affected mangroves and other coastal habitats.

- Marine-based tourism also leads to environment degradation through the construction of hotels, beach clubs and marinas involving infilling, dredging and resuspension of contaminated silts.
- Oil pollution threatens the coral reefs, where boating activities for tourism and fishing are unregulated. Most of the shallow water coral reef habitats of Sri Lanka, Maldives and India were severely damaged as a result of bleaching.
- Coastal erosion is a coastal hazard that is accelerated due to poor land use practices and deforestation in the catchment areas, coral mining for the manufacture of cement and sand-removal.
- Other hazards, like the occurrences of algal blooms along coastlines, and saline water intrusion into ground water are also increasing at alarming rate.

8. RESEARCH PRIORITIES

New estimates of areas and rates of major land-use/cover conversions have greatly narrowed down uncertainties. A number of more subtle land changes still need to be better quantified at national and regional scale. This is particularly the case for anthropogenic changes that strongly interact with natural environmental variability and therefore require longitudinal data over a long time period for a reliable assessment.

Different assumptions about human-environment relationships and temporality lead to varying explanations and interpretations of the causes and significance of environmental changes. A systematic analysis of local scale land-use change studies, conducted over a range of timescales, helps to uncover general principles to provide an explanation and prediction of new land-use changes.

Improved understanding of processes of land-use change should lead to emphasis on the potential for ecological restoration through land management (Victor and Ausubel, 2000) to mitigate human impact on the environment for the present and future scenarios. To fulfill this evolution of the new research questions, methods and scientific paradigm have to be developed.

Improved understanding of the complex dynamic processes underlying land-use change will allow more reliable projections and more realistic scenarios of future changes. Institutional and technological innovations may lead to negative feedback loops that decrease the rate of change or even reverse land-use/cover change trends. The relative strength of amplifying and attenuating feedback can be influenced by policies. The analysis of interaction, coherence, or conflict between social and biophysical responses to changes in both ecosystem services and earth system processes caused by land changes is still a largely unfocussed area which deserves to be given priority.

The three dimensions of land-use change models, namely, space, time and human decision; along with two distinct attributes of each, i.e., scale and complexity, provide the foundation for comparing and reviewing land-use change models. Hence, models and synthesis of land transformation in South Asia should be developed to deal with the complex situation prevailing here.

Land-use changes are now increasingly analyzed as part of the system interactions rather than interpreting deviations from a predisturbance state as problematic, leading to co-evolution of natural and social systems. Throughout their history, human societies have coevolved with their environment through change, instability, and mutual adaptation. The coupled human-environment systems should therefore be considered as a whole when we assess sustainability and vulnerability.

9. CONCLUSIONS

Significant progress in the quantification and understanding of land-use/cover changes has been achieved over the last decade. Land transformations in South Asia are complex and resulted in measurable consequences across various categories of land-use/cover with spatial variability. However, many data sources are inconsistent and suffer from scaling problems. Much remains to be learned, however, before we can fully assess and project the future role of land transformation in the functioning of the earth system particularly role of South Asia and identify conditions for sustainable land use.

REFERENCES

Adeel, Z. and Pomeroy, R., 2002. Assessment and management of mangrove ecosystems in developing countries. *Trees*, **16(2-3),** 235-238.

Agarwal, C., Glen, M.G., Morgan, J.G., Tom, P.E. and Charles, M.S., 2002. A Review and Assessment of Land-Use Change Models: Dynamics of Space, Time, and Human Choice. CIPEC Collaborative Report No. 1. USFS Publication GTR-NE-297. USDA Forest Service, Northeastern Forest Research Station, USA.

Ahmed, A., Rajan, K.S. and Shibasaki, R., 2001. Modeling agricultural land use change in Bangladesh: Farmer as an agent of change. Paper presented at the 22nd Asian Conference on Remote Sensing, 5-9, November, Singapore.

Alcamo, J., 1994. *IMAGE 2.0: Integrated Modeling of Global Climate Change.* Kluwer Academic Publishers, Dordrecht.

Amacher, G.S., Hyde, W.F. and Kanel, K.R., 1996. Household fuelwood demand and supply in Nepal's Terai and Mid-hills: Choice between cash outlays and labor opportunity. *World Development*, **24,** 1725-1736.

Barbier, E.B., 2001. The economics of tropical deforestation and land use: An introduction to the special issue. *Land Economics*, **77(2),** 155-171.

Baskota, K. and Sharma, B., 1995. Economic and natural resource conditions in the districts of Bagmati zone and their implications for the environment: an adaptive simulation model. MEI. ICIMOD, Nepal. Series No. 95/9.

Biksham, Gujja and Stich, A.F., 1996. What price prawn? Shrimp aquaculture's impact in Asia. *Environment*, Washington.

Bluffstone, R.A., 1995. The effect of labor market performance on deforestation in developing countries under open access: An example from rural Nepal. *Journal of Environmental Economics and Management*, **29,** 42-63.

Bose, Vasudeva Gupta and Sinha, 1997. Teri Report No. 97/ED/52.

Bouwman, A.F., 1995. Compilation of a Global Inventory of Emissions of Nitrous Oxide. Ph.D. Thesis. Agricultural University, Wageningen.

Briassoulis, H., 2000. Analysis of Land Use Change: Theoretical and Modeling Approaches, http://www.rri.wvu.edu/WebBook/Briassoulis/contents.htm.

Brown, K. and Pearce, D.W., 1994. The Causes of Tropical Deforestation: The Economic and Statistical Analysis of Factors Giving Rise to the Loss of the Tropical Forests. University College London Press, London.

CDIAC, 2001. Carbon Dioxide Information Analysis Center, Oak Ridge National Laboratory, Oak Ridge, Tennessee, USA. http://cdiac.esd.ornl.gov/epubs/ndp/ndp050/ndp050.html.

Chomitz, K.M. and Gray, D.A., 1996. Roads, land use, and deforestation: A spatial model applied to Belize. *World Bank Econ. Rev.*, **10(3),** 487-512.

CLUE, 2006. A spatially explicit model for the analysis of Land Use Change and its Effects. Department of Environmental Sciences, Wageningen University, Wageningen, The Netherlands.

Das, D.C., 1977. Soil conservation practices and erosion control in India – A case study. *FAO Soils Bulletin*, **33,** 11-50.

de Bie, C.A.J.M. 2000. Comparative performance analysis of agro-systems. ITC dissertation no. 75, 232p; Available at http://www.itc.nl/education/larus/landuse/

Deichmann, U., Balk, D. and Yetman, G., 2001. Transforming population data for interdisciplinary usages: From census to grid. Work. Pap., Cent. Int. Earth Sci. Inf. Netw., Columbia Univ.

Doos, B.R., 2002. Population growth and loss of arable land. *Glob. Environ. Change: Hum. Policy Dimens.*, **12(4),** 303-311.

Edwards, Peter, 2000. Aquaculture, poverty impacts and livelihoods, ODI Natural Resource Perspectives, Number 56, June 2000, http://www.odi.org.uk/nrp/56.html.

Erika Lepers, Eric F. Lambin, Anthony C. Janetos, Ruth Defries, Frederic Achard, Navin Ramankutty and Robert J. Scholes, 2005. A Synthesis of Information on Rapid Land-cover Change for the Period 1981-2000, *BioScience*, **55(2)**, 115.

FAO, 1996. Global climate change and agricultural production. Direct and indirect effects of changing hydrological, pedological and plant physiological processes. FAO, Rome, Italy. http://www.fao.org/docrep/W5183E/w5183e00.HTM

FAO, 1999. State of the world's forests. Rome, Italy.

FAO, 2000. ftp://ftp.fao.org/agl/agll/docs/landglos.pdf.

FAO, 2001. Global forest resources assessment 2000 (FRA 2000): main report, FAO For. Pap. 140, FAO, Rome.

FAO, 2003. State of forestry in Asia and the Pacific – 2003: Status, changes and trends, Asia-Pacific Forestry Commission, Rap Publication 2003/22.

FAO/UNEP, 1999. Terminology for integrated resources planning and management. FAO, 2000. Land cover classification system (LCCS), 179 p.

Fisher, M.J., Rao, I.M., Ayarza, M.A., Lascano, C.E., Sanz, J.I., Thomas, R.J. and Vera, R.R., 1994. Carbon storage by introduced deep-rooted grasses in the South American savannas. *Nature*, **371,** 236-238.

Food and Agriculture Organization of the United Nations (FAO), 2006. FAOSTAT Online Statistical Service. Rome: FAO. Available online at: http://faostat.fao.org.

Gautam, A.P., Webb, E.L., Shivakoti, G.P. and Zoebisch, M.A., 2003. Land use dynamics and landscape change pattern in a mountain watershed in Nepal. *Agriculture, Ecosystems and Environment*, **99,** 83-96.

Goldewijk, K.K. and Ramankutty, N., 2004. Land cover change over the last three centuries due to human activities: Assessing the differences between two new global data sets. *GeoJournal*, **61,** 335-344.

Goldewijk, K.K., 2001. Estimating global land use change over the past 300 years: The HYDE database. *Glob. Biogeochem. Cycles*, **15(2),** 417-434.

Gr¨ubler, A., 1994. Technology. *In:* Changes in Land Use and Land Cover: A Global Perspective, W.B. Meyer, B.L. Turner (eds), 287-328. Cambridge, UK: Cambridge Univ. Press.

Hinrichsen, D., 1990. Our common seas: Coasts in crisis. London: Earthscan Publications.

Hossain, Md Shahadat, 2001. Biological aspects of the coastal and marine environment of Bangladesh. *Ocean & Coastal Management*, **44,** 261-282.

Houghton, R.A. and Hackler, J.L., 2001. Carbon Flux to the Atmosphere from Land-Use Changes: 1850 to 1990. ORNL/CDIAC-131, NDP-050/R1.

Humphries, S., 1998. Milk cows, migrants, and land markets: Unravelling the complexities of forest to pasture conversion in Northern Honduras. *Econ. Dev. Cult. Change*, **47(1),** 95-124.

INSA, CAS, USNAS, 2001. Indian National Science Academy, Chinese Academy of Sciences and US National Academy of Sciences. (Expert Panel: Wolman, M.G., George, P.S., Kulkarni, S., Vashishtha, P.S., Shidong, Z., Qiguo, Z., Yi, Z., Long J.F., Rosenzweig, C., Solecki, W.D.) Growing Populations, Changing Landscapes: Studies from India, China, and the United States. Washington, DC: Natl. Acad. 324 pp.

IPCC, 2000. Land Use, Land-Use Change, and Forestry. A Special Report of the Intergovernmental Panel on Climate Change, http://www.gcrio.org/OnLnDoc/pdf/lulucf_spm.pdf.

Kagatsume, M. and Kitamura, T., 1999. A Model for Land Use/Cover Change – the Kitamura & Kagatsume Model. *In:* Land Use for Global Environmental Conservation (LU/GEC) – Final Report of the LU/GEC First Phase (1995-1997) ed. K. Otsubo, pp.1-13, CGER.

Kaimowitz, D. and Angelsen, A., 1998. Economic Models of Tropical Deforestation: A Review. Centre for International Forestry Research, Jakarta, Indonesia.

Kauppi, P., Mielikäinen, K. and Kuusela, K., 1992. Biomass and carbon budget of European forests, 1971 to 1990. *Science*, **256,** 70-74.

Kohlin, G. and Parks, P.J., 2001. Spatial variability and disincentives to harvest: deforestation and fuelwood collection in South Asia. *Land Economics*, **27,** 206-218.

Lambin, E.F., Turner II, B.L., Geist, H.J., Agbola, S., Angelsen, A., Bruce, J.W., Coomes, O., Dirzo, R., Fischer, G., Folke, C., George, P.S., Homewood, K., Imbernon, J., Leemans, R., Li, X., Moran, E.F., Mortimore, M., Ramakrishnan, P.S., Richards, J.F., Skånes, H., Steffen, W., Stone, G.D., Svedin, U., Veldkamp, T., Vogel, C., Xu, J., 2001. The Causes of Land-Use and Land-Cover Change: Moving Beyond the Myths. *Global Environmental Change*, **2001/4**.

Lambin, E.F., Baulies, X., Bockstael, N., Fischer, G. and Krug, T., 1999. Land-use and land-cover change (LUCC): Implementation strategy. IGBP Rep. 48, IHDP Rep. 10, Int. Geosph.-Biosph. Program., Int. Hum. Dimens. Glob. Environ. Change Program., Stockholm/Bonn.

Lambin, E.F., 1994. Modelling deforestation processes. Trees Series B: Research Report No. 1. Brussels, Luxembourg: ECSCEC-EAEC.

Lambin, E.F., Geist, H.J. and Lepers, E., 2003. Dynamics of Land-Use and Land-Cover Change. *In:* Tropical Regions. *Annu. Rev. Environ. Resour.*, **28,** 205-241.

Leemans, R., Lambin, E.F., McCalla, A., Nelson, J., Pingali, P. and Watson, B., 2003. Drivers of change in ecosystems and their services. *In:* Ecosystems and Human Well-Being: A Framework for Assessment, (eds.) Mooney, H., Cropper, A. and Reid, W. Washington, DC. Island Press.

Leemans, R., 1996. Incorporating land-use change in Earth system models illustrated by IMAGE 2. *In:* Global Change and Terrestrial Ecosystems. B. Walker and W. Steffen (eds.). Cambridge University Press, Cambridge (in press).

Meyer, W.B. and Turner II, B.L., 1992. Human population growth and global land use/land cover change. *Annu. Rev. Ecol. Syst.*, **23,** 39-61.

MoEF, 1999. National Forestry Action Programme. Ministry of Environment and Forests, Govt. of India, New Delhi, Vol. I, pp. 79.

Mukherjee, B.K., Das, D.C., Shamsher Singh, Prasad, C.S., Samuel, J.C., 1985. Statistics: soil and water conservation – watershed management, land resources and land reclamation. Soil and Water Conservation Division, Ministry of Agriculture and Rural Development, GoI, New Delhi. p. 21.

OECD, 1996. Guidelines for aid agencies for improved conservation and sustainable use of tropical and subtropical wetlands. Organisation for Economic Co-operation and Development, Paris, France.

Ostrom, E., 1990. Governing the Commons: The Evolution of Institutions for Collective Action. Cambridge, UK: Cambridge Univ. Press. 280 pp.

Puigdef´abregas, J., 1998. Ecological impacts of global change on drylands and their implications for desertification. *Land Degrad. Dev.*, **9(5),** 393-406.

Rai, S.C. and Sharma, P., 2004. Carbon flux and land-use/cover change in a Himalayan watershed. *Current Science*, **86(12),** 1594-1596.

Ramankutty, N., Foley, J.A., 1999. Estimating historical changes in global land cover: croplands from 1700 to 1992. *Glob. Biogeochem. Cycles*, **13(4),** 997-1027.

Ramankutty, N., Foley, J.A. and Olejniczak, N.J., 2002. People on the Land: Changes in global population and croplands during the 20th century. *Ambio*, **31(3),** 251-257.

Richards, P., 1990. Land transformation. *In:* The Earth as Transformed by Human Action: Global and Regional Changes in the Biosphere over the Past 300 Years. (eds.) B.L. Turner, W.C. Clark, R.W. Kates, J.F. Richards, J. Mathews and W.B. Meyer, 163-78. Cambridge, UK: Cambridge Univ. Press.

Robinson, J., 1994. Land-Use and Land-Cover Projections: Report of working group C. *In:* Changes in Land Use and Land Cover: A global perspective. (eds.) W. Meyer and B.L. Turner II. Cambridge University Press, pp. 73-92.

SACEP, 2001. State of the Environment – South Asia 2001. http://www.sacep.org/default.htm.

SAGE, 1999. http://www.sage.wisc.edu:16080/iamdata/datasets.php.

Sah, B.P., 1996. Watershed degradation and its socio-economic impacts using RS and GIS: A case study of Trijuga watershed, Nepal. Dissertation (M.Sc.) NR 96-20. AIT, Bangkok, Thailand.

Sahay, K.B., 2000. Problem of livestock population. Delhi: *The Tribune*, 11 April 2000.

Sankhayan, P.L., Gurung, N., Sitaula, B.K. and Hofstad, O., 2003. Bioeconomic modelling of and use and forest degradation at watershed level in Nepal. *Agriculture, Ecosystems and Environment*, **94,** 105-116.

Scott, D.A. and Poole, C.M., 1989. A status overview of Asian wetlands. No. 53. Asian Wetland Bureau, Kuala Lumpur, Malaysia.

Sehgal, J., Abrol, I.P., 1994. Soil degradation in India: Status and impact. New Delhi: Oxford and IBH. 80 pp.

SFR, 1987. The State of Forest Report. Forest Survey of India, MoEF, Dehradun.

Shajaat Ali, A.M., 2006. Rice to shrimp: Land use/land cover changes and soil degradation in Southwestern Bangladesh. *Land Use Policy*, **23,** 421-435.

Skole, D. and Tucker, C., 1993. Tropical deforestation and habitat fragmentation in the Amazon: Satellite data from 1978 to 1988. *Science*, **260,** 1905-1910.

State of the Environment – South Asia, 2001. http://www.sacep.org/default.htm.

Thapa, G.B. and Weber, K.E., 1995. Status and management of Watersheds in the upper Pokhara valley, Nepal. *Environmental Management*, **19,** 497-513.

Thenkabail, P.S., Schull, M. and Turral, H., 2005. Ganges and Indus river basin land use/land cover (LULC) and irrigated area mapping using continuous streams of MODIS data. *Remote Sensing of Environment*, **95,** 317-341.

Tilman, D., 1999. Global environmental impacts of agricultural expansion: The need for sustainable and efficient practices. *Proc. Natl. Acad. Sci. USA*, **96(11),** 5995-6000.

Tiwari, P.C., 2000. Land-use changes in Himalaya and their impact on the plains ecosystem: Need for sustainable land use. *Land Use Policy*, **17,** 101-111.

Turner, B.L., 2002. Toward integrated land-change science: Advances in 1.5 decades of sustained international research on land-use and land cover change. *In:* Steffen, W., Jäger, J., Carson, D.J., Bradhsaw, C. (eds). Challenges of a Changing Earth. Berlin: Springer, pp. 21-26.

U.N., 2002. World Urbanization Prospects: The 2001 Revision (ESA/P/WP.173) New York: UN Publication. 328pp. http://www.un.org/esa/population/ publications/ wup2001/

Upadhyay, K.P., 1995. Shifting Cultivation in Bhutan: A Gradual Approach to Modifying Land Use Patterns. Food and Agriculture Organization of the United Nations, Rome, Italy.

Upadhyay, T.P., Solberg, B. and Sankhayan, P.L., 2006. Use of models to analyse land-use changes, forest/soil degradation and carbon sequestration with special reference to Himalayan region: A review and analysis. *Forest Policy and Economics*, **9,** 349-371.

Veldkamp, T. and Lambin, E.F., 2001. Predicting land-use change. Agriculture, Ecosystems and Environment. Special issue.

Veldkamp, A. and Lambin, E.F., 2001. Predicting land-use change: editorial. *Agriculture, Ecosystems and Environment*, **85,** 1-6.

Victor, D.G. and Ausubel, J.H., 2000. Restoring the forests. *Foreign Aff.*, **79(6),** 127-144.

World Bank, 2000. World Development Indicators Report. The World Bank. http://www.worldbank.org.

WRI, 2000. Regional Reefs at Risk: South East Asia. http://www.wri.org/marine/

Yeung, Y., 2001. Coastal Megacities in Asia: Transformation, sustainability and management. *Ocean & Coastal Management*, **44,** 319-333.

Young, A., 1999. Is there really spare land? A critique of estimates of available cultivable land in developing countries. *Environ. Dev. Sustain.*, **1,** 3-18.

5 Atmospheric Composition Change and Air Quality

A. Jayaraman, G. Beig, U.C. Kulshrestha, T. Lahiri, M.R. Ray, S.K. Satheesh, C. Sharma and C. Venkataraman

1. INTRODUCTION

Rapid economic growth and increasing energy demand, being witnessed in the South Asian region in the recent years, have its impact on atmospheric composition change and air quality. The Indian Ocean Experiment conducted during the late nineties has shown the spread of aerosol haze over most of the tropical Indian Ocean during winter months and an increase in the concentration of pollutant gases such as CO and ozone. The reported wintertime clear sky aerosol radiative forcing at the surface was in the range of 23 to 27 W/m^2 over the polluted ocean region and a sufficiently large atmospheric absorption, in the range of 16 to 20 W/m^2. Presence of significant amount of soot particles is found responsible for the large atmospheric forcing. Subsequent field measurements made within the Indian mainland showed high concentration of soot particles in the range of 10 to 15% of the total dry aerosol mass concentration over urban locations. Larger atmospheric forcing is found over land region than over ocean regions as a result of the higher soot concentration as well as higher surface reflectance that amplifies the effect. As the arid and semi-arid regions and the surrounding ocean region are a major source of natural aerosol particles such as sand-derived dust and sea salt particles, it is necessary to delineate the aerosol forcing caused due to natural aerosols. In a parallel study, it was demonstrated that over the tropical Indian Ocean the forcing due to natural aerosols often exceeds the anthropogenic forcing by as much as 1.5 times during summer months.

For future climate impact assessment, estimation emissions for various atmospheric constituents is very essential. A detailed, spatially resolved ($0.25° \times 0.25°$) emissions inventory of SO_2 over India shows that the annual emission is about 4.16 Tg SO_2 y^{-1}, of which fossil fuel combustion dominates the emission while the contribution from biofuel combustion is only about 4%. Coal combustion in the thermal power plants was the single largest contributor to SO_2 emissions with balance emissions coming mostly from diesel use in road transport and fuel oil combustion. As far as carbonaceous aerosols are concerned, of the about 450 Gg y^{-1} BC emission, over 50% come from the biofuel combustion in cooking stoves, followed by crop waste burning and fossil fuel combustion. Another study shows the emissions from S. Asia, including Bangladesh, Bhutan, India, Nepal, Pakistan and Sri Lanka as 881 Gg y^{-1} for BC and 4535 Gg y^{-1} for OC in which the Indian emission accounted for 683 Gg y^{-1} and 3487 Gg y^{-1} for BC and OC respectively. An analysis of trends in aerosol emissions from India during the 1981-2000 period shows a growth rate of 2% yr^{-1} for biomass burning emissions and about 9% yr^{-1} for fossil fuel burning.

Consistency remains to be achieved among various studies and estimates of aerosol emissions. Particularly in the case of BC, S. Asian emissions from energy use estimates are believed to be too low to explain measurements from the Indian Ocean Experiment. Differences in emission estimates primarily arise from the use of differing emission factors for S. Asian sources, whose measurement remains a pressing need to reduce uncertainties. Regional atmospheric modelling studies, which use newly available emissions databases and integrate model predictions with in-situ aerosol measurements and satellite data, will further the understanding of aerosol emissions and effects in the S. Asian region.

The distribution of tropospheric ozone and its precursors over the Asian region simulated using a 3-D chemistry transport model with improved national emission inventories of pollutants show that the impact of Indian emissions on ozone over the surrounding South Asian region is of the order of 1-7 ppb at the surface, 5-15 ppb in the lower troposphere and in the range of 15-25 ppb in the upper troposphere during the summer monsoon season. The background level of ozone over the Indian region, major part of which can be interpreted as the long-range transport from the surrounding Asian region, is estimated to be around 9-25 ppb for July, at the boundary layer, which is almost 50% of the total abundance. The impact is found to be the maximum in the northeast part of India. Increasing chemical emissions of pollutants in the Indian tropical region during the 1990s have had a substantial impact on the concentration of tropospheric ozone at the regional scale in the boundary layer and a moderate effect in the free troposphere. Model calculations suggest that increases in ozone concentration in the 1990s were typically 2–9 ppbv (5-20%) in the boundary layer during July, and 3–5% in the free troposphere. The magnitude of the variations in the free troposphere during July is higher than during the post-monsoon months. Perturbations in CO concentrations for July amount to 4-8% in 1990s in the free troposphere but were relatively larger in the boundary layer. The variation is found to be significantly dependent on season, which are estimated to be larger during the monsoon season extending up to the free troposphere, thereby highlighting the role played by photochemistry, convection and long-range transport over this part of the globe.

The changing oxidising capacity of the tropical atmosphere is of great concern. The increased emissions of NO_x and non-methane hydrocarbons from combustion, along with higher penetration of UV radiation tend to increase the ozone and OH concentrations in the tropical-free troposphere. On the contrary, the increased emission of CO from biomass and fossil fuel burning tends to reduce the OH concentrations. Also an increase in methane emission due to increase in natural gas usage, cattle and paddy field also contribute to a reduction in the global OH concentration. Model estimates on the decrease in OH concentration since the pre-industrial era differ anywhere between 5 and 30%. Future changes in the oxidising or the cleansing capacity of our atmosphere is going to depend critically on the future anthropogenic emissions. It is strongly believed that pollutants enter into the stratosphere mainly through the tropical tropopause. A detailed understanding of the processes involved in the transport and transformation of short-lived species is very critical if we have to assess the future anthropogenic influence on atmospheric composition. In situ observational techniques need to be strengthened for studying the atmospheric chemistry over the tropics. Apart from improvements in the traditional balloon soundings, aircraft observations are to be strengthened. Integration of various observational techniques and new data analysis techniques need to be strengthened for our improvement in the understanding of the changes taking place in the atmospheric composition.

With rapid industrialization air quality has become a serious concern in the South Asian region. In most of the Indian cities the concentrations of suspended particulate matter (SPM) have been found to be exceeding the National Ambient Air Quality Standards' (NAAQS) limit while concentrations of other pollutants such as oxides of nitrogen (NO_x) and sulphur dioxide (SO_2) in most Indian cities have been found to be below the respective NAAQS limits. The situation is more or less similar in other cities of South Asia. In Colombo, in over-thickly populated areas the PM_{10} concentration is higher than the recommended levels. In the Pakistan cities of Lahore and Karachi while the PM_{10} and TSP values are found to exceed the standard limit the concentrations of gaseous pollutants like SO_2, NO_x, O_3, CO and Pb in the ambient air have been found to be below or around the national standard values. The monthly averaged PM_{10} concentrations measured at different locations in Kathmandu during 2002-03 shows that the PM_{10} concentrations remain much above the national standard during November to June period and at lower values during June to October. At locations like Lahore the reason for the observed high concentration of SPM is ascribed to low rainfall and in-situ soil characteristics, which favours high atmospheric loading of natural particles. However in most of the urban areas of South

Asia, transport and industrial sectors are considered to be the major contributors to the deteriorating air quality.

In most of the cities, the number of vehicles has been going up very sharply. Policy interventions have been initiated in South Asian cities in the transport sector for regulating/minimizing the transport related emissions of different trace gases and particulate matter. Since 1996 there have been several policy interventions implemented in India's transport sector which included phasing out of lead from gasoline, reduction of sulphur content in diesel, tightening of mass emission standards for new vehicles etc. The results of these policy interventions have been by and large positive as reflected by the long-term air quality monitoring data of different cities in India. In Pakistan, CNG is increasingly used in vehicles, in Kathmandu electrical vehicles have been introduced, lead has been phased out in Dhaka, and vehicle emission norms have been improved in Colombo. These measures and their strict implementation are likely to improve the ambient air quality in South Asian cities in the near future.

Because of the high levels of SPM and relatively less gaseous pollutants levels, the acid rain is still not a problem in most of the regions in S. Asia. In general the nature of rain water in India is alkaline, having a pH value >5.6. At present, at most of the sites where detailed measurements are conducted there is no threat of acid rain. The reason for the alkalinity is the abundance of loose soil, which is rich in Ca carbonate. The pH of the soil in most parts of India is also very high as compared to the pH of soil in acidified regions of the world. The acidity generated by the oxidation of gases like SO_2 is neutralized by soil-derived particles. In some areas NH_3 is found to be the most important neutralizer. Acid rain events are observed under limited conditions such as, if it is raining continuously for several hours, if the soil of the region is acidic itself, or in the industrialized areas. Results from the Indian Ocean Experiment revealed that the situation over Indian Ocean is very different as compared to Indian continental sites. The pH of rainwater over Indian Ocean has been observed between 3.8 and 5.6, while at continental sites it is between 5.6 and 7. The acidic nature of rainwater over Indian Ocean is due to insignificant influence of soil dust and the dominance of anthropogenic sulphate contributed by long-range transport. A systematic network of sites needs to be set up to monitor future scenario of acidification in the region. Deposition modelling coupled with above measurement network are necessary to estimate the future trends.

Several studies have documented the adverse effects of air pollution on the respiratory and general health of children and adults of both rural and urban regions. In India besides vehicular and industrial pollution, indoor air pollution from biomass fuel usage is a major problem. Rural women rely primarily on biomass fuels for cooking food and in that process they are exposed to an array of pollutants including potential carcinogens. Chronic inhalation of biomass smoke may lead to stillbirth, low birth weight and diseases in early infancy and increases the risk of acute respiratory infections like pneumonia and tuberculosis. Particulate pollutants appear to be responsible for the adverse health as they can cross the lung-blood barrier and circulates inside the body causing a systemic damage. Studies have shown that air pollution often causes alterations in body's immunity, enhancing the susceptibility to infection. Acute respiratory infections is one of the main killers of children in developing countries, causing four million child deaths per year surpassing the number of deaths due to diarrhoea. Chronic exposure to particulate air pollution increases the risk of atherosclerosis, stroke and myocardial infarction among Indians at a relatively younger age. Over-expression of platelet P-selectin in biofuel users indicates high risk of cardiovascular diseases, while genotoxic changes are risk factor for several diseases including cancer.

In the South Asian region it is the increasing concentration of suspended particulate matter which poses a serious threat to the atmospheric composition and air quality than compared to gaseous pollutants. While the decrease in the oxidising capacity of the tropical troposphere is becoming a serious issue in the near future, presently it is the large atmospheric radiative forcing caused by absorbing aerosols and its possible effect on the regional climate change which is a major concern. But, the magnitude of the threat posed by air pollution to the health of both urban and rural population, particularly the children of the region, is of very serious and immediate concern.

2. NATURAL AEROSOLS

Aerosols are produced both due to natural or anthropogenic activities. Natural aerosols account for ~70% of the global aerosol loading. The main contributors are sea salt, dust and natural sulphates. The abundance of these shows significant variability from region to region and season to season. Recent investigations have shown that a proportion of dust is due to anthropogenic activities (Satheesh and Moorthy, 2005). Similarly, a proportion of anthropogenic soot originates from natural forest fires (Crutzen and Andreae, 1990). It is difficult to separate the anthropogenic components of dust from natural or natural components of soot from anthropogenic. Besides, away from the source regions, both natural and anthropogenic components mix together (Jacobson, 2000; Chandra et al., 2004) and on a global scale it is almost impossible to exactly apportion the natural and anthropogenic shares of the total aerosol (Satheesh and Moorthy, 2005). Nevertheless, several investigations and coordinated field campaigns have been carried out to discriminate impact of anthropogenic aerosols on climate. In recent years, there has been a substantial increase in interest in the influence of anthropogenic aerosols on the climate through both direct and indirect radiative effects. Several extensive investigations and coordinated field campaigns have been carried out to assess the impact of anthropogenic aerosols on climate. However, studies of natural aerosols are few compared to those of anthropogenic aerosols, despite the importance of the former.

Several experiments and simulations have attempted to quantify the radiative impacts of natural aerosols, particularly sea salt, dust and oceanic sulphate, yet large uncertainties persist in these estimates especially due to fewer data over oceans, especially on sea salt aerosols. At high wind speeds, it is difficult to make accurate measurements over oceans. For land-derived dust particles there is inadequate understanding of their optical/radiative properties and large regional differences exist depending on the soil type at the source region.

A study over the tropical Indian Ocean has demonstrated that oceanic winds have a significant and overlooked effect on Earth's heat balance by increasing sea-salt aerosols (Satheesh and Lubin, 2003). It is shown that as sea-surface wind speed increases from 0 to 15 m s^{-1}, the magnitude of aerosol forcing at the TOA is enhanced by ~6 W m^{-2} (i.e., larger negative value) (Satheesh, 2002; Satheesh and Lubin, 2003). The magnitude of composite aerosol forcing at the TOA observed over the tropical Indian Ocean was about 10 Wm^{-2} (Satheesh and Ramanathan, 2000; Satheesh et al., 2002) while modulation in forcing by sea salt aerosols could be quite significant.

There are a number of investigations available in the literature regarding the transport of dust aerosols from continents to ocean and vice versa (Prospero et al., 2002; Zender et al., 2003). Some of these authors found the existence of Saharan dust even over the remote areas of the Atlantic Ocean and Arabian Sea (Satheesh and Srinivasan, 2002; Li et al., 2005). The South Asian region has a unique weather pattern on account of the Indian monsoon and the associated winds that reverse direction seasonally. Recent experiments conducted over the oceanic regions adjacent to the Indian sub-continent such as Indian Ocean Experiment (INDOEX) have revealed the presence of anthropogenic aerosol haze during January to March (Ramanathan et al., 2001). Chemical analysis of aerosols over the tropical Indian Ocean have shown that during summer months natural aerosols contribute more than 50% to composite aerosol optical depth (Satheesh and Srinivasan, 2002; Satheesh et al., 2002). The radiative forcing due to Arabian/Saharan aerosols (mostly natural) during April-May period was comparable and often exceeded (as much as 1.5 times) the forcing due to anthropogenic aerosols during January to March period (Satheesh and Moorthy, 2005).

To accurately predict the impact of dust aerosols on climate, the spatial and temporal distribution of dust is essential (Deepshikha et al., 2005). However, regional characteristics of soil dust production, transport and removal processes are poorly understood. Many global models do not accurately simulate regional distribution of dust due to their low grid resolution and inaccuracy of dust source function. More extensive measurements of the dust optical properties, along with the vertical distribution of the dust layer, are needed to reduce the uncertainty of the climate response to dust aerosols. Similarly, there are very few data on sea salt aerosols where wind speeds are high. In such conditions accurate measurements are extremely difficult. Thus the data on the global distributions of two major natural aerosol types (sea salt and mineral dust) are not adequate.

3. ANTHROPOGENIC AEROSOL EMISSIONS FROM SOUTH ASIA

Aerosol emission inventories currently in use for climate assessment (Cooke et al., 1999; Arndt et al., 1997; Cooke and Wilson, 1996; Liousse et al., 1996; Akimoto and Narita, 1994; Penner et al., 1993; Spiro et al., 1992) are guided by a need for uniformity but have limitations in capturing regional emissions because of their use of national average fuel consumption, limited sectoral detail in technology, fuel related information and emission factors, especially for inefficient combustion sources. The increasing importance of comprehensive assessments of regional climate change has prompted efforts to develop emissions databases on a national (Habib et al., 2004a, b; Reddy and Venkataraman, 2002a, b), regional (Streets et al., 2003a, b), and global scale (Bond et al., 2004), which address region-specific characteristics. A detailed, spatially resolved ($0.25° \times 0.25°$) emissions inventory of SO_2, PM2.5 particles and its constituents such as black carbon (BC), organic matter (OM), inorganic oxidized matter (IOM) has recently become available for India (Reddy and Venkataraman, 2002a, b), which uses sector-wise technology, fuel-composition and emissions data for fossil fuel combustion. A new methodology based on food consumption statistics to estimate biofuel use and related emissions from India has also been developed (Habib et al., 2004a).

These studies estimate Indian annual SO_2 emissions at about 4.16 Tg SO_2 y^{-1} dominated by fossil fuel combustion with a negligible contribution (4%) from combustion of biofuels, largely dried cattle manure. Coal combustion in the thermal power plants was the single largest contributor to SO_2 emissions, with balance emissions mostly from diesel use in road transport and fuel oil combustion in industrial sector. Zero SO_2 control was assumed in these estimates. Carbonaceous aerosols from India are primarily emitted from biofuel combustion and open biomass burning (Fig. 1). BC emissions of 450 Gg y^{-1} are over 50% from biofuel combustion in cooking stoves, with important contributions from crop waste burning and fossil fuel combustion, in particular diesel transport. OM emissions (organic carbon corrected for associated hydrogen and oxygen using a factor of 1.3) are estimated at 1190 Gg y^{-1} with significant contributions from open biomass burning (crop waste and forest fires), followed by biofuel and fossil fuel combustion. Among fossil fuel sources, brick kilns using a mixture of powdered coal and locally available biomass are the largest emitters of OM from their low-temperature and inefficient combustion. Coal combustion is the major source of IOM

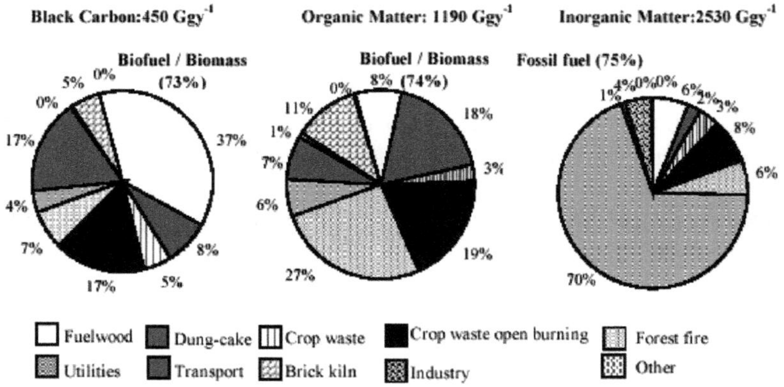

Fig. 1: Emissions of black carbon (BC), organic matter (OM) and inorganic oxidised matter (IOM) from India for 2000, and relative contribution of biofuels (fuelwood, dung-cake, crop waste), biomass open burning (crop waste, forest) and fossil fuel use sectors (electric utilities, transport, industry and "other" including domestic and small-scale industries).

emissions of 2530 Gg y^{-1} primarily consisting of fly-ash, followed by open biomass burning. Fly-ash from coal-fired utilities is of specific importance in S. Asia coal used in this sector. Installed particle control devices were assumed functional 50% of operation time in estimating IOM emissions. Emissions from S. Asia, including Bangladesh, Bhutan, India, Nepal, Pakistan and Sri Lanka have been estimated as 881 Gg y^{-1} of BC and 4535 Gg y^{-1} OC, in which Indian emissions accounted for 683 Gg y^{-1} BC and 3487 Gg y^{-1} OC (Streets et al., 2003a, b). Another recent estimate for India (Bond et al., 2004) is 597 Gg y^{-1} of BC and 2817 Gg y^{-1} of OC, with totals given for Asia, but not S. Asia. There have been assertions in the literature of the possible adulteration of petrol (or gasoline) in India (Dickerson et al., 2002), potentially leading to larger emissions. This is believed to occur from the diversion of kerosene, subsidised for use as a domestic fuel, and its use for adulteration at retail outlets (petrol pumps) of the distribution system. Indian oil refineries maintain fuel specifications at the refinery, but the distribution system is not well monitored. While suitable methods for fuel testing at the pump are being developed to establish the extent of the problem, it has not been possible to quantify at this time the percentage of petrol/diesel suffering adulteration in India. Excluding emission changes from fuel adulteration, differences among the studies arise from differences in their choice of mean and upper bounds in emission factor for various fuels and sources.

An analysis of trends in aerosol emissions from India during 1981-2000 (Habib et al., 2004b) was made projecting biofuel use as a function of national population and crop waste open burning from annual crop production for 1981-2000. Climatological mean forest burning was derived by combining the source term with fire counts from ATSR (Along-Track Scanning

Radiometer aboard satellite ERS-2) (Reddy and Boucher, 2004) averaged for the period 1996-2000. Fossil fuel use trends were taken from CMIE (2001) for coal, low-speed and high-speed diesel, petrol (motor gasoline), fuel (furnace) oil, and kerosene along with emission factors based on technology and fuel consumption (Reddy and Venkataraman, 2002a), to derive emission trends from fossil fuels. These projections (Fig. 2) show an increase in annual emissions of BC from 270 in 1981 to 450 Gg in 2000, of OM from 800 in 1981 to 1190 Gg in 2000 and of IOM from 1100 in 1981 to 2530 Gg in 2000. The increase in emissions from biomass burning of about 2% yr^{-1} is reflected in the growth rate in user population of biofuels (i.e. the rural population) and the increase in crop waste burnt as a function of growth in crop production during this period. The increase in emissions from fossil fuel burning of about 9% yr^{-1} reflects an increase in industrial fossil fuel use during this twenty-year period.

Consistency remains to be achieved among various studies and estimates of aerosol emissions from S. Asia. Especially in the case of BC, S. Asian emissions from energy use estimates (Bond et al., 2004; Habib et al, 2004a, b; Streets et al., 2003a, b; Reddy and Venkataraman, 2002a, b), are believed to be too low to explain measurements from the Indian Ocean Experiment (Dickerson et al., 2002). However, a recent modelling study (Reddy et al., 2004) suggests that S. Asian BC emissions may not be significantly underestimated in relation to emissions of scattering aerosol constituents and precursors. Differences in emission estimates primarily arise from the use of differing emission factors for S. Asian sources, whose measurement remains a pressing need to reduce uncertainties in atmospheric emissions from the region. Regional atmospheric modelling studies, which use these newly available emissions databases and integrate model predictions with in-situ aerosol measurements and satellite remote sensing of aerosols, will further the understanding of aerosol emissions and effects in the S. Asian region.

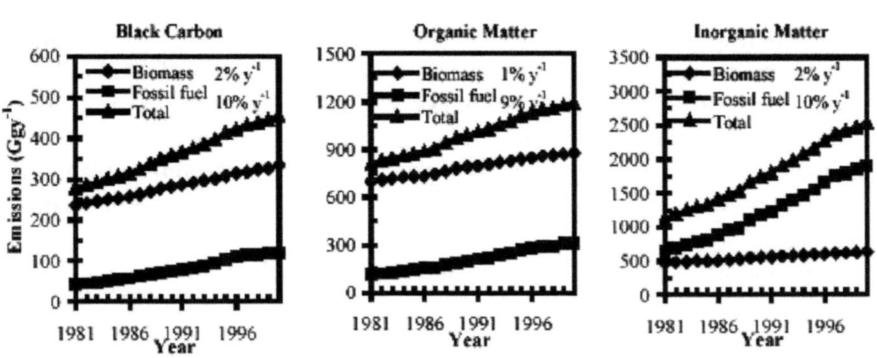

Fig. 2: Annual average emissions during 1981-2000 of BC, OM and IOM from India and contributions from fossil fuel combustion and biofuel/open biomass burning.

4. DIRECT AND INDIRECT AEROSOL RADIATIVE FORCINGS

Determination of the radiative effects of aerosols is currently one of the challenging problems in climate research. Aerosols influence the earth's radiative balance directly by scattering incoming short-wave radiation back to space and indirectly through their influence on cloud properties (Jones et al., 1994; Meehl et al., 1996). The indirect effect is considered to be one of the largest uncertainties in current global climate models (GCM's) (Meehl et al., 1996). Accurately estimating the indirect radiative effect of aerosols requires understanding of the role of aerosols in influencing cloud properties. The estimation of the indirect radiative effect of aerosols requires understanding of the role of aerosols in influencing cloud properties. Several investigations have focused on the determination of indirect effect, but most of them were confined to the anthropogenic sulphate aerosols. While there exist a few studies on the direct radiative forcing, studies on indirect effect of aerosols are rather a few over South Asian regions.

During a comprehensive aerosol field campaign, simultaneous measurements were made of aerosol spectral optical depths, black carbon (BC) mass concentration, total and size segregated aerosol mass concentrations over an urban continental location, Bangalore (13°N, 77°E, 960 m msl), in India (Babu et al., 2002). This study has revealed large amounts of BC, both in absolute terms and fraction of total mass (~11%) implying a significantly low single scatter albedo. The surface radiative forcing as reported by these investigators were as high as –23 W m^{-2} and top of the atmosphere forcing was +5 W m^{-2} and lead to a net atmospheric absorption which translates to an atmospheric heating of 0.8 K/day (Babu et al., 2002). Later, Pandithurai et al. (2004), using measurements of aerosol radiative properties and radiative fluxes, estimated aerosol radiative forcing at a tropical urban site in India, Pune. These studies have reported that during the dry season of 2001 and 2002 (November-April), surface, top of the atmosphere and atmospheric radiative forcing were –33, 0 and +33 W m^{-2}, respectively. The estimated single scattering albedo was ~0.81. In 2004, extensive measurements of aerosol black carbon and aerosol optical depths were made during December, at Kanpur, an urban continental location in northern India (Tripathi et al., 2005). High BC concentration is found both in absolute terms (6-20 $\mu g\ m^{-3}$) and mass fraction (~10%) yielding low single scattering albedo (~0.78). The estimated surface forcing was as high as –62 W m^{-2} and atmospheric absorption was +71 W m^{-2} which lead to a lower atmospheric heating of ~1.8 K/day.

As far as oceanic regions adjacent to the Indian sub-continent are concerned, there were no aerosol measurements prior to 1995. Aerosol observations commenced over the Arabian Sea in 1995 with regular measurements of aerosol spectral optical depths using a multi-wavelength radiometer (MWR) from a tiny island location, Minicoy (8.3°N, 73.0°E), ~400 km due west of the southern tip of the Indian peninsula (Moorthy and

Satheesh, 2000). Subsequently, several cruise-based experiments were conducted as part of various experimental campaigns (Jayaraman et al., 1998). Recently, Satheesh et al. (2005) using aerosol data collected using ship-borne and island platforms (for eight years from 1995 to 2002) along with MODIS (onboard TERRA satellite) data have evolved a comprehensive characterization of the spatial and temporal variation in the physical, chemical, and radiative properties of aerosols over the Arabian Sea. Over northern Arabian Sea (regions lying north of 12°N), AODs do not show significant latitudinal variations; the average aerosol optical depth for this region was 0.29±0.12 during winter monsoon season (WMS; November to March) and 0.47±0.14 during summer monsoon season (SMS; April/May to September). The single scattering albedo did not show significant seasonal variations (remains within ~0.93 to 0.98 throughout the year). During WMS (SMS), top of the atmosphere diurnally averaged aerosol forcing (diurnally averaged) remains around –6.1 (–14.3) W m^{-2} over northern Arabian Sea up to around 12°N and decreases southwards till it attains a value of –3.8 (–3.4) W m^{-2} at the equator. The surface forcing remains around –16.2 (–15.2) W m^{-2} over northern Arabian Sea up to 12°N and decreases southwards to a value of –5.5 (–3.5) W m^{-2} at equator. Comparable values of radiative forcing were reported by Ramachandran (2005). Recently, during the second phase of the Arabian Sea Monsoon Experiment (ARMEX-II), extensive measurements of spectral aerosol optical depth, mass concentration, and mass size distribution of ambient aerosols as well as mass concentration of aerosol black carbon (BC) were made during monsoon transition period over the Arabian Sea (Babu et al., 2005; Moorthy et al., 2005). The estimated radiative forcing was –27 W m^{-2} at the surface and –12 W m^{-2} at the top of the atmosphere (TOA) which are consistent with long-term averages reported by Satheesh et al. (2005) and Ramachandran (2005).

Prior to 2001, measurements of aerosols over the Bay of Bengal were non-existent. In recent years, there have been a few efforts to understand aerosols over this region as part of various experimental campaigns (Satheesh, 2001; Moorthy et al., 2003; Ganguly et al., 2005). The Bay of Bengal region (BoB) is a rather small oceanic region, surrounded by landmasses with distinct natural and anthropogenic activities. These, along with the seasonally changing air mass types, make this region rather complex from the perspective of aerosols and thus their radiative impact. Measurements of aerosol properties over the Bay of Bengal region have revealed that the aerosol optical depths (AODs) and black carbon (BC) mass fraction (MF) over northern BoB reaches their maximum value during April/May (AOD ~0.48 ± 0.06; BCMF ~6%) and minimum during October/November (AOD ~0.19 ± 0.02; BCMF ~3%) in contrast to the seasonal pattern reported over Arabian Sea by earlier investigations (Vinoj et al., 2004). Over equatorial Indian Ocean south of BoB, AODs were low (~0.11 ± 0.03) and seasonal variations were not that significant. The aerosol surface radiative forcing (0.2 to 40 μm) over northern BoB is in the range of –9 to –30 W m^{-2} whereas that over southern BoB is in the range of –3 to –12 W m^{-2}. The corresponding atmospheric forcing was in the range of +6 to +20 W m^{-2} and +1 to +6 W m^{-2} respectively.

It is shown that surface wind has a significant role in changing the chemical composition of aerosols over the sea and hence the forcing (Satheesh and Lubin, 2003). Aerosol short wave, long-wave and net forcing as a function of wind speed is shown in Fig. 3 (data from Satheesh and Lubin, 2003) (lower panel). The experiments by Satheesh and Lubin (2003) over the tropical Indian Ocean suggest that sea-salt aerosols generated by even moderate winds reduce solar heating by nearly half and concluded that radiative effects of sea-salt aerosols are significant and should be included in models of regional and global climate. Estimates of sea-salt aerosol radiative

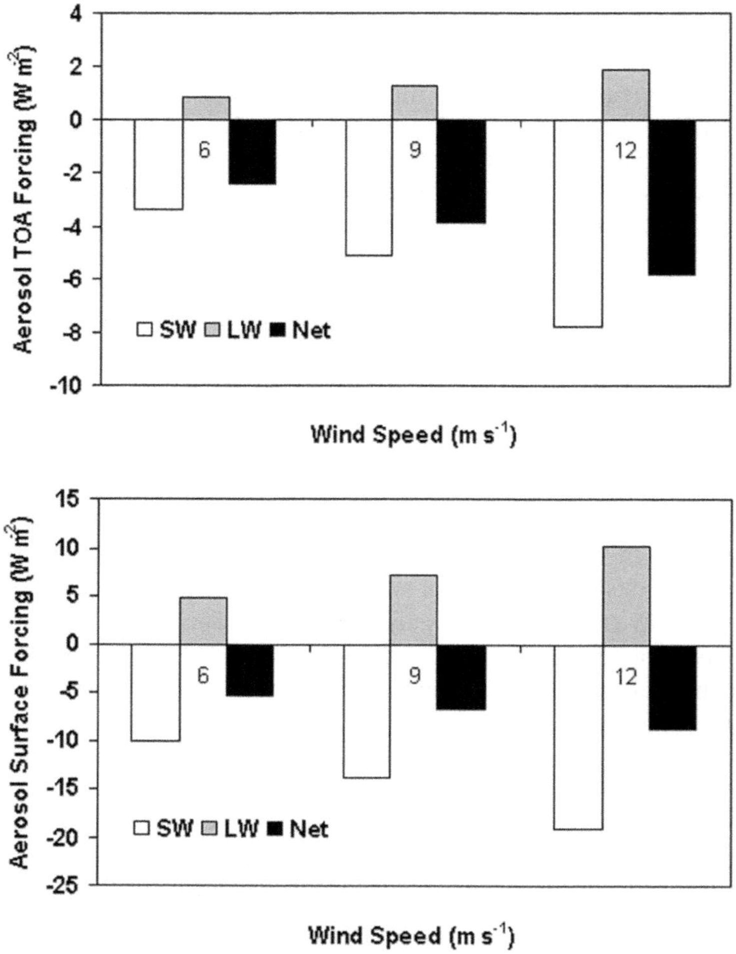

Fig. 3: Aerosol forcing at different wind speeds. The open, gray and black bars represent short wave, long wave and net radiative forcing (from Satheesh and Lubin, 2003).

forcing by Winter and Chylek (1997) showed that at low wind speed, the sea salt radiative forcing is in the range of –0.6 to –2 W m^{-2} and at higher wind speeds this can be as high as –1.5 W m^{-2} to –4 W m^{-2}. This negative forcing by naturally occurring sea-salt aerosol is quite significant when we consider the fact that forcing caused by projected doubling of CO_2 is about +4 W m^{-2}. The forcing caused by the increase in CO_2 since the advent of the industrial era is about +1.46 W m^{-2} (Charlson et al., 1992; Winter and Chylek, 1997). The data on sea-salt aerosols where wind speeds are high is very few. Moreover, in such conditions the measurements are difficult. Thus there is a considerable uncertainty in sea-salt aerosol radiative forcing (Gong et al., 2002; Kinne et al., 2003). Observations over the tropical Indian Ocean have shown that radiative forcing due to sea-salt aerosol is –1.36 ± 0.46 W m^{-2} and that due to dust and soot are respectively –0.72 ± 0.3 W m^{-2} and +0.64 ± 0.38 W m^{-2}. The radiative forcing due to sulphate (natural and anthropogenic) aerosol was –6.4 W m^{-2}.

While there exist a few studies on the direct radiative forcing, studies on indirect effect of aerosols are rather a few over South Asian regions. Based on direct measurements of aerosols, cloud droplet concentration and supersaturation over the tropical Indian Ocean, Ramanathan et al. (2001) derived empirical relations between aerosol number and various parameters such as cloud drop number, cloud drop effective radius, cloud optical depth and so on. Investigations have revealed that sea-salt number concentration over the ocean is a function of wind speed (Moorthy et al., 1997). The estimates of sea salt direct and indirect effects over the Indian Ocean were –2 ± 1 W m^{-2} and –7 ± 4 W m^{-2} respectively (Vinoj and Satheesh, 2004). This is quite large compared to anthropogenic aerosol forcing reported over this region (–5 ± 2.5 W m^{-2}) (Ramanathan et al., 2001). Thus, clearly the direct and indirect effects of sea-salt aerosols have a significant role in offsetting the positive forcing by absorbing aerosols and greenhouse gases (GHGs). The direct and indirect forcing due to sea-salt aerosols compared with anthropogenic forcing over the Indian Ocean is shown in Fig. 4 (lower panel). A comparison of radiative forcing due to various aerosol species along with that due to well-mixed greenhouse gases is shown in Fig. 4 (upper panel). The magnitude of indirect radiative forcing (and uncertainty) due to sea-salt aerosols is several-fold more than the direct radiative forcing of sea-salt aerosols. The large magnitude and variability in both direct and indirect forcing due to sea-salt aerosols emphasises the importance of natural aerosols.

5. EMISSIONS OF TRACE GASES

The quantification of the increasing contribution of atmospheric pollutants including nitrogen oxide (NO_x), carbon monoxide (CO), and volatile organic compounds (VOC) to global environmental changes has received much

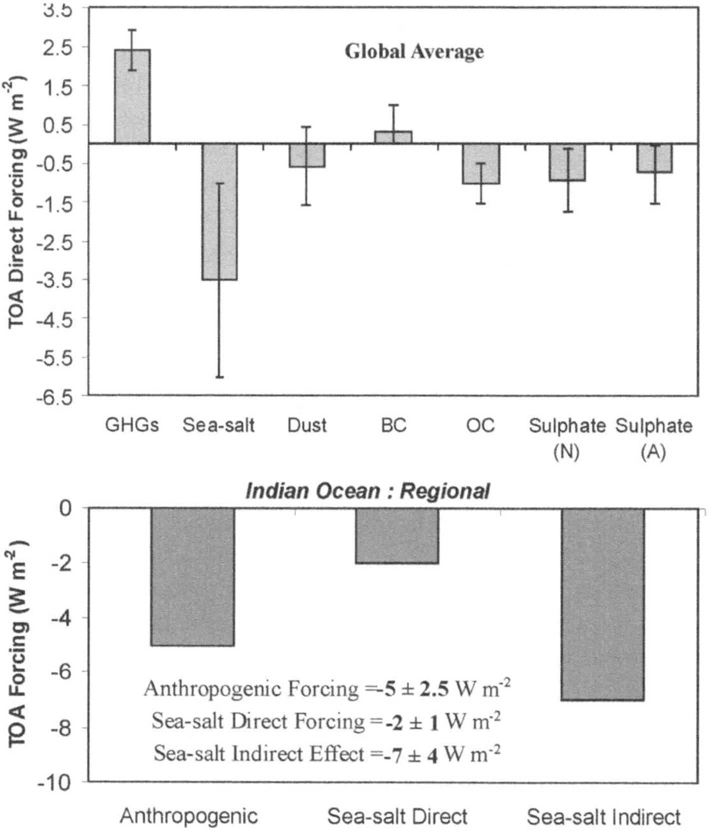

Fig. 4: Comparison of greenhouse gas forcing with that of aerosol forcing due to various species (upper panel) and natural versus anthropogenic forcing over tropical Indian Ocean (lower panel) (from Satheesh and Moorthy, 2005).

attention in recent years (IPCC, 2001). The anthropogenic emissions of these trace gases and their impact on the tropospheric distribution of ozone and its precursors are not only confined locally but likely to influence other regions of the globe depending on the prevailing meteorological conditions and long range transport. However, the emissions of pollutants are region specific and systematic inventory data is rather poorly defined in the models, especially over the Asian region. The estimates of anthropogenic emissions of these trace gases and their impact on the tropospheric distribution of ozone on long time scales is of great significance in that context. It is estimated that by year 2010 over four billion people will be living in eastern and southern Asia. India dominates this region in terms of population growth and has a high economic potential. Industrial activities as well as transportation are growing rapidly in this region. At current growth rates, Asian energy demand will be doubling every 12 years. The major source of energy in rural India (where 72% of the population resides) is bio-fuel from wood, kerosene and coal combustion. This gives rise to considerable CO emissions along with emissions of other source gases including NO_x and VOCs (volatile organic compounds). The CO emission emitted as a result of bio-fuel use dominates the sources of CO emissions in India (Dalvi et al., 2005). Although India's contribution to global CO emissions is merely 6%, there has been a rapid growth in recent years (Mitra and Sharma, 2002; Mitra and Bhattacharya, 1998) in contrast with the globally declining trend in CO level reported, for example, by Novelli (1998).

In many Asian countries, coal burning is one of the major sources of energy (Shrestha and Bhattacharya, 1991), which results in large increases in nitrogen oxides (NO_x) emissions. Global chemical transport models (e.g. Prather et al., 2003; Brasseur et al., 2005) suggest that surface ozone concentration could increase by as much as 25-30 ppbv in India between years 2000 and 2100, assuming an economic growth described by the IPCC/SRES A2 scenario. Saraf and Beig (2004) reported on the basis of ozonosonde observations made at a few Indian locations that surface ozone concentrations have already increased significantly over the last 20 years. Several modelling studies that have assessed the trend in global tropospheric ozone and its precursors in response to changing emission loads (Saraf et al., 2003; Granier and Brasseur, 2003) have not discussed specifically the long-term trends and their impact in the Indian tropical region. A major source of uncertainty is the lack of systematic grided inventory of precursor gases over the Indian region. Changing levels of ozone have implications for human health and vegetation growth while increasing NO_x level can adversely affect the quality of ground water and contribute to eutrophication processes.

The attention of scientific community is therefore fast focussing to explore the impact of long-range transport from South Asian region to the other part of globe due to the above mentioned factors. In this direction, Indian Ocean Experiment (INDOEX) studied the natural and anthropogenic climate forcing by aerosols and feedbacks on regional and global climate (Crutzen and Ramanathan et al., 2001). Several observational and modelling studies related to tropospheric ozone were also conducted over the Indian Ocean region (Saraf et al., 2003; deLaat and Lelieveld, 2000; deLatt et al., 1999), which indicated the role of convective transport in bringing the continental pollutants from south Asia towards the pristine marine environment. Recently there are several modelling studies, which examined the tropospheric ozone and its precursors and the effective radiative forcing on the climate as a result of changing emissions scenario on global scale (Brasseur et al., 2005; Lamarque et al., 2005; Doherty et al., 2005). However, the studies related to impact of emissions focussing on the Asian region in detail and their contributions on the surrounding geographical regions are not numerous.

The distribution of ozone and its precursors in the Indian region are studied using the global chemical transport model MOZART-2 (Horowitz et al., 2003). MOZART-2 simulations are driven by assimilated meteorological data from the European Center for Medium Range Weather Forecast (ECMWF) with a standard resolution of ~1.8 degrees in longitude and latitude, and with 31 vertical levels from the surface to approximately 10 hPa pressure. The model accounts for the distribution of 63 chemical species, 168 reactions including 33 photolysis reactions. It considers surface emissions of several chemical compounds (CO, NO_x, VOC, etc.). The emissions due to fossil fuel combustion, agricultural burning, bio-fuel, etc. are taken from the recent estimates by the POET project (Granier et al., personal communication, 2003). The biomass burning emissions are deduced from satellite observations of fires (Granier and Lamarqus, personal communication, 2005). Over the Indian subcontinent, recent high-resolution emission estimates from the Indian national inventory (Dalvi et al., 2005) which account for the rapid temporal variability and small-scale geographical variations (e.g. hotspots) is found to be different from other global inventories like EDGAR (Olivier et al., 1996) and POET. The Indian national inventory considers the discrete mixture of large and modern urban populations co-existing with traditional rural and agriculture dominated populations. Such diverse social differences are major factors that are not present in most developed countries (Garg and Shukla, 2002).

In the adopted Indian emission inventory, the CO emission for 2001 from bio-fuel sources (wood and cow dung burning) is around 34,282 Gg, which represents almost 50% of total CO emissions in India. The total CO emissions from all sources over the Indian region for 1991 and the absolute change (Gg) from 1990 to 2001 (Dalvi et al., 2005) are shown in Fig. 5. The largest emissions are found to occur in the Western region of India (West Bengal) where the corresponding changes in 1990s are of the order of 100 Gg. The location of emission hot spots seems to correlate with the densely populated rural areas, emphasizing the contribution of rural bio-fuel emissions through fuel wood burning, etc. The emissions and corresponding changes in 1990s are smallest in the states of Jammu and Kashmir, some North East Indian states and small Union Territories. These states with undulating topography and surrounded by the Tibetan and Himalayan mountain regions have low population density, limited agriculture and limited vehicular proliferation. The total CO emission in India (59.3 Tg in 1991 and 69.4 Tg in 2001) has increased by 16% during the 1990s, whereas the world's CO concentration has declined during the same period (Novelli et al., 1998). The total surface NO_x emissions in the Indian region for 1991 and 2001 are estimated to be 3.5 Tg NO_2/yr and 4.3 Tg NO_2/yr respectively, an increase of about 22% during the decade. The maximum contribution is from coal (1.54 Tg NO_2 in 1991), which amounts to about 45% of total NO_x emissions in India. The second largest contribution is provided by fossil fuel combustion in 2001, which has increased dramatically (78%) during 1990s due to rapid increase in vehicular traffic. The location of emission hot spots (Fig. 5)

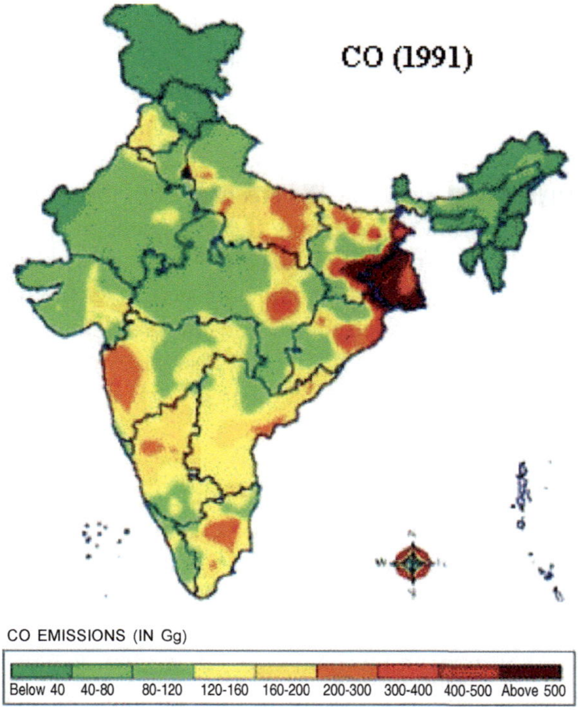

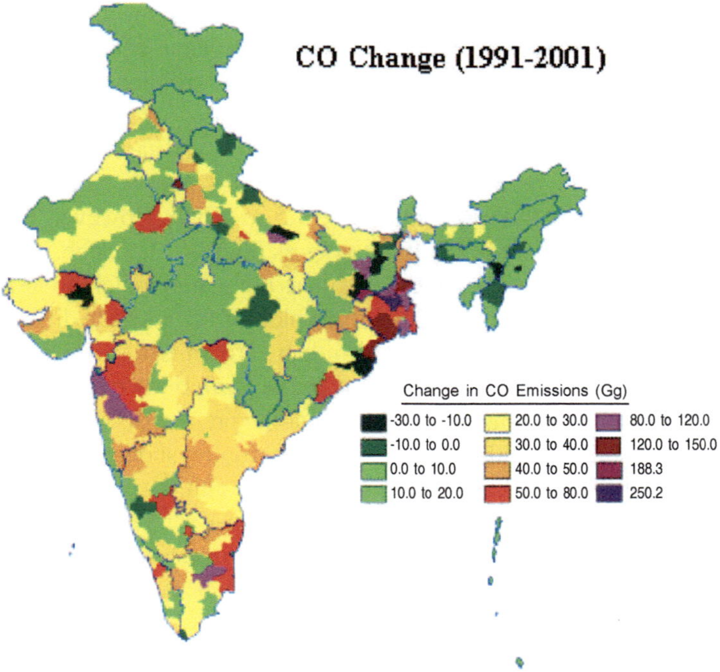

Fig. 5: Grided emission inventory of CO (in Gg NO_2) for the year 1991 and its departure from 2001 (change from 1991 to 2001).

correlates with the location of mega cities, emphasizing the contributions from transport sector and thermal power plants.

Two different scenarios are considered to simulate the distribution of ozone and its precursors over the South Asian region. In the first scenario (reference scenario) the model is run with normal surface emissions for all pollutants for 2001. In scenario 2 (India zero scenario), the emissions from India are turned off and for other regions they are kept as in the case of scenario 1. Hence, the difference of scenario 1 and 2 will provide the information related to the influence of Indian emissions over the surrounding South Asian region. Results obtained for the zero India simulations (scenario 2) provide us with background distribution of these gases over India, which has major contribution from long range transport from distance sources in the S. Asian region.

Figure 6 shows the model-simulated distribution of the volume-mixing ratio of ozone (ppb), CO (ppb) and NO_x (ppt) at the surface over the S. Asian region with normal background emission inventory scenario for July 2001. The ozone volume mixing ratio is found to be as low as 15-25 ppb near the coastal region of south-west India which increases towards north-east direction due to strong south-westerly monsoon winds. The concentration becomes highest (60-80 ppb) in the Indo-Gangetic region of India, Bhutan, Nepal and Bangaladesh and is moderate (35-50 ppb) in the northeastern China. In many other parts of China, ozone shows very high value. In western India, values are found to be of the order of 40-60 ppb excepting a few regions in Arabian countries like that of Qatar, Iran and Iraq where the value exceeds more than 80 ppb. The trend of variation and gradient in CO and NO_x is found to be almost similar to that of ozone. The CO level is found to be in the range of 120-280 ppb in most parts of India excepting Indo-Gangetic plane where it increases to 300-450 ppb. The volume-mixing ratio of NO_x over India is found to be 1.5 to 4 ppb. However, in some parts of China, Iraq and part of Delhi, it goes as high as 10-17 ppb. The NO_x level is found to be relatively high over India as compared to neighbouring regions.

Figure 7 shows the latitude-altitude plot for the absolute difference (ppbv) between the normal emissions and Indian zero emissions for ozone over the South Asian region. The impacts of Indian emissions over the surrounding region for the fixed longitude (78.6^{o}E) for the pre-monsoon (May) and intense monsoon (July) months are shown. Top figure shows the impact in the boundary layer (1.2 km) during monsoon month of July. The high-density white patch represents the ozone distribution over India from the original normal emissions for this region. This patch in the upper map seems to be pushed to north-eastward for the month of July because the monsoon current prevails over the Indian Peninsula. The change in ozone concentration over most part of South Asian region can be interpreted as the influence of the transport of Indian surface emissions over other parts of South Asia. Near the boundary layer, the change is found to be of the order of 2-7 ppb in

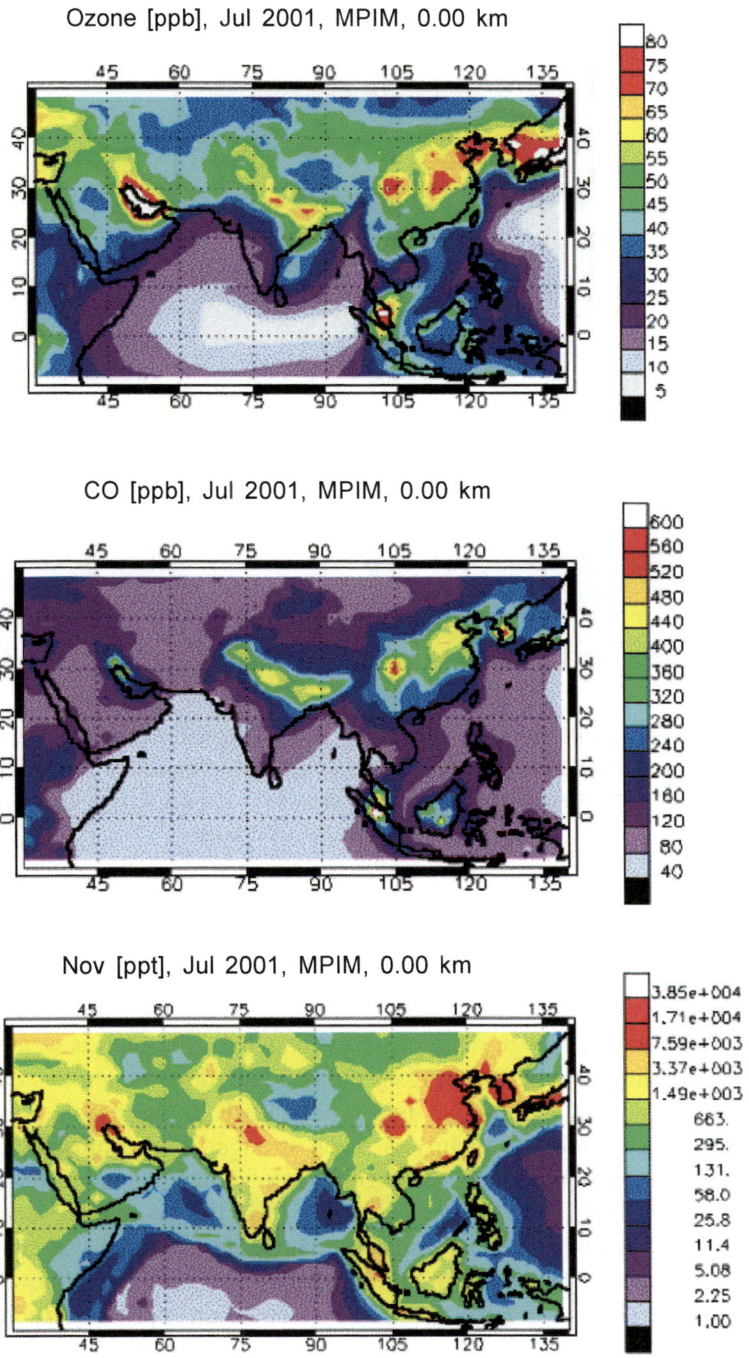

Fig. 6: Model simulated distribution of the volume-mixing ratio of ozone (ppb), CO (ppb) and NO_x (ppt) at the surface over the South Asian region with normal background emission inventory scenario for July 2001.

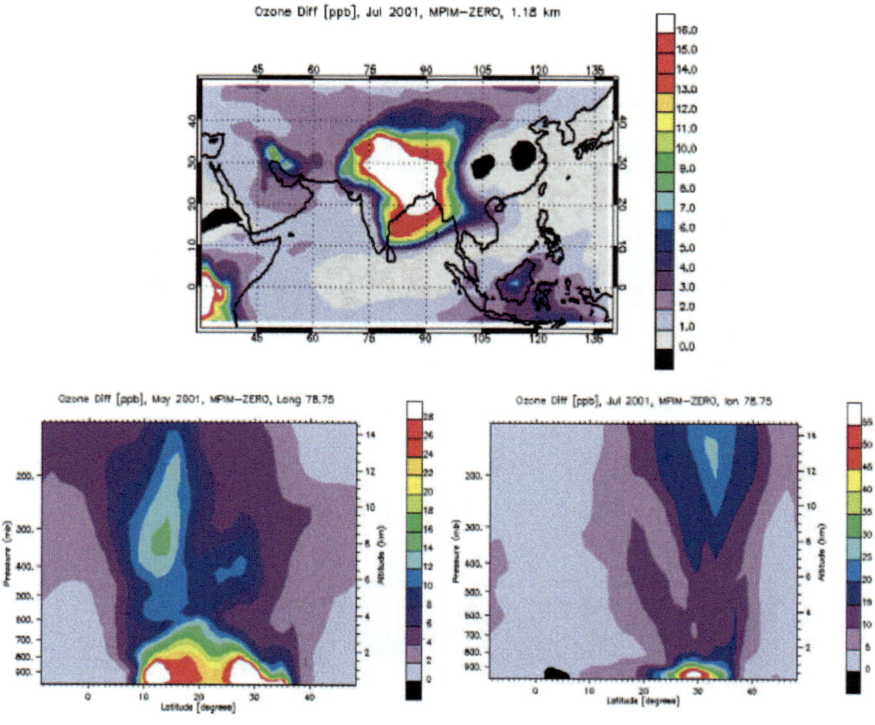

Fig. 7: Absolute difference in ozone concentration (ppbv) estimated for normal emissions and for Indian zero emissions showing the impact of Indian chemical emissions over the South Asian. Top figure shows the impact at 1.18 km and bottom figures are for the surface level.

eastern India and 1-4 ppb in western India for the month of July. Southwest monsoon is active over almost whole of India in July and the marine wind brings the moisture over the region. Monsoon trough connecting the Rajasthan and the Bay of Bengal is established and the wind pattern becomes south-westerly below the trough region whereas above the trough it is weak south-easterly. In the lower panel of Fig. 7, the two latitude-altitude plots of ozone are for the months of May and July. They clearly show that the advection towards eastern side during July is much stronger than compared to May. During the monsoon period, the magnitude of vertical spread in the tropics centred over 30°N is of the order of 5-15 ppb in the lower troposphere and 15-25 ppb in the upper troposphere. The magnitude of similar vertical spread is almost half for the month of May and it is centred over 10°N.

Figure 8 shows the level of ozone concentration (ppb) over India close to the boundary layer (1.2 km) during the month of May (pre-monsoon) and July (monsoon) when emissions over the Indian region are switched off in the model simulations. The diffused transport of ozone from other northern and western part over the Indian region is seen. The background level of

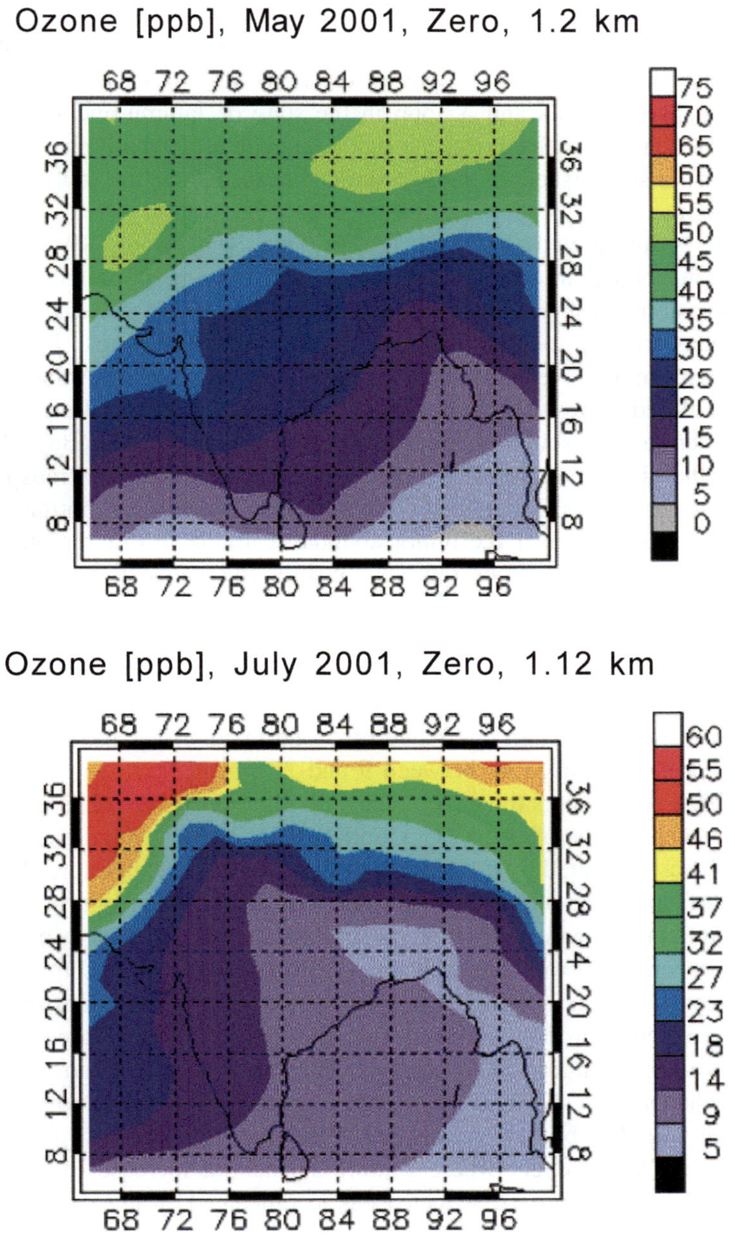

Fig. 8: Influence of South Asian emissions of secondary pollutants over the Indian tropical region for pre-monsoon (May) and monsoon months (July) in 2001.

ozone, which can be interpreted as the ozone transported from the distance sources from the neighbouring regions depict some wavy nature for the month of May, spreading over 64°E to 100°E where the volume mixing ratio is found to be 15-45 ppvb as against the normal range of 35-70 ppbv (when reference emissions for 2001 are used over India). The May plot shows an increase in ozone from 15 ppbv in the northwest to around 30 ppb in the central part of India and 35 ppb in the state of Jammu and Kashmir. The specific features noticed over the Indo-Gangetic region and other mega cities are not distinguished in both months because they are related to local emissions.

It is interpreted that over the Arabian Sea, strong SW to westerly components prevail during this period, which is bringing pollutants from the northeastern coast area of Africa to Arabian sea. Also, the weak northerly to northeasterly components prevailing over Gulf countries, Pakistan and Afghanistan and over more northern regions meet over the Arabian Sea and are modified to westerly. Thus it can be stated that ozone is getting transported over central and southern India, Arabian Sea and over the Bay of Bengal from the above mentioned regions. On the other hand, the transport from India to the Bay of Bengal and other adjoining western and northern regions could be nil. However the minor abundance found in this region is due to transportation from other regions. In July the convection is strong and Southwest monsoon becomes vigorous over almost whole of India and air is advected towards Northeast direction, which is seen in Fig. 8, where the upper push towards Northern India is stronger in July as compared to May. The magnitude of zone abundance has however reduced in July as compared to May and the background value of ozone in July over most part of India at 1.2 km is found to be between 9 and 25 ppb.

Figure 9 presents the geographical distribution of the relative changes in the concentrations of ozone, CO, and NO_x in 2001 over India as compared to 1991 as simulated by the MOZART model. These changes are shown near the surface for the month of July when the convective activity is maximum. The amount of ozone and its precursors in July are expected to be relatively low as compared to other seasons due to pronounced washout effects and surface deposition of soluble chemical species involved in the ozone budget. It is noticed from Fig. 9 that the general pattern in the ozone changes is followed by changes in CO and NO_x but the hotspots are mostly linked to NO_x changes. As shown in Fig. 6, the increase in the ozone volume-mixing ratio is found to be 6-10% (2-4 ppbv) during 1990s in most parts of India except in West Bengal where a more substantial increase of 5-8 ppbv is found. This maximum ozone increase is accompanied by an equally significant increase in the NO_x level of 0.5-1.5 ppbv in this part of India. The absolute change in CO appears to be relatively high (40-60 ppbv) near the Indo-Gangetic plains because the abundance of CO is found to be very high in this region, which corresponds to about 10-20% variation. The high CO abundance in this region is also reported by observations carried out over this region by Ali et al. (2004). The increase in CO is relatively limited near the coastal regions, and, in particular, becomes negligible in the southern coastal regions of India. The pattern of changes in ozone in the southern part

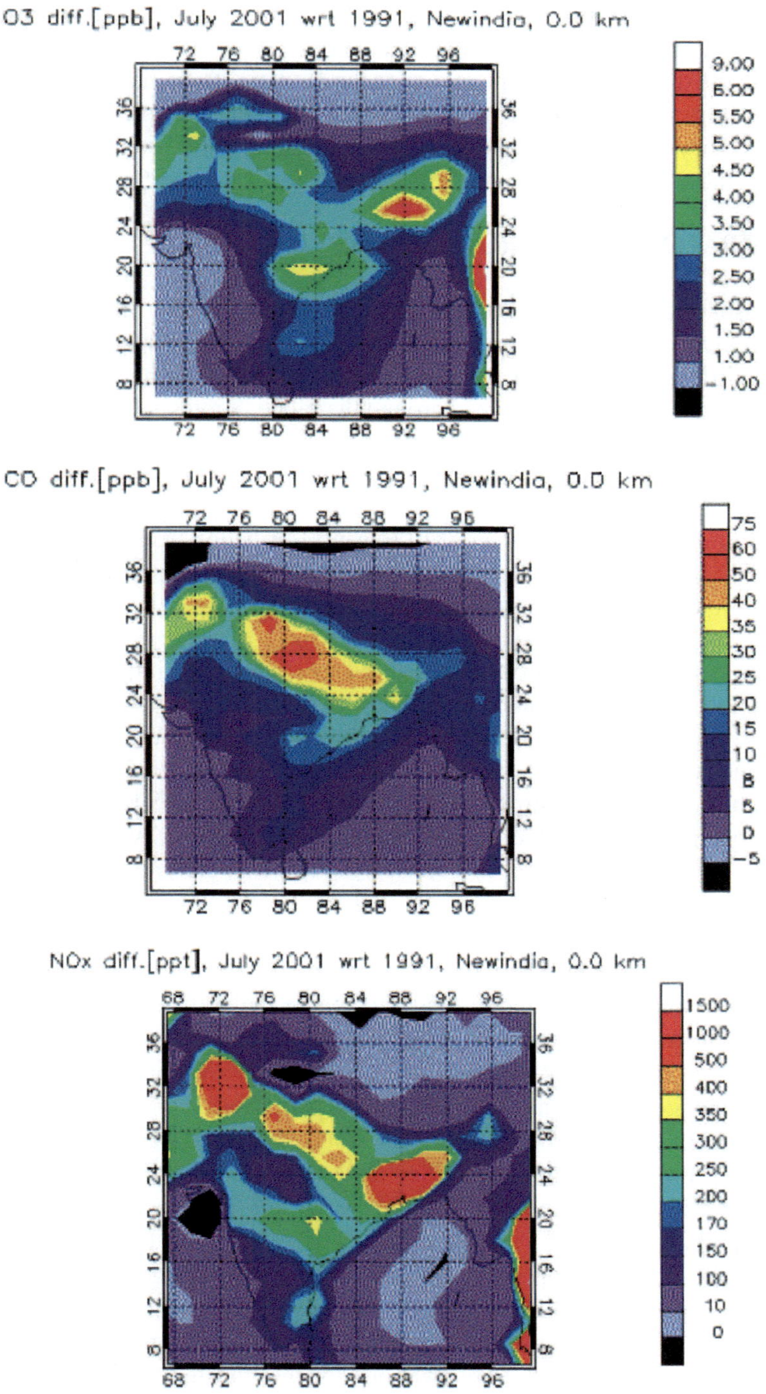

Fig. 9: Model simulated change (ppb) in ozone, CO, and NO_x, from 1991 to 2001 (July) at the surface over the Indian region.

of India and near the Indian Ocean follows the variation in NO_x (Fig. 9). Saraf and Beig (2004) have estimated the trend over the three Indian stations based on the past 30 years of ozonosonde data. They have identified an increasing trend in ozone in the troposphere. For example at Trivandrum (8°N), they reported a surface trend of around 1.4% per year, which is comparable to the present model predicted results of 1.0% per year.

6. STRATOSPHERIC-TROPOSPHERIC EXCHANGE

Atmospheric transport is responsible for carrying energy and matter from one region of the Earth to another. It enables chemical species originating from different regions to mix and interact and maintains a link between various sources and sinks spread around the globe. Stratospheric-Tropospheric exchange (STE) is a key process in ozone chemistry. For example, the man-made pollutants released in the troposphere enter into the stratosphere and initiate catalytic reactions, which are responsible for the stratospheric ozone depletion, while it is the downward transport of ozone from the stratosphere that significantly contributes to the ozone budget in the troposphere. Satellite data on water vapour distribution is often used as the effective dynamical tracer for identifying the regions of STE (e.g. Cooper et al., 1998). For example specific humidity value is used to identify the intrusion of dry air from the stratosphere and such feature is found often associated with enhancement in ozone concentration. It is also the downward transport, which is responsible for the ultimate removal of a variety of stratospheric species including those that destroy ozone. However, the role of STE in altering the tropospheric ozone budget is yet to be fully quantified (WMO, 1999).

One important difference between the troposphere and the stratosphere is the timescales for the vertical mixing. Vertical transport of air and hence the vertical mixing of chemical species in the troposphere occur in timescales in the range of few hours to a day assisted by the strong convective updrafts present in this region while in the stratosphere the vertical mixing takes few months to a year. Because of this large difference in the vertical mixing timescales, the troposphere and the stratosphere are sometime treated as individual compartments and the stratospheric-tropospheric exchange is described in terms of the rate at which tropospheric species enter into or are removed from the stratosphere. Chemical constituents that have long lifetime such as CFCs, N_2O and CH_4 exhibit nearly uniform mixing ratios in the troposphere and in the stratosphere their highest value of mixing ratio is found just above the tropopause. Over the tropical tropopause the high mixing ratio region is found to extend deeper like an air plume from the troposphere, which is indicative of the STE. There are evidences from both theoretical calculations and from field observations for the transfer of trace gases from the upper troposphere to the lower stratosphere (Bregman et al., 1997; Fisher et al., 2000). The vertical mixing is however limited to only the lowest part of the stratosphere. The material after entering the stratosphere is more readily transported horizontally along the surface of constant potential temperature.

One of the important questions related to climate change and involving STE process is: How the equatorial stratospheric chemistry gets modified due to anthropogenic gases originating from the rapidly industrializing countries in the equatorial region? The ability of the atmosphere to remove a chemical species injected into it is determined by both chemical and physical processes. For example, some pollutants are removed from the atmosphere more rapidly either because they are chemically very reactive or they dissolve quickly in water and get scavenged with rain. The chemically stable and water insoluble species such as chlorofluorocarbons can remain in the troposphere for a long duration, of decades and even centuries, and can become chemically active in a suitable environment, far away from their source regions, such as the case of CFCs which deplete ozone in the polar stratosphere. The chemical cleansing capacity of the lower atmosphere largely depends on the amount of hydroxyl (OH) radical, which is often referred to as the "detergent" of the atmosphere. Almost all hydrogen-containing gases, whether of natural or anthropogenic origin, are oxidised by OH. Also, hydrogen peroxide (H_2O_2) is an oxidising agent, which converts sulphur dioxide to sulphuric acid within rain and cloud droplets, and is responsible for causing the acid rain. Thus the oxidation capacity of the atmosphere is linked to the global environment problem. In this context the question, whether the cleansing or the oxidising capacity of the atmosphere has changed due to anthropogenic activities becomes important.

There are reasons to believe that the oxidising capacity of the atmosphere could have changed in the recent time. For example, the increased emissions of NO_x and non-methane hydrocarbons from combustion, along with higher penetration of UV radiation tend to increase the ozone and OH concentrations in the tropical troposphere. On the contrary, the increased biomass and fossil fuel burning in the recent times have resulted in an increase in the emission of CO, which tends to reduce the OH concentrations. Likewise, the increase in methane emission due to increase in natural gas usage, cattle and paddy field also contribute to a reduction in the global OH concentration. In order to make a quantitative estimate on the decrease in global OH concentration values, which is responsible for reducing the oxidising capacity of the atmosphere, a long-term monitoring of tropospheric OH is necessary both at regional and global scales. In the absence of that, estimates are made using model calculations and observational data on the abundance of species that are destroyed by OH. Though there is a consensus between various estimates that there is a considerable decrease in the OH concentration since the pre-industrial era, the absolute value differs. For example, Crutzen and Zimmermann (1991) estimate a range of 5 to 10% for the decrease while Roelofs et al. (1997) estimate about 22%. In an earlier one dimensional model study, Thompson (1992) predicted decrease ranging between 10 and

30%. The factors that control the change in OH concentration such as the increased CO, hydrocarbons, NO_x emissions, increased tropospheric UV radiation due to stratospheric ozone depletion, partially offset each other on a global scale (Brasseur et al., 2003). However at shorter spatial and temporal scales there could be larger differences. Future changes in the cleansing capacity of our atmosphere is going to depend critically on the future anthropogenic emissions.

Another important aspect in the realm of STE and global change is the radiative and chemical effects of aerosols in the equatorial middle atmosphere. Apart from altering the solar radiation budget in the lower atmosphere, aerosols also physically and chemically interact with the gaseous constituents of the atmosphere and can induce changes in the thermal and dynamical properties of the atmosphere. Aerosols are formed in the atmosphere by two mechanisms, one by direct injection of particles from sources located mainly at the earth's surface and the other by nucleation processes happening within the atmosphere. The directly injected particles, except soot, are mostly coarse in nature and do not reach higher altitudes. Conversely, production of particles due to gas-to-particle conversion could occur up to very high altitudes, and as the freshly nucleated particles are in the sub-micron size range they are readily transported to long distances. Therefore study of the factors controlling the nucleation processes in the equatorial upper troposphere and lower stratosphere becomes important. Also the low temperature at the equatorial tropopause helps in a variety of heterogeneous nucleation processes involving ice particles, sulphates etc.

Meteorological processes can affect the microphysical properties of aerosols such as their size distribution and chemical composition. For example, in convective systems when particles are transported upwards, a separation between water soluble and insoluble particles occurs as the precipitating clouds remove the soluble components of the aerosols. The insoluble components of the aerosols that penetrate above the clouds contribute to the heterogeneous nucleation processes in the free troposphere (Hegg et al., 1990; Clarke et al., 1999). The important aerosol precursor gases are sulphur dioxide (SO_2) emitted from volcanoes and industries, dimethylsuphide (DMS) from oceans, oxides of nitrogen (NO_x), ammonia, and volatile organic compounds. Volcanic eruptions are one of the major sources of SO_2 into the atmosphere and explosive eruptions can eject the material directly into the free troposphere and into the stratosphere. Majority of the land-based active volcanoes (about two third) are located in the northern hemisphere, while Indonesia, a tropical country, south of equator has also many of them. The removal process for the material injected into the free troposphere is slower and hence the volcanic sulphur species has longer residence time leading to more efficient long-range transport. In the equatorial stratosphere the volcanic aerosol layer, such as that formed after the Mt. Pinatubo eruption lasted for about three years (Jayaraman et al., 1995).

In the boundary layer the concentration of sulphate particles are mainly controlled by SO_2 emitted by anthropogenic activities, while in the free troposphere and stratosphere, volcanic eruptions and high flying aircrafts are important in controlling the sulphate burden. However more studies are required to estimate the transformation rates of SO_2 to sulphate in warm and moist conditions (Heintzenberg et al., 2003). Soot particles can also find its way to free troposphere, as it is insoluble in water and less scavenged by cloud precipitation. However, contrary to sulphate particles, which are good scatterers of light, soot absorbs radiation and contributes to atmospheric heating. INDOEX studied the cross-equatorial flow of pollutants from the northern hemisphere into the ITCZ and its impact on the radiative forcing (Ramanathan et al., 2001). Lidar and air-borne measurements of aerosols and condensation nuclei (CN) concentrations across the ITCZ (Muller et al., 2001; Heymsfield and McFarquhar, 2001) revealed multiple layers of absorbing aerosols well above the marine boundary layer and a modification in the number concentration and droplet size of the CN particles in the polluted side of the ITCZ.

It is strongly believed that pollutants enter into the stratosphere mainly through the tropical tropopause. A detailed understanding of the processes involved in the transformation and transport of short-lived species across the ITCZ is very critical if we have to assess the future changes of the global atmosphere due to man-made activities. In situ observational techniques need to be strengthened for studying the atmospheric chemistry over the tropics. Apart from improvements in the traditional balloon soundings, aircraft observations are also needed to be strengthened. It is also necessary to assess the large horizontal variability in atmospheric constituents and dynamics over the equatorial region, which strongly influences the vertical transition. Integration of various observational techniques is necessary for this purpose. New data analysis techniques need to be developed for the true understanding of the three-dimensional structure of the atmosphere. Establishment of permanent observatories at key locations over the equatorial region is very important to this study.

7. AIR QUALITY IN SOUTH ASIA

Air quality is a serious issue in South Asia as almost all the urban centres in this region exhibit poorer air quality compared to international as well as respective national ambient air quality standards. In view of the serious consequences of poor air quality especially on human health, and mounted public pressure the policy makers are forced to push through the required policy interventions.

In most of the Indian cities, the concentration of suspended particulate matter (SPM) is found to exceed the National Ambient Air Quality Standards' (NAAQS) limit during the last few years (Central Pollution Control Board, New Delhi, www.cpcb.nic.in) which has caused concern in all the

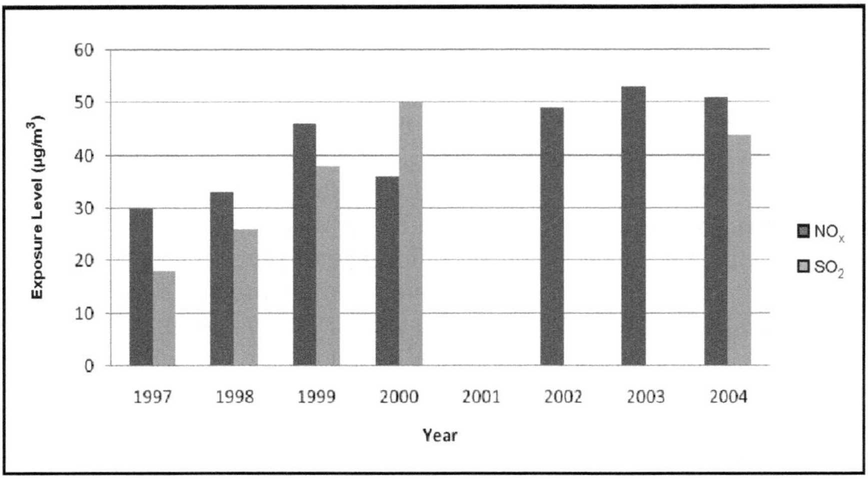

Fig. 10: Variation in annual average NO_x and SO_2 concentrations from 1997 to 2004 observed over Colombo.

stakeholders. However, the concentrations of other pollutants like oxides of nitrogen (NO_x) and sulfur dioxide (SO_2) in most of the Indian cities are found to be below the respective NAAQS limits. The situation is not very different in other cities of South Asia. In Colombo, the concentrations of PM_{10} at thickly populated areas, during 1998 to 2002, show higher values than the recommended level. The concentrations of NO_x and SO_2 in Colombo during 1997 to 2004 period show an increase in ambient concentrations (Fig. 10). In the Pakistan cities of Lahore and Karachi, the concentrations of most of the pollutants like SO_2, NO_x, O_3, CO and Pb in the ambient air have been found to be below or around the national standard values but the TSP (Total Suspended Particulate) matter and PM_{10} values are observed to be much higher than the permissible levels (Figs 11 and 12 respectively). In Kathmandu also, the SPM and PM_{10} concentrations in the ambient air have been found to exceed the standard limits (Fig. 13). The monthly average PM_{10} concentrations measured at different locations in Kathmandu during 2002-03 show that the value remains much above the national standard during November to June period but lower values during June to October period. In Dhaka, the studies have revealed that PM_{10} concentrations in the urban areas during November to January period hovers at >100 $\mu g/m^3$ compared to other period when it hovers at $\leq$ 50 $\mu g/m^3$. Various reasons have been ascribed to the observed increase in SPM and PM_{10} (and also other pollutants) in the ambient air of urban areas of South Asia, which include emissions from point, line and other sources, long-range and short-range transportation and meteorological conditions.

According to a report (by JICA and Pak-EPA),which investigated the high SPM concentration observed in March 2000 in Lahore (where the

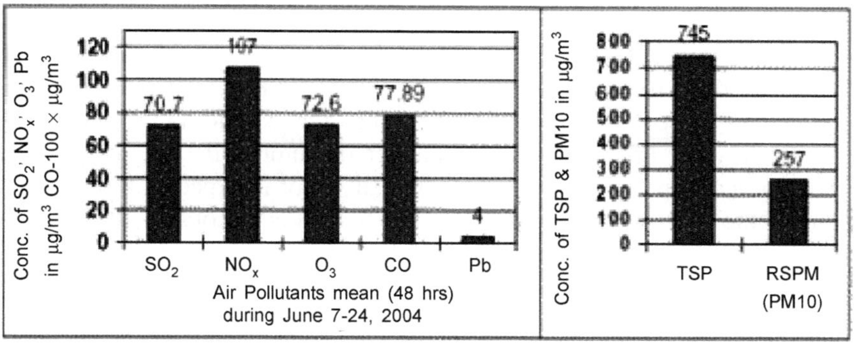

Fig. 11: Ambient air quality data showing different pollutant levels over Lahore.

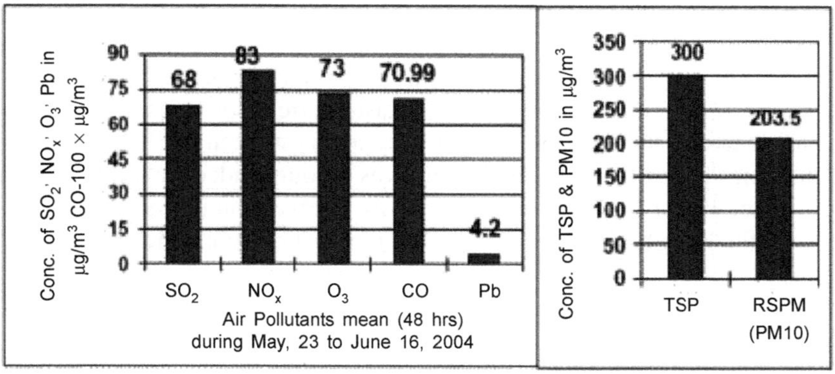

Fig. 12: Ambient air quality data showing different pollutant levels over Karachi.

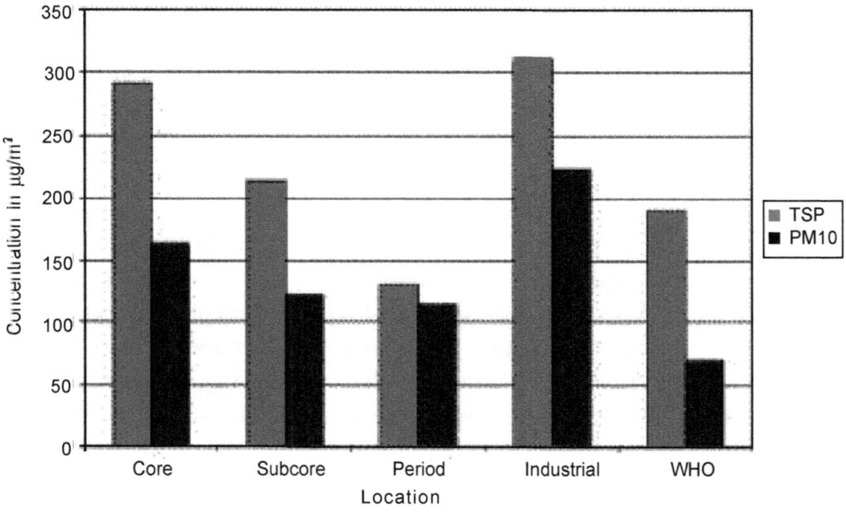

Fig. 13: TSP and PM10 concentrations obtained at different locations in Kathmandu. For comparison the permissible limits proposed by WHO are also shown.

maximum value reached about 1500 μg/m^3), very low rainfall coupled with in-situ soil characteristics favoured high loading of natural SPM. However in most of the urban areas of South Asia, transport and industrial sectors are considered to be the major contributors to the deteriorating ambient air quality. In most of the cities, the number of automobiles has been found growing very rapidly. Though the industrial sector is expected to contribute more to the ambient loading of gases like SO_2, it is not observed in most of the South Asian cities but the transport sector has been identified as the most important sector affecting the ambient air quality. In view of this realization, the policy interventions have been initiated in South Asian cities in the transport sector for regulating/minimizing the transport related emissions of different trace gases and particulate matter. Though other sectors have also received policy interventions they are not to the extent witnessed by the transport sector.

Delhi has become an example in recent years where most drastic policy interventions in terms of fuel switching, vehicular technology up-gradation, traffic management, road and other infrastructure improvement have been implemented in India. Delhi's experience has now been followed for most of the cities in India in a gradual manner as recommended by the Mashelkar Committee Report on 'Road map for policy intervention in transport sector'. Since 1996, there has been several policy interventions implemented in India's transport sector which were aimed to reduce pollutants emissions which included phasing out of lead from gasoline, reduction of sulphur contents in diesel, tightening of mass emission standards for new vehicles, introduction of Compressed Natural Gas (CNG) as fuel substitute for most of the commercial vehicles plying within the city etc. Delhi has witnessed complete conversion of commercial vehicle fleet into CNG mode while in other cities of India gradual introduction of CNG is being done along with introduction of other cleaner fuels like LPG.

The results of these policy interventions have been by and large positive as reflected by the long-term air quality monitoring data of different cities in India. Table 1 gives a comparison of per cent change in different air quality parameters observed in Delhi from 2003 to 2004. Almost all the species like SO_2, NO_2, SPM, RSPM and CO show a decrease by 2 to 5% at traffic junction monitoring points in Delhi during 2004 compared to 2003. The results are encouraging all the more because of the fact that number of vehicles has been increasing very fast during this period and also the increased traffic congestions have increased travel time with a consequence of increased emissions.

Other cities in South Asia are also witnessing increased policy interventions in the transport sector. In Pakistan, CNG is being increasingly used in vehicles; in Kathmandu electrical vehicles (three wheelers) have been introduced; lead has been phased out in Dhaka; and vehicle emission norms have been improved in Colombo. These measures are likely to help in improving the ambient air quality in South Asia but, probably, much more is required to be done in a more comprehensive manner.

Table 1: Comparison of the ambient air quality in Delhi for the 2003 and 2004 periods

Parameter	*Percent increase/decrease in 2004 with respect to 2003*	
	Location	*Increase /Decrease*
Sulphur dioxide (SO_2)	Residential	0%
	Industrial	+11%
	Traffic intersection	–11%
Nitrogen dioxide (NO_2)	Residential	+5%
	Industrial	+17%
	Traffic intersection	–5%
Suspended Particulate Matter (SPM)	Residential	+5%
	Industrial	–4%
	Traffic intersection	–2%
Respirable Suspended Particulate Matter (RSPM)	Residential	+3%
	Industrial	–4%
	Traffic intersection	–7%
Carbon monoxide	Traffic intersection	–9%

8. ACID RAIN SCENARIO IN INDIA

Acid rain occurrence in Europe, North America and East Asia has been observed as common phenomenon (Galloway et al., 1987; Rodhe et al., 1995; Hara et al., 1995). In these regions very low pH of rainwater has been reported. But the precipitation studies carried out in India at several sites have revealed that in general, the nature of rain water is alkaline having pH >5.6. A number of studies has been reported by several workers (Jain et al., 2000; Khemani et al., 1989; Kulshrestha et al., 1996, 2003; Rao et al., 1990; Saxena et al., 1991). A few systematic network studies like Indo-Swedish Network on precipitation chemistry in Indian and Nepal (Parashar et al., 1996) and BAPMoN network (Mukhopadyay et al., 1992) have also reported the alkaline pH of rainwater in India. Recently, in a review report Kulshrestha et al. (2005) have compiled precipitation studies from around 100 sites in India and found pH of rainwater in most of the sites as alkaline. The possible reason for alkaline feature is abundance of calcium in rainwater that neutralizes the acidity. $CaCO_3$ is contributed by loose soil dust into the atmosphere. The acidity generated by the oxidation of gases like SO_2 is neutralized by soil-derived particles. Mostly, calcium concentration in the soil is reported the highest among other soluble chemical constituents, while in some areas NH_3 is found to be the most important neutralizer. It is reported that most of SO_2

in India is not contributing to sulphuric acid rather it forms calcium sulphate with $CaCO_3$ in the atmosphere which is scavenged by the rain water without increasing the acidity of rain water (Kulshrestha et al., 2003).

Acid rain events have been observed under limited conditions e.g., if it is raining continuously for several hours, if the soil of the region is itself acidic, or near highly industrialized areas. Over Indian Ocean, acid rain occurrence has been reported during Indian Ocean Experiment (Kulshrestha et al., 1999; Granat et al., 2002) where nss SO_4 levels were very high as compared to nss Ca. Results from Indian Ocean Experiment (INDOEX) revealed that the situation over the Indian Ocean is very different as compared to Indian continental sites. The pH of rain water over Indian Ocean has been observed between 3.8 and 5.6, while at continental sites it lies between 5.6 and 7 and sometimes even more than 7. The acidic nature of rain water over Indian Ocean is due to insignificant influence of soil dust and the dominance of anthropogenic sulphate from long-range transport. In the present scenario, only certain pockets in areas like central-east, northeast and southwest regions are observed to have acidic deposition where further systematic investigations are needed (Kulshrestha et al., 2005). Interestingly, at continental sites, with the present 5% annual increase in sulphur no increasing trend in acidity is reported.

Table 2 gives the mean concentration of different chemical constituents and pH of rainwater found at rural, sub-urban, urban and industrial sites in India. Stations located within the cities are termed as urban while that located on the outskirts of the city are termed as suburban. Sites without any important anthropogenic emissions in the neighbourhood are denoted as rural and are thought to be representative for the surrounding countryside. Sites that are located in industrial areas are termed as industrial sites. All sites show alkaline nature of rain water (pH >5.6). It is to be noted that at rural sites, the concentration of HCO_3 is the highest. The concentrations of nss SO_4 increase from rural to industrial site while HCO_3 decrease from rural to industrial sites indicating crustal influence at rural site. pH decreases from rural to industrial sites indicating some influence of industrial SO_2. However, it remains alkaline as the soil-dust that is rich in $CaCO_3$ neutralizes the acidity caused by oxides of nitrogen and sulphur.

Table 2: Mean chemical composition (μeql^{-1}) and pH studied at four categories of stations in India

	Na^+	NH_4^+	K^+	Mg^{2+}	Ca^{2+}	Cl^-	NO_3^-	nss SO_4^{2-}	HCO_3^- (calc)	pH
Rural	55	12	15	40	105	49	25	21	54	6.5
Sub-urban	79	15	11	51	121	82	15	21	40	6.7
Urban	76	22	11	36	105	80	36	34	37	6.4
Industrial	38	26	8	28	89	38	21	85	33	6.1

It is well known that Sweden is one of the most affected countries by acid rain problem. As reported by Granat (1990), rainwater at all the sites of Swedish network had pH below 5.6. Figures 14a and 14b show variation of pH and SO_4 at different sites in India and Sweden. These figures show that in Sweden as SO_4 increases, pH decreases while at Indian sites, even at higher SO_4 levels, the pH of rain water is observed to be higher. It indicates that in Sweden, the sulphate is primarily contributed from sulphuric acid but in India its pathway could be different. As reported by Kulshrestha et al. (2003), the possible mechanism of SO_2 removal is its adsorption onto the

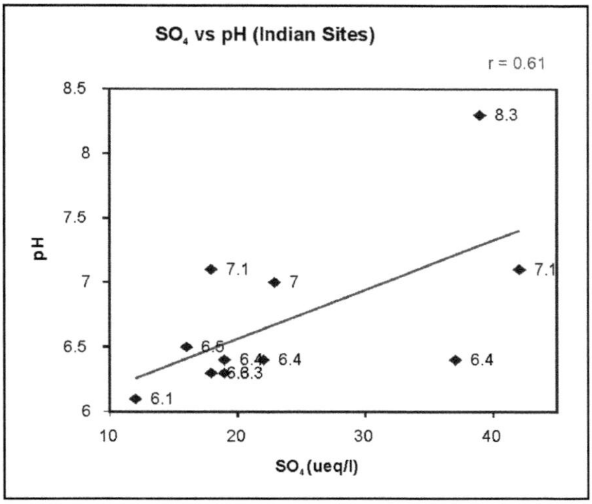

Fig. 14a: Variation of SO_4 with rain water pH obtained at different sites in India.

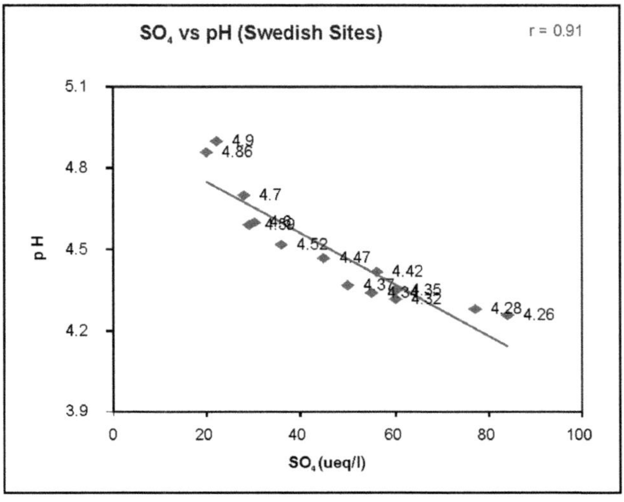

Fig. 14b: Variation of SO_4 with rain water pH obtained at different Swedish sites.

Table 3: Soil and their pH in India (Source: Sehgal, 1996)

Category	*pH range*
Alluvial soil	8.3-8.6 and 9.2-10.6
Black soil	8.8-9.4
Red soil	6.0-6.9
Laterite soil	5.4-5.8
Salt affected soil	9.3-10.3
Desert soil	8.0-8.3
Forest soil	4.6-5.6 and 6.6-6.7

$CaCO_3$ dominated dust particles that forms calcium sulphate, which does not lower the pH.

The soils of India are highly alkaline (Table 3) which contribute alkaline suspended dust into the atmosphere during dry weather conditions. Even during the monsoon period (June-September), the rain is often intermittent at several locations and the soil dust is resuspended again into the atmosphere. Estimates of dust emissions have been reported by Tegen and Fung (1995). The dust so emitted preferably scavenges SO_2 in the presence of sufficient moisture forming calcium sulphate. Thar desert is a major source of $CaCO_3$ dust into the atmosphere. It also spreads to other parts of the country during 'Aandhi' (dust storm) events and by the westerly wind prevailing during summer season. Due to high $CaCO_3$ loadings, SO_2 is scavenged significantly. This may be one of the reasons why increasing SO_2 levels have not been reported at most of the sites. Anthropogenic sources have influence in certain parts like central-east part, which is a mining area with several thermal power plants. In the regions like Kerala (southwest India) and northeast, acidity is reported due to natural vegetation. These areas need to be further investigated (Kulshrestha et al., 2003).

In contrast to continental sites in India, situation over Indian Ocean is very much different where suspended soil dust is insignificant but nss SO_4 concentrations are very high as observed during Indian Ocean Experiment (INDOEX) (Kulshrestha et al., 1999; Granat et al., 2002). Lower Ca levels found over the Indian Ocean region are not sufficient to neutralize the aerosols and acidity of rainwater becomes very high. Figures 15 and 16 show the variation of nss Ca and nss SO_4 aerosols measured over the Indian Ocean during INDOEX. The nss Ca particles which are predominantly crustal origin are coarse in size mode and are transported close to coast but higher nss SO_4 observed up to 5° N indicates that SO_2 is transported to long distances and is transformed into sulphuric acid resulting in acid rain over the Indian Ocean. The source of SO_2 over Indian Ocean includes India, Middle East, Africa, Bay of Bengal and Australian regions as found by the air back trajectory analysis by Granat et al. (2002).

A systematic close network of sites with quality assurance of data is required to be set up to monitor future scenario of acidification in India. This

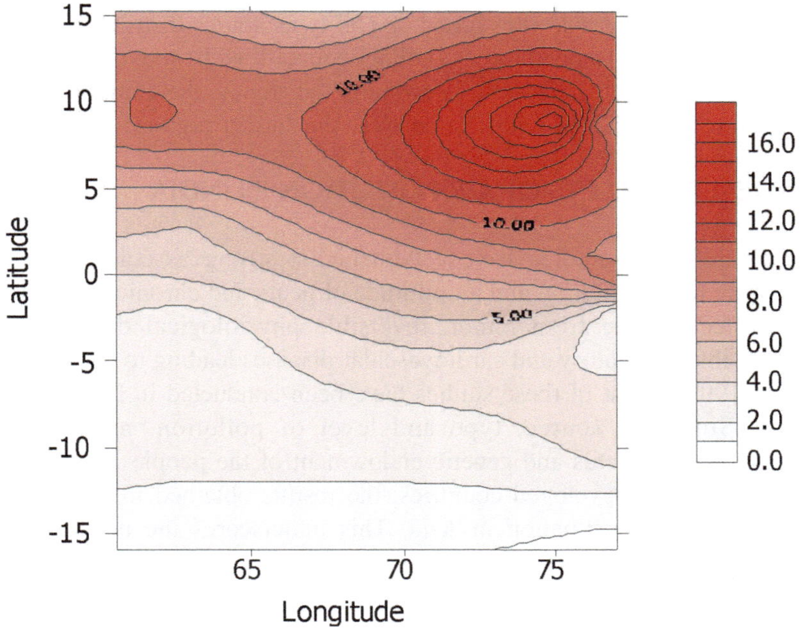

Fig. 15: Variation in nss SO_4 aerosol concentration (μg m^{-3}) obtained over the Indian Ocean during winter months.

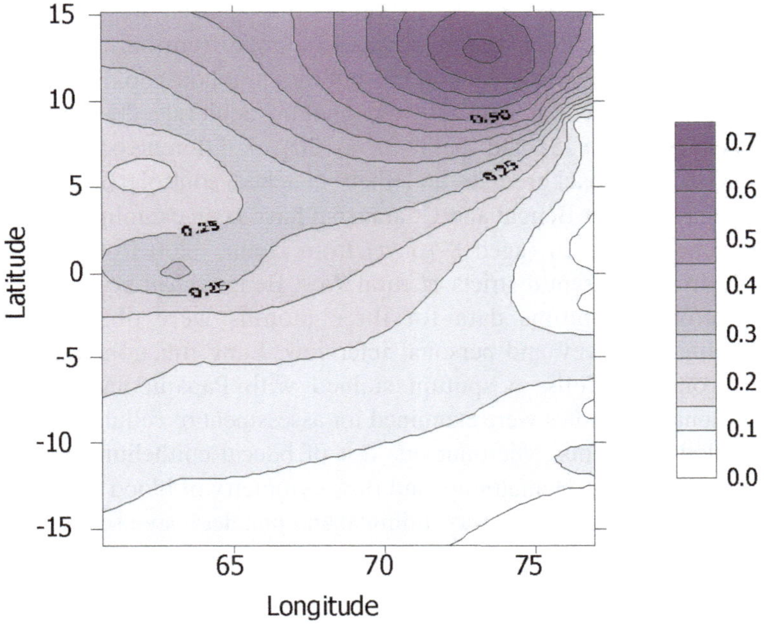

Fig. 16: Variation in nss Ca aerosol concentration (μg m^{-3}) obtained over the Indian Ocean during winter months.

network should include sites based on soil type, source distribution and land use pattern etc. Deposition modelling coupled with above measurement network needs be carried out to estimate the trends, their implications and the future potential of acidification over the Indian region.

9. HEALTH EFFECTS OF AIR POLLUTION IN INDIA

Epidemiological studies have established a strong association between exposure to air pollution and a multitude of acute and chronic health effects. The effects range from minor, reversible physiological disturbances to irreversible respiratory and cardiovascular diseases leading to death (Holgate et al., 1999). Most of these studies have been conducted in Europe and the USA. Since the source, type and level of pollution, nutritional and immunological status and genetic endowment of the people are different in developing and developed countries, the results obtained in the West may not represent the situation in Asia. This underscores the need for health studies of pollution-exposed population in this part of the world.

Millions of urban Indians are exposed to some of the highest pollutant levels in the world. Kolkata and Delhi are among the highly polluted cities in India where the concentration of respirable particulate matter (PM_{10}) over the last ten years remained far above the National Ambient Air Quality Standards. Considering these, a comprehensive, ten-year study on health effects of chronic exposure to air pollution of Kolkata and Delhi has been carried out. In addition, health impact of smoke from burning of biomass fuel used for domestic cooking in most of the rural households in the country is being investigated. In this study non-smoking adult (age 20-65 yr) residents of Kolkata (n = 835) and Delhi (n = 720) of different occupational and socio-economic backgrounds and their matched controls (n = 300) from rural areas of West Bengal and Uttaranchal have been examined. In addition, 3500 school children (aged 8-16 yr) from Delhi, 3200 from Kolkata and 12,000 from different districts of rural West Bengal have been investigated. Respiratory symptoms data for three months were obtained through questionnaire survey and personal interview. Lung function was evaluated by spirometry. Cells in sputum stained with Papanicolau method and monoclonal antibodies were examined for assessment of cellular lung reaction to inhaled pollutants. Micronucleus test of buccal epithelium was used for genotoxicity assay. Hematology and flow cytometry of blood cells were used for assessment of inflammatory, immune and platelet responses. Air pollution data were obtained from state and central pollution control boards. Appropriate statistical procedures were used for data analysis and interpretation.

The three-month respiratory symptom data shows that a majority of the citizens of Kolkata and Delhi have respiratory symptoms indicating presence of underlying respiratory illness (Fig. 17). Upper respiratory symptoms (URS) like common cold, running or stuffy nose, sinusitis and sore throat were present in 58% adults examined in Delhi and 74% in Kolkata compared with 34% in controls. Lower respiratory symptoms (LRS) including dry cough, wheeze and chest discomfort were prevalent in 60% in Delhi, 68% in Kolkata in contrast to 31% in controls. Women and persons from lower socio-economic background suffered most in both the cities. URS were present in 70% children in Kolkata, 68% in Delhi compared with 41% in controls. Similarly, LRS were present in 65% children in Kolkata, 60% in Delhi and 32% in controls. In both cases, girls were affected more than the boys. Like respiratory symptoms, lung function impairment was more prevalent in both adults and children of Kolkata and Delhi compared with respective controls (Figs 17 and 18). URS, LRS and lung function impairment were positively correlated with the concentrations of PM_{10} in ambient air.

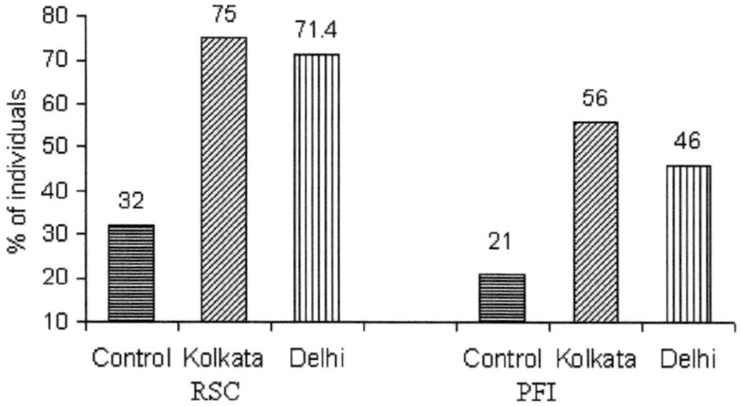

Fig. 17: Prevalence of respiratory symptom complex (RSC) and pulmonary function impairment (PFI) among non-smoking adults of Kolkata and Delhi.

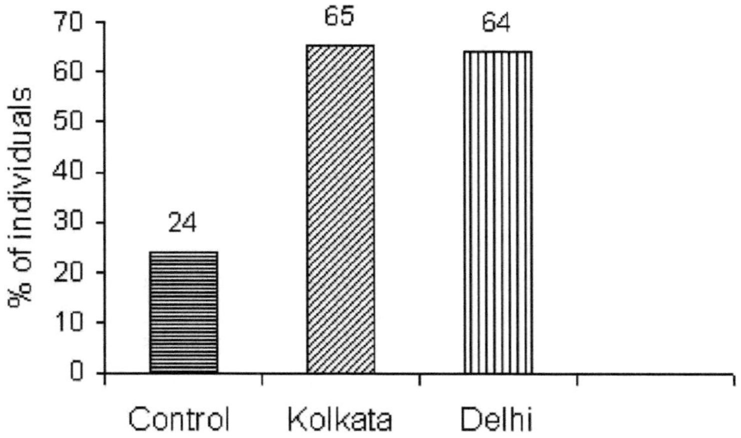

Fig. 18: Prevalence of lung function impairment in children of Kolkata and Delhi.

Marked increase in the number of alveolar macrophages (AMs) and inflammatory cells like neutrophils, eosinophils and lymphocytes was recorded in the sputum of residents of Kolkata and Delhi compared with rural controls. The AM of urban group produced tissue-damaging enzyme elastase and inflammatory mediators like tumor necrosis factor-alpha and nitric oxide in excess. The urban people had several times more iron-laden macrophages in sputum than that of controls, suggesting, among others, covert pulmonary haemorrhage.

Sputum neutrophilia, an indication of infection, inflammation, airway obstruction and lung damage, was recorded in 36% individuals of Kolkata, 31% of Delhi against 22% of controls. Increased sputum eosinophil count observed in 22% adult non-smokers of Kolkata, 29% of Delhi compared with 8% of controls is associated with bronchitis, wheeze and hypersensitivity of the lung as in case of asthma. Apoptosis i.e. programmed cell death was greatly increased in AM of urban group indicating injury to these frontline defense cells in the lung. Squamous metaplasia of respiratory epithelial cells were recorded in 48% examined individuals of Kolkata and 38% of Delhi in contrast to 17% of controls. The change indicates faulty repair of bronchial cells following damage.

Air pollution was found to alter body's immune response. A fall in CD4+ T-helper cells and corresponding rise in CD8+ T-cytotoxic cells have been recorded in urban subjects as well as in rural women who use biofuels for cooking. Air-borne pollutants contain many mutagens and carcinogens. We found elevated micronucleus frequency in buccal epithelium of the urban group, indicating greater prevalence of chromosomal damage. Similarly, p^{53}, the tumor suppressor gene, was found mutated in bronchial epithelial cells of 4% of Kolkatans against 0.5% in rural controls. The findings suggest genetic changes in exposed tissues following chronic inhalation of polluted air.

The present and earlier studies (Lahiri et al., 2000a, b, 2002; Basu et al., 2001; Roy et al., 2001; Ray et al., 2003) have documented a series of adverse effects of air pollution on the respiratory and general health of children and adults of rural and urban India. Besides vehicular and industrial pollution, indoor air pollution from biomass fuel use is a major problem in India. Rural women rely primarily on biomass fuels for cooking food. In the process they are exposed to an array of pollutants including potential carcinogens equivalent to smoking several packs of cigarettes per day (Smith and Liu, 1994). Chronic inhalation of biomass smoke may lead to stillbirth, low birth weight and diseases in early infancy (Saksena and Smith, 2003), and increases the risk of acute respiratory infections (ARI) like pneumonia (Smith et al., 2000) and tuberculosis (Mishra et al., 1999). Particulate pollutants, especially the ultra-fine ones appear to be responsible for the adverse health outcome, because they can cross the lung-blood barrier and circulate inside the body causing systemic damage (Nemmar et al., 2002).

As shown in this study, air pollution often causes alterations in body's immunity, enhancing the susceptibility to infection (Johnson et al., 1990). ARI is one of the main killers of children in developing countries, causing four million child deaths per year surpassing the number of deaths due to diarrhoea (Murray and Lopez, 1996).

Chronic exposure to particulate air pollution increases the risk of atherosclerosis, stroke and myocardial infarction among Indians at a relatively younger age (Cropper et al., 1997). Over-expression of platelet P-selectin in biofuel users in our study indicates high risk of cardiovascular diseases, while genotoxic changes are risk factor for several diseases including cancer. In essence, the studies point to the magnitude of the threat posed by air pollution to the health of both urban and rural population of the country.

REFERENCES

Akimoto, H., Narita, H., 1994. Distribution of SO_2, NO_x and CO_2 emissions from fuel combustion and industrial activities in Asia with 1° × 1° resolution. *Atmospheric Environment*, **28,** 213-215.

Ali, K., Momin, G.A., Safai, P.D., Chate, D.M. and Rao, P.S.P., 2004. Surface ozone measurements over Himalayan region and Delhi, North India. *Ind. J. Radio and Space Phys.*, **33,** 391-398.

Arndt, R.L., Carmichael, G.R., Streets, D.G. and Bhatti, N., 1997. Sulphur dioxide emissions and sectoral contribution to sulphur deposition in Asia. *Atmospheric Environment*, **31,** 1553-1582.

Babu, S.S., Satheesh, S.K. and Krishna Moorthy, K., 2002. Enhanced aerosol radiative forcing due to aerosol black carbon at an urban site in India. *Geophys. Res. Lett.*, **29,** 10.1029/2002GL015826.

Basu, C., Ray, M.R. and Lahiri, T., 2001. Traffic-related air pollution in Calcutta associated with increased respiratory symptoms and impaired alveolar macrophage activity. *J. Environ. Poll*, **8,** 187-195.

Bond, T.C., Streets, D.G., Yarber, K.F., Nelson, S.M., Woo, J.H. and Klimont, Z., 2004. A technology-based global inventory of black and organic carbon emissions from combustion. *J. Geophys. Research*, **109**, D14203, doi: 10.1029/2003JD003697.

Brasseur, G.P., Hauglustaine, D.A., Walters, S., Rasch, P.J., Muller, J.-F., Granier, C. and Tie, X.X., 1998. MOZART, A global chemical transport model for ozone and related chemical tracers 1. Model description. *J. Geophys. Res.*, **103,** 28,265-28,289.

Brasseur, G.P. et al., 2003. An integrated view of the causes and impacts of anthropogenic changes. *In:* Atmospheric Chemistry in a Changing World, G.P. Brasseur, R.G. Prinn and A.P. Pszenny (eds.), Springer, Germany.

Brasseur, G.P., Schultz, M., Granier, C., Saunois, M., Diehl, T., Botzet, M., Roeckner, E. and Walters, S. 2006. Impact of climate change on the future chemical composition of the global troposphere. *J. Climate*, **19**, 3932-3951.

Bregman, A. et al., 1997. In situ trace gas and particle measurements in the summer lower stratosphere during STREAM II: Implications for O_3 production. *J. Atmos. Che.*, **26(3),** 275-310.

Chandra, S. et al., 2004. Can the state of mixing of black carbon aerosols explain the mystery of 'excess' atmospheric absorption? *Geophys. Res. Lett.*, **31,** doi: 10. 1029/2004GL020662.

Charlson, R.J. et al., 1992. Cimate Forcing by Anthropogenic Aerosols. *Science*, **255,** 423-430.

Clarke, A.D., Kapustin, V.N., Eisele, F.L., Weber, R.J. and McMurry, P.H., 1999. Particle production near marine clouds: Sulfuric acid and prediction from classical binary nucleation. *Geophys. Res. Lett.*, **26,** 2425-2428.

CMIE, 2001. Economic intelligence service: Energy – 2001. Centre for Monitoring of Indian Economy Private Limited, New Delhi.

Cooke, W.F., Wilson, J.J.N., 1996. A black carbon aerosol model. *Journal of Geophysical Research*, **101(D14),** 19395-19409, doi: 10.1029/96JD00671.

Cooke, W.F., Liousse, C., Cachier, H., Feichter, J., 1999. Construction of a 1° × 1° fossil fuel emission dataset for carbonaceous aerosol and implementation and radiative impact in the ECHAM-4 model. *Journal of Geophysical Research*, **104(D18),** 22137-22162, doi: 10.1029/1999JD900187.

Cooper, O.R., et al., 1998. The influence of springtime weather systems on vertical ozone distributions over three north American sites. *J. Geophys. Res.*, **103,** 22091-22103.

Cropper, M.L., Simon, N.B. and Alberini, A., 1997. The health effects of air pollution in Delhi. The World Bank Development Research Group, December, 1997.

Crutzen, P.J. and Andreae, M.O., 1990. Biomass burning in the tropics: Impact on atmospheric chemistry and biogeochemical cycles. *Science*, **250,** 1669-1678.

Crutzen, P.J. and Zimmerman, P.H., 1991. The changing photochemistry of the troposphere. *Tellus*, **43A,** 136-151.

Crutzen, P. and Ramanathan, V., 2001. The Indian Ocean Experiment: Foreword, INDOEX Special Issue. *J. Geophys. Res.*, **106,** 28,369-28,371.

Dalvi, M., Beig, G., Patil, U., Kaginalkar, A. Sharma, C. and Mitra, A.P., 2006. A GIS based methodology for Gridding of Large Scale Emission Inventories: Application to Carbon-Monoxide Emissions over Indian Region. *Atmos. Environ.*, **40**, 2995-3007.

Deepshikha, S. et al., 2005. Regional distribution of absorbing efficiency of dust aerosols over India and adjacent continents. *Geophys. Res. Lett.*, **32,** doi : 10. 1029/2004GL022091.

de Laat, A.T.J., Zachariasse, M., Roelofs, G.J., Velthoven, P. Van, Dickerson, R.R., Rhoads, K.P., Oltmans, S.J. and Lelieveld, J., 1999. Tropospheric ozone distribution over the Indian Ocean during Spring 1995 evaluated with a chemistry-climate model. *J. Geophys. Res.*, **104,** 13,881-13,894.

de Laat, A.T.J. and Lelieveld, J., 2000. Diurnal ozone cycle in the tropical and subtropical marine boundary layer. *J. Geophys. Res.*, **105(D9),** 11,547-11,559.

Dickerson, R.R., Andreae, M.O., Campos, T., Mayol-Bracero, O.L., Neusuess, C. and Streets, D.G., 2002. Analysis of black carbon and carbon monoxide observed over the Indian Ocean: Implications for emissions and photochemistry. *Journal of Geophysical Research*, **107(D19),** 8017, doi: 10.1029/2001JD000501.

Doherty, R.M., Stevenson, D.S., Collins, W.J. and Sanderson, M.G., 2005. Influence of convective transport on tropospheric ozone and its precursors in a chemistry-climate model. *Atmos. Chem. Phys. Discuss.*, **5,** 3747-3771.

Fisher, H. et al., 2000. Tracer correlations in the northern high latitude lower most stratosphere: Influence of cross-tropopause mass exchange. *Geophys. Res. Lett.*, **27(1),** 97-100.

Ganguly, D., Jayaraman, A. and Gadhavi, H. 2005. In situ ship cruise measurements of mass concentration and size distribution of aerosols over Bay of Bengal and their radiative impacts. *J. Geophys. Res.*, **110,** D06205, doi: 10.1029/2004JD005325.

Garg, A. and Shukla, P.R., 2002. Emissions inventory of India. Tata McGraw-Hill Publishing Company Limited, New Delhi.

Gong, S.L., Barrie, L.A. and Lazare, M. 2002. Canadian Aerosol Module (CAM): A size-segregated simulation of atmospheric aerosol processes for climate and air quality models, 2, Global sea-salt aerosol and its budgets. *J. Geophys. Res.*, **107(D24),** 4779, doi: 10.1029/2001JD002004.

Granat, L., 1990. Luft-och nederbordskemiska stationsnatet inom PMK. Report 3942, Meteorological Institute at Stockholm University, Stockholm.

Granat, L., Das, S.N., Tharkur, R.S. and Rodhe, H., 2001. Atmospheric deposition in rural areas of India- Net and potential acidity. *Water, Air and Soil Pollution*, **130,** 469-474.

Granat, L., Norman, N., Leck, C., Kulshrestha, U.C. and Rodhe, H., 2002. Wet scavenging of sulfur and other compounds during INDOEX. *J. Geophysical Research*, **107(D19),** 8025, doi: 1011029/2001JD000499.

Granier, C. and Brasseur, G.P., 2003. The impact of road traffic on global tropospheric ozone. *Geophys. Res. Lett.*, **30,** 1086, doi: 10.1029/2002GL015972.

Habib, G., Venkataraman, C., Shrivastava, M., Banerjee, R., Stehr, J.W. and Dickerson, R.R., 2004. New methodology for estimating biofuel consumption for cooking: Atmospheric emissions of black carbon and sulfur dioxide from India. *Global Biogeochemical Cycles*, **18**, p. GB3007.

Habib, G., Venkataraman, C., Chiapello, I., Ramachandran, S., Boucher, O. and Reddy, M.S., 2006. Seasonal and interannual variability in absorbing aerosols over India derived from TOMS: Relationship to regional meteorology and emissions. *Atmospheric Environment*, **40(11)**, 1909-1921.

Hara, H., Kitamura, M., Mori, A., Noguchi, I., Ohizumi, T., Sinya, S., Takuchi, T. and Deguchi, T., 1995. *Water, Air and Soil Pollution*, **85,** 2307-2312.

Hegg, D.A., Radke, L.F. and Hobbs, P.V., 1990. Particle production associated with marine clouds. *J. Geophys. Res.*, **95,** 13917-13926.

Heintzenberg, J., Raes, F. and Schwartz, S.E., 2003. Tropospheric aerosols. *In:* Atmospheric Chemistry in a Changing World, G.P. Brasseur, R.G. Prinn and A.P. Pszenny (eds.), Springer, Germany.

Heymsfield, A.J. and McFarquhar, G.M., 2001. Microphysics of INDOEX clean and polluted trade cumulus clouds. *J. Geophys. Res.*, **106,** 28643-28674.

Holgate, T.H., Samet, J., Koren, H.S., Maynard, R.L., 1999. Air Pollution and Health. Academic Press, San Diego.

Horowitz, L.W., Walters, S., Mauzerall, D.L., Emmons, L.K., Rasch, P.J., Granier, C., Tie, X.X., Lamarque, J.-F., Schultz, M.G., Tyndall, G.S., Orlando, J.J. and Brasseur, G.P., 2003. A global simulation of tropospheric ozone and related tracers: Description and evaluation of MOZART, version 2. *J. Geophys. Res.*, **108,** 4784, doi: 10.1029/2002JD002853.

Intergovernmental Panel on Climate Change (IPCC), 2001. The Climate Change, 2001, The Scientific Basis, Edited by J.T. Houghton et al., Cambridge Univ. Press, New York.

Jacobson, Mark Z., 2001. Strong Radiative heating due to the mixing state of black carbon in atmospheric aerosols. *Nature*, **409,** 695-697.

Jain, M., Kulshrestha, U.C., Sarkar, A.K. and Parashar, D.C., 2000. Influence of crustal aerosols on wet deposition at urban and rural sites in India. *Atmospheric Environment*, **34,** 5129-5137.

Jayaraman, A., Ramachandran, S., Acharya, Y.B. and Subbaraya, B.H., 1995. Pinatubo volcanic aerosol layer decay observed at Ahmedabad (23N) India using Nd:YAG backscatter lidar. *J. Geophysical Research*, **100,** 23209-23214.

Jayaraman, A., Lubin, D., Ramachandran, S., Ramanathan, V., Woodbridge, E., Collins, W.D. and Zalpuri, K.S., 1998. Direct observations of aerosol radiative forcing over the tropical Indian Ocean during the Jan-Feb 1996 Pre-INDOEX cruise. *J. Geophys. Res.*, **103(D12),** 13827-13836.

Jayaraman, A., 2001. Aerosol radiative forcing over the tropical Indian Ocean. *Proc. Ind. Nat. Sci. Acd.*, **67,** 385-394.

Johnson, J.D., Houchens, D.P., Kluwe, W.M., Craig, D.K., Fisher, G.L., 1990. Effects of mainstream and environmental tobacco smoke on the immune system in animals and humans: A review. *Crit Rev Toxicol*, **20,** 369-395.

Jones, A. and Slingo, A., 1996. Predicting cloud-droplet effective radius and indirect sulphate aerosol forcing using a general circulation model. *Quarterly Journal of the Royal Meteorological Society*, **122(535),** 1573-1595 Part A.

Khemani, L.T., 1989. Physical and chemical characteristics of atmospheric aerosols. *In:* Air Pollution Control, Vol 2, Encyclopedia of Environ. Control, Tech., Gulf Publ. Co. USA.

Kinne, S. et al., 2003. Monthly averages of aerosol properties: A global comparison among models, satellite data, and AERONET ground data. *J. Geophys. Res.*, **108(D20),** 4634, doi: 10.1029/2001JD001253.

Krishnamurti, T.N., Jha, B., Prospero, J.M., Jayaraman, A. and Ramanathan, V., 1998. Aerosol and pollutant transport over the tropical Indian Ocean during the 1996 northeast monsoon and the impact on radiative forcing. *Tellus*, **B50,** 521-542.

Kulshrestha, U.C., Sarkar, A.K., Srivastava, S.S. and Parashar, D.C., 1995. Wet-only and bulk deposition studies at New Delhi (India). *Water, Air and Soil Pollution*, **85,** 2137-2142.

Kulshrestha, U.C., Sarkar, A.K., Srivastava, S.S. and Parashar, D.C., 1996. Investigation into atmospheric deposition through precipitation studies at New Delhi (India). *Atmospheric Environment*, **30,** 4149-4154.

Kulshrestha, U.C., Kulshrestha, M.J., Sekar, R., Sastry, G.S.R. and Vairamani, M., 2003a. Chemical characteristics of rain water at an urban site of south-central India. *Atmos. Environ.*, **37,** 3019-3026.

Kulshrestha, U.C., Granat, L. and Rodhe, H., 2003b. Precipitation chemistry studies in India – A search for regional patterns. Report CM-99, Department of Meteorology, Stockholm University, Sweden.

Kulshrestha, M.J., Kulshrestha, U.C., Parashar, D.C. and Vairamani, M., 2003c. Estimation of SO_4 contribution by dry deposition of SO_2 onto the dust particles in India. *Atmospheric Environment*, **37(22),** 3057-3063.

Kulshrestha, U.C., Granat, L., Engardt, M. and Rodhe, H., 2005. Review of precipitation monitoring studies in India - A search for regional patterns. *Atmospheric Environment*, 7403-7419.

Lahiri, T., Ray, M.R., Mukherjee, S., Basu, C. and Lahiri, P., 2000a. Marked increase in sputum alveolar macrophages in residents of Calcutta: Possible exposure effect of severe air pollution. *Curr Sci*, **78,** 399-404.

Lahiri, T., Basu, C., Ganguly, S., Ray, M.R. and Lahiri, P., 2000b. Air pollution in Calcutta elicits adverse pulmonary reaction in children. *Indian J Med Res*, **112,** 21-26.

Lahiri, T., Ray, M.R. and Lahiri, P., 2002. Health Effects of Air Pollution: A study in Kolkata. Department of Environment, Government of West Bengal.

Lamarque, J.-F., Hess, P., Emmons, L., Buja, L., Washington, W. and Granier, C., 2005. Tropospheric ozone evolution between 1890 and 1990. *J. Geophys. Res.*, **110,** D08304, doi: 10.1029/2004JD005537.

Lelieveld, J., Crutzen, P.J., Ramanathan, V., Andreae, M.O., Brenninkmeijer, C.A.M., Campos, T., Cass, G.R., Dickerson, R.R., Fisher, H., De Gouw, J.A., Hansel, A., Jefferson, A., Kley, D., de Laat, A.T.J., Lal, S., Lawrence, M.G., Lobert, J.M., Mayol-Bracero, O.L., Mitra, A.P., Novakov, T., Oltmans, S.J., Prather, K.A., Reiner, T., Rodhe, H., Scheeren, H.A., Sikka, D. and Williams, J., 2001. The Indian Ocean Experiment: Widespread air pollution from South and Sotheast Asia. *Science*, **291(5506),** 1031-1036.

Li, F. and Ramanathan, V., 2002. Winter to summer monsoon variation of aerosol optical depth over the tropical Indian Ocean. *J. Geophys. Res.*, **107(D16),** 4284, doi: 10.1029/2001JD000949, AAC 2-1~ AAC 2-13.

Liousse, C., Penner, J.E., Chuang, C., Walton, J.J., Eddleman, H. and Cachier, H., 1996. A global three-dimensional model study of carbonaceous aerosols. *Journal of Geophysical Research*, **101(D14),** 19411-19422, doi: 10.1029/95JD03426.

Meehl, G.A. et al., 1996. Climate change from increased CO_2 and direct and indirect effects of aerosols. *Geophys. Res. Lett.*, **23(25),** 3755-3758.

Mishra, V., Retherford, R.D. and Smith, K.R., 1999. Biomass cooking fuels and prevalence of tuberculosis in India. *Int J Infect Dis*, **3,** 119-129.

Mitra, A.P. and Bhattacharya, S., 1998. Greenhouse Gas Emissions in India for 1990 Using IPCC Standard Methodology. *Scientific Report #11*. Center on Global Change, NPL, New Delhi, India.

Mitra, A.P. and Sharma, C., 2002. Indian aerosols: Present status. *Chemosphere*, **49,** 1175-1190.

Moorthy, K.K., Satheesh, S.K. and Murthy, B.V.K. 1997. Investigations of marine aerosols over the tropical Indian Ocean, J. Geophys. Res., 102, 18,827-18,842.

Moorthy, K.K. and Satheesh, S.K., 2000. Characteristics of aerosols over a remote island, Minicoy in the Arabian Sea, Optical properties and retrieved size characteristics. *Quarterly Journal of the Royal Meteorological Society*, **126,** 81-109.

Mukhopadhyay, B., Datar, S.V. and Srivastava, H.N., 1992. Precipitation chemistry over the Indian region. *Mausam*, **43,** 249-258.

Murray, C.J.L. and Lopez, A.D. (eds.), 1996. Global burden of disease. Harvard University Press, Cambridge, MA.

Nemmar, A., Hoet, P.H., Vanquickenborne, B., Dinsdale, D., Thomeer, M., Hoylaerts, M.F., Vanbilloen, H., Mortelmans, L. and Nemery, B., 2002. Passage of inhaled particles into the blood circulation in humans. *Circulation*, **105,** 411-414.

Novelli, P.C., Lang, P.M. and Masarie, K.A., 1998. Distributions and recent changes of carbon monoxide in the lower troposphere. *J. Geophys. Res.*, **103,** 19015-19033.

Olivier, J.G.J., Bouwman, A.F., van der Maas, C.W.M., Berdowski, J.J.M., Veldt, C., Bloos, J.P.J., Visschedijk, A.J.H., Zanfeld, P.Y.J. and Haverlag, J.L., 1996. Description of EDGAR Version 2.0: "A Set of Global Emission Inventories of Greenhouse Gases and Oxon-depleting Substances for all Anthropogenic and most Natural Sources on a per Country Basis and on a 1 × 1 grid". RIVM Report 771060 002, RIVM, Bilthoven, the Netherlands.

Pandithurai, G., Pinker, R.T., Takamura, T. et al., 2004. Aerosol radiative forcing over a tropical urban site in India. *Geophys. Res. Lett.*, **31(12),** Art. No. L12107.

Parashar, D.C., Granat, L., Kulshrestha, U.C., Pillai, A.G., Naik, M.S., Momin, G.A., Rao, P.S.P., Safai, P.D., Khemani, L.T., Naqvi, S.W.A., Narvekar, P.V., Thapa, K.B. and Rodhe, H., 1996. Chemical composition of precipitation in India and Nepal – A preliminary report on an Indo-Swedish Project on Atmospheric Chemistry. Report CM 90, Department of Meteorology, Stockholm University.

Penner, J.E., Eddleman, H. and Novakov, T., 1993. Towards the development of a global inventory for black carbon emissions. *Atmospheric Environment*, **27A,** 1277-1295.

Prather, M. et al., 2003. Fresh air in the 21st Century? *Geophys. Res. Lett.*, **30,** 1100-1104.

Prospero, J.M., Ginoux, P., Torres, O., Nicholson, S. and Gill, T., 2002. Environmental Characterization of Global Sources of Atmospheric Soil Dust Identified with the NIMBUS7 Total Ozone Mapping Spectrometer (TOMS) Absorbing Aerosol Product. *Rev. Geophys.*, **40**(1), 1002, doi: 10.1029/2000RG000095.

Rajeev, K., Ramanathan, V. and Meywerk, J., 2000. Regional aerosol distribution and its long-range transport over the Indian Ocean, *J. Geophy. Res.*, **105,** 2029-2043.

Rajeev, K. and Ramanathan, V., 2001. Direct observations of clear-sky aerosol radiative forcing from space during the Indian Ocean Experiment. *J. Geophys. Res.*, **106,** 17,221-17,236.

Ramachandran, S., 2005. Premonsoon shortwave aerosol radiative forcings over the Arabian Sea and tropical Indian Ocean: Yearly and monthly mean variabilities. *J. Geophys. Res.*, **110(D7),** D07207.

Ramanathan, V. et al., 2001. The Indian Ocean Experiment: An integrated assessment of the climate forcing and effects of the great Indo-Asian haze. *J. Geophys. Res.*, **106,** 28371-28398.

Ramanathan, V. and Crutzen, P.J., 2001. Concept paper on Asian Brown clouds; Air pollution in the Indo-Asia-Pacific region: Impact on Climate & Environment. *United Nations Environment Programme/Environment Assessment Programme* (available at http://www.abc-asia.ucsd. edu/ABCconceptFinal23May01.pdf).

Ramanathan, V. et al., 2005. Atmospheric brown clouds: Impacts on South Asian climate and hydrological cycle. *Proc. Nat. Sci. Acad.*, **102,** 5326-5333.

Rao, P.S.P., Momin, G.A., Safai, P.D., Pillai, A.G., Naik, M.S. and Khemani, L.T., 1990. Impact of Ca and SO_4 on pH of rain water in rural environments in India. *Indian J Environmental Protection*, **10,** 941-943.

Ray, M.R., Basu, C. and Lahiri, T. 2003. Haematological changes and up-regulation of P-selectin expression in circulating platelets of Indian women chronically exposed to emissions from biomas fuels. *Comp Clin Path*, **12,** 151-154.

Reddy, M.S. and Venkataraman, C., 2002a. A $0.25^{\circ} \times 0.25^{\circ}$ inventory of aerosol and sulfur dioxide emissions from India. I: Fossil fuel combustion. *Atmospheric Environment*, **36,** 677-697.

Reddy, M.S. and Venkataraman, C., 2002b. A $0.25^{\circ} \times 0.25^{\circ}$ inventory of aerosol and sulfur dioxide emissions from India. II: Biomass combustion. *Atmospheric Environment*, **36,** 699-712.

Reddy, M.S. and Boucher, O., 2004. A study of global cycle of carbonaceous aerosols in the LMDZT general circulation model. *Journal of Geophysical Research*, **109**, D14202, doi: 10.1029/2003 JD004048.

Reddy, M.S., Boucher, O., Venkataraman, C., Verma, S., Bellouin, N., and Pham, M., 2004. GCM estimates of aerosol transport and radiative forcing during INDOEX. *Journal of Geophysical Research*, **109**, D16205, doi: 10.1029/2004 JD004557.

Rodhe, H., Langner, J., Gallardo, L. and Kjellström, E., 1995. Global scale transport of acidifying pollutants. *Water, Air and Soil Pollution*, **85,** 37-50.

Roelofs, G.J., Lelieveld, J. and van Dorland R., 1997. A three-dimensional chemistry/ general circulation transport model simulation of anthropogenically derived ozone in the troposphere and its radiative climate forcing. *J. Geophys. Res.*, **102,** 23,389-23,401.

Roy, S., Ray, M.R., Basu, C., Lahiri, P. and Lahiri, T., 2001. Abundance of siderophages in sputum. Indicator of adverse lung reaction to air pollution. *Acta Cytol*, **45,** 958-964.

Saksena, S. and Smith, K.R., 2003. Indoor air pollution. *In:* Air pollution and health in rapidly developing countries. McGranahan, G., Murray, F. (eds.), pp. 129-145, Earthscan Publications, London.

Saraf, N., Beig, G. and Schultz, M., 2003. Tropospheric distribution of ozone and its precursors over the tropical Indian ocean. *J. Geophys. Res.*, **108,** 4636, doi: 10.1029/2003JD003521.

Saraf, N. and Beig, G., 2004. Long-Term Trends in Tropospheric Ozone Over the Indian Tropical Region. *Geophys. Res. Lett.*, **31,** L05101, doi: 10.1029/ 2003GL018516.

Satheesh, S.K. and Ramanathan, V., 2000. Large differences in tropical aerosol forcing at the top of the atmosphere and earth's surface. *Nature*, **405,** 60-63.

Satheesh, S.K., 2002. Aerosol Radiative Forcing over Tropical Indian Ocean: Modulation by Sea-surface winds. *Current Science*, 10th February.

Satheesh, S.K. and Srinivasan, J., 2002. Enhanced aerosol loading over Arabian Sea during premonsoon season: Natural or Anthropogenic? *Geophys. Res. Lett.*, 10.1029/2002GL 015687.

Satheesh, S.K. et. al., 2002. Physical, Chemical, and Radiative Properties of Indian Ocean Aerosols. *J. Geophys. Res.*, 10.1029/2002JD002463.

Satheesh, S.K. and Lubin, D., 2003. Short wave versus long wave radiative forcing by Indian Ocean aerosols: Role of sea-surface winds. *Geophys. Res., Lett.*, **30.**

Satheesh, S.K. and Krishna Moorthy, K., 2005. Radiative Effects of Natural Aerosols: A review. *Atmospheric Environment.*

Saxena, A., Sharma, S., Kulshrestha, U.C. and Srivastava, S.S., 1991. Factors affecting alkaline nature of rain water in Agra (India). *Environmental Pollution*, **74,** 129-138.

Sehgal, J., 1996. Pedology – Concepts and Applications. Kalyani Publishers, New Delhi.

Shrestha, R.M. and Bhattacharya, S.C., 1991. Coal use in electricity generation in Asia and emission control options. *Proceedings from the 3rd annual conference on acid rain and emissions in Asia*, 18-21 Nov., Bangkok, Thailand.

Smith, K.R., Samet, J.M., Romieu, I. and Bruce, N., 2000. Indoor air pollution in developing countries and lower respiratory tract infections in children. *Thorax*, **55,** 518-532.

Smith, K.R. and Liu, Y., 1994. Indoor air pollution in developing countries. *In:* Samet, J.M. (ed.), Epidemiology of lung cancer. Lung Biology in Health and Disease Series. **74,** 151-184.

Spiro, P.A., Jacob, D.J. and Logan, J.A., 1992. Global inventory of sulphur emissions with $1^\circ \times 1^\circ$ resolution. *Journal of Geophysical Research*, **97(D5),** 6023-6036.

Streets, D.G., Bond, T.C., Carmichael, G.R., Fernandes, S.D., Fu, Q., He, D., Klimont, Z., Nelson, S.M., Tsai, N.Y., Wang, M.Q., Woo, J.H. and Yarber, K.F., 2003. An inventory of gaseous and primary aerosol emissions in Asia in the year 2000. *Journal of Geophysical Research*, **108(D21),** 8809, doi: 10.1029/2002JD003093.

Tegen, I. and Fung, I., 1995. Contribution to the atmospheric mineral load from land surface modification. *J. Geophys. Res.*, **100,** 18,707-18,726.

Thomson, A.M., 1992. The oxidizing capacity of the earth's atmosphere: Probable past and future changes. *Science*, **256,** 1157-1168.

Tripathi, S.N., Dey, S., Tare, V. et al., 2005. Aerosol black carbon radiative forcing at an industrial city in northern India. *Geophys. Res. Lett.*, **32(8),** Art. No. L08802.

Winter, B. and Chylek, P., 1997. Contribution of sea salt aerosol to the planetary clear-sky albedo. *Tellus-B*, **49(1),** 72-79.

WMO, 1999. Scientific assessment of ozone depletion: 1998. Global Ozone Research and Monitoring Project, Geneva, Switzerland.

Zender, C.S., Bian, H. and Newman, D., 2003. Mineral Dust Entrainment and Deposition (DEAD) model: Description and 1990s dust climatology. *J. Geophys. Res.*, **108(D14)**, 4416, doi: 10.1029/2002JD002775.

Appendix

INDOEX AND ABC

During winter months, from December to April, the northeast trade wind transports gaseous and particulate pollutants from the Indian subcontinent to over the tropical Indian Ocean resulting into formation of a layer of haze extending up to about 3 km in height and spreading over most of the north Indian Ocean and South and Southeast Asia. The Indian Ocean Experiment (INDOEX) studied this haze through multi-platform observations including ships, station observations, aircrafts, satellites and balloons and integrated the results with models to derive the aerosol radiative forcing over the region (Ramanathan et al., 2001: hereinafter referred to as R1). Large latitude gradient in aerosol amount from coastal India to the pristine ocean (Fig. A.1) with aerosol optical depth value varying from about 0.6 near the coast to as large as 0.2 over the equatorial Indian Ocean shows the significance of the transport and transformation of the particles (Jayaraman et al., 1998; Krishnamurti et al., 1998).

The aerosols found are a composite of several inorganic and carbonaceous species, including black carbon clusters and fly ash. The high concentration of black carbon, of the order of 14% in the fine particle mass (Lelieveld et al., 2001) contributes to absorption of solar radiation which gives a brownish tinge to the haze. The north to south dispersion of particles is, however, limited by the Inter Tropical Convergence Zone (ITCZ) where the surface trade winds from the two hemispheres meet, which is also a region of strong convective activity. ITCZ exhibits annual oscillation with respect to equator, and for the whole globe, the maximum excursion is found over the Indian Ocean, from about 10°S in January to about 20°N in July. Yearly variation in the strength of the ITCZ and its progression greatly influence the spatial and temporal distribution of aerosol loading over the Indian Ocean (Rajeev et al., 2000; Li and Ramanathan, 2002). Based on the chemical composition it is estimated that about 80% (±10%) of the aerosols found in the northern Indian Ocean during winter is anthropogenic in nature (R1). Agricultural burning and the biofuel use are found to be the main cause for the observed high CO concentration (Lelieveld et al., 2001). The chemical properties of the particles determine their refractive index and hence the scattering and absorbing efficiency of aerosols. The single scattering albedo obtained both through observations and model fittings was between 0.85 and 0.9 along coastal India and north Indian Ocean.

The major outcome of the INDOEX study is estimation of the impact of the haze in radiative forcing. Different estimates of the direct aerosol forcing under clear sky condition reveal a large reduction in the surface reaching radiation, in the range of 23 to 27 W/m^2 over the polluted region with sufficiently large atmospheric absorption, in the range of 16 to 20 W/m^2 (Satheesh and Ramanathan, 2000; Jayaraman, 2001; Rajeev and

Ramanathan, 2001). The regionally averaged aerosol forcing for the North Indian Ocean is summarized in Fig. A.2. For clear skies the top of the atmosphere (TOA) forcing is –7 ± 1 W/m^2 and cloudy skies it ranges from 0 to –4 W/m^2. The uncertainty in the forcing value estimation comes mainly from the uncertainty in the prescription of the cloud fraction and the vertical profile of aerosols. Taking the mid value of –2 ± 2 W/m^2 for the cloudy sky TOA forcing, it is 10 times less than the surface forcing while the ratio of surface to TOA for the clear sky is only about three. The large atmospheric forcing (difference between the surface and TOA forcing) is mainly due to

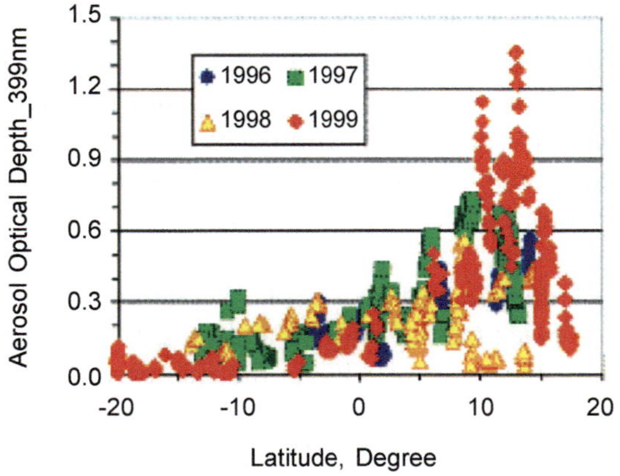

Fig. A.1: Winter time latitudinal variation in aerosol optical depth from coastal India to the pristine Indian Ocean south of the Inter Tropical Convergence Zone (ITCZ). During NH winter the ITCZ is between 10 and 15° S. Measurements were made using sun-photometer from the Indian Research vessel Sagar Kanya.

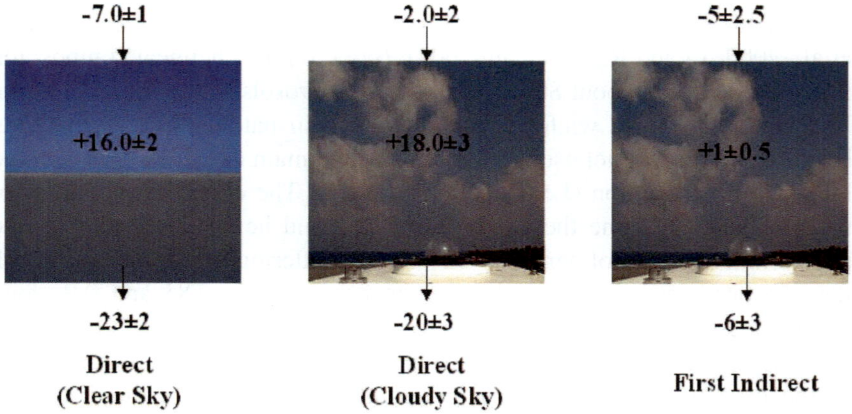

Fig. A.2: Aerosol radiative forcing (both natural and anthropogenic) for the North Indian Ocean Region (0° to 20°N and 40° to 100°E). The values on the top, middle and bottom represent forcing at the top of the atmosphere, within the atmosphere and at the surface respectively (Ramanathan et al., 2001).

absorption of radiation by soot as well as water vapour absorption of radiation scattered by aerosols (R1). C-130 observation through the polluted clouds demonstrated the first indirect effect of more cloud drops and smaller effective drop radii than compared to pristine clouds (Heymsfield and McFarquhar, 2001). Inserting this in the radiative transfer model, the regionally averaged radiative forcing for the first indirect effect is estimated as -5 ± 2 W/m^2 for the TOA and 1 ± 0.5 W/m^2 and -6 ± 3 W/m^2 for the atmosphere and surface respectively (R1). The significant finding is that clouds decrease the TOA direct forcing by 5 W/m^2 with respect to clear sky forcing while enhance the –ve forcing by the same amount through first indirect effect. This is mainly because the absorbing aerosols are located within the cloud as well as above the clouds (R1).

The INDOEX findings have led to a new major international effort, called the Atmospheric Brown Cloud (ABC) Project (Ramanathan and Crutzen, 2001). The project, to be coordinated by the United Nations Environment Program (UNEP) has its first focus on the Asia and the Pacific regions. The goal is to address the major environmental and climate challenges facing this region in the coming decades, due to the rising levels of air pollutants. The major components of the ABC project are observations, regional impacts and assessment modelling to understand the air pollution issue in a broader context with the end goal of helping policymakers to arrive at informed decisions. Compared to INDOEX, ABC broadens the scope of the research by considering the entire Asian region and the adjacent tropical Indian Ocean and western Pacific Ocean. It will begin with a new framework (Fig. A.3), which recognizes both the regional and the global nature of the anthropogenic pollutants and their impact on regional and global scales.

The underlying principles of Project ABC include promoting regional capacity building and facilitating interactions between scientific and policy making process. The specific objectives of the ABC are: 1. To develop the science and capacity to study the issue of aerosols in the region; 2. To assess the impacts of Atmospheric Brown Clouds on climate change and water budget, health, ecosystem and agriculture under one common framework; and 3. To raise awareness on the issue and promote actions for mitigation.

In order to achieve the objectives the project will have major components on observations, regional impacts modelling, and awareness and mitigation. ABC is in the process of establishing an integrated network of surface climate observatories across the entire Asia-Pacific region. These strategically placed observatories will monitor the chemical composition, radiative effects and transport of atmospheric aerosols and related atmospheric pollutants.

The observatories will also provide critical information for calibration of satellite observations and baseline measurements for long-term analysis of changes in regional pollutant gases and aerosol characteristics, in addition to enhancing regional capacity building. Each surface observatory will be

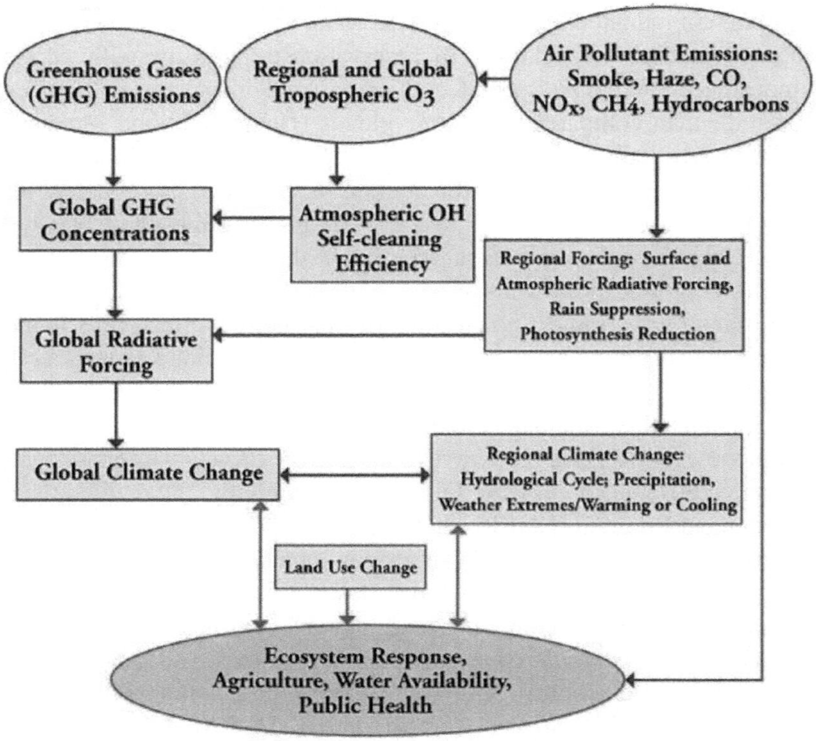

Fig. A.3: The new framework calls for a regional focus. Air pollution and aerosols exert a large regional radiative forcing, which can have direct impacts on regional climate. Both regional forcing and regional climate change can aggregate to influence global climate change (Ramanathan and Crutzen, 2001).

equipped with radiation, aerosol and chemistry instruments, to be operated with renewable power (wind and solar) and use electricity generated from fossil fuels only as backup. Regional scale source-receptor models are to be used in conjunction with the data from observatories and satellites to identify the relative contribution of the various Asian regions to the observed aerosol loading. Direct short-wave and long-wave aerosol radiative forcing at the surface and top of the atmosphere, estimated from aerosol data in conjunction with remote radiometric measurements will be related to the regional aerosol emission sources.

Coupled ocean atmosphere model simulation of the decreases in surface solar radiation, changes in surface and atmosphere temperatures over land and sea, and decrease in monsoon rainfall are similar to the trends observed during the period 1930-2000 over South Asia (Ramanathan et al., 2005). The rainfall disruption is surprisingly large, and is characterized mainly by increasing precipitation over Bay of Bengal and drying of areas northwest of India. More reliable estimates of the aerosol effect on climate will be derived from the data collected through the observatories. Similarly, numerical experiments to be conducted using sophisticated mathematical crop models for different scenarios with respect to water and nutrient intakes to quantify the response of potential crop yields (initially rice, wheat and sugarcane) to atmospheric pollution. This is to be supported with in-situ and satellite observations to monitor crop health and water stress. The results will facilitate better strategies to help optimize resources such as water and land, given the complex feedback effects between atmospheric pollution and food production. For the study of health effects, epidemiologists will be engaged to work with pathologists to make case studies of deaths caused by atmospheric pollution. Physical and chemical properties of the haze, seasonal and spatial coverage of the pollution events, will be used as environmental background for these studies.

6 Global Warming, Changes in Hydrological Cycle and Availability of Water in South Asia

M. Monirul Qader Mirza and Ahsan Uddin Ahmed

1. INTRODUCTION

Various regions of South Asia experience high climate variability, both spatially and temporally. The hydrological regime of major parts of the region is predominantly influenced by monsoon, which brings 70-80% of total annual rainfall during early June to September. The post-monsoon months become dry and there is hardly any appreciable rainfall during winter months (December to February). Moreover, the western parts of Bangladesh and India generally receive significantly lower amounts of rainfall compared to the eastern parts of both the countries, which is a manifestation of high spatial distribution of rainfall. Rainfall in Nepal is also higher in the eastern part compared to the western region. Topographically, in the Terai region with flatter topography along the Indian border, rainfall is very high. In Pakistan, the Punjab and Kashmir regions in the north experience the highest precipitation whereas in the southern region it is one-fifth of the north. In a warmer climate in future, the overall pattern of rainfall/precipitation is expected to change spatially as well as temporally. Water resources of the South Asia region, however, are highly sensitive to climate variability and change. Therefore, an anticipated change in climate system – as a consequence of global warming and subsequent sea level rise – could considerably affect both the hydrological cycle as well as distribution, which in turn would affect the lives and livelihoods of hundreds of millions of inhabitants.

2. MONSOON VARIABILITY AND HYDROLOGICAL CYCLE IN SOUTH ASIA

Monsoon is the most important climatic phenomenon in South Asia. Economic, social and cultural lives of hundreds of millions of South Asians are intertwined with the onset and departure of monsoon. Until recently, South Asian economy has always been dependent on subsistent agricultural practices, where rainfed cropping played a significant role. While a monsoon with normal rainfall brings joy to the South Asians, especially to the farming communities, an abnormal monsoon (either dry or extremely) can cause destruction of crops and hardships and may lead to nutritional hazards and famine. Onset and retreat dates of monsoon (Fig. 1) and the amount of rainfall it brings in, determine crop calendar, river characteristics, sediment supply and moisture availability in South Asia.

Munot and Kothawale (2000) identified a large scale temporal variability and intensity of daily rainfall variations during monsoon over India. It was found that, on an average, the monsoon was active for 103 days over northeast (NE-India) India, for 75-78 days over central-northeast (CNE-India) India and so on for all the regions. Pant (2003) identified temporal scale of monsoon variability and associated hydrological features (Table 1). Floods and droughts and their frequencies are associated with interannual and decadal and century scale variability, respectively. Average daily normal rainfall (ADNRF) is at a maximum of 14.7 mm over NE-India and at a minimum of 4.6 mm over NW-India (Munot and Kothawale, 2000). Increase of greenhouse gas concentrations would likely to intensify the Asian summer monsoon and its variability. The intensification of monsoon resulted mainly from an enhanced land-sea contrast and a northward shift of the convergence zone. They simulated gradual increase of the monsoon variability from the year 2030 onwards. The intensification seemed to be connected with the corresponding

Table 1: Temporal scales of monsoon variability

Scale	*Intraseasonal*	*Interannual*	*Decadal and century*	*Millennium and longer*
Features	Active and break-monsoon phases; 30-50 day oscillations	Droughts and floods	Changes in the frequency of droughts and floods	Changes in the areal extents of monsoons
Factors	Atmospheric variability; tropical-midlatitude interactions; soil moisture; sea-surface temperatures	Atmospheric interactions; El Nino Southern Oscillation; Top layers of tropical oceans; snow cover; Land surface characteristics	Monsoon circulation variations; Deep ocean involvement; Greenhouse gases increase; Human activities; Biospheric changes; Volcanic dust	Global climate excursions; Ice ages; Warm episode; Sun-earth geometry

Source: Pant, 2003.

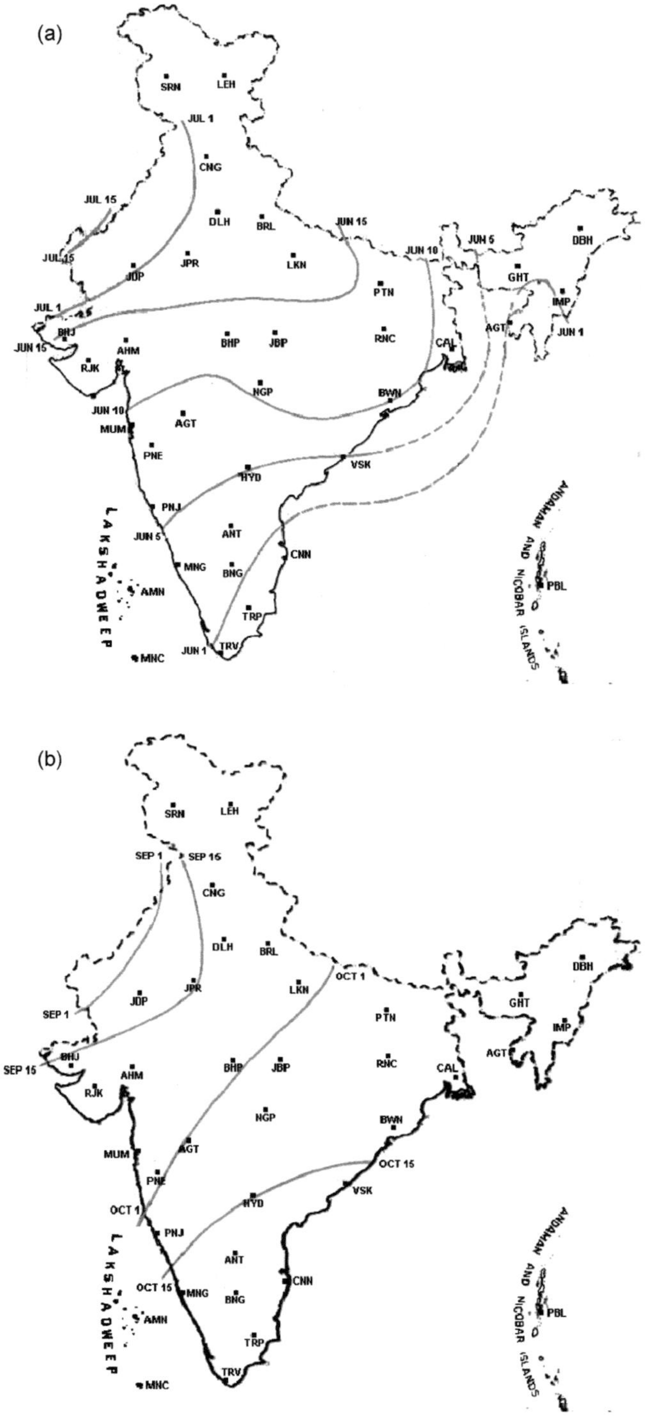

Fig. 1: Onset (a) and retreat (b) dates of monsoon.
Source: India Meteorological Department.

increase of the sea surface temperature variability over the tropical Pacific. Lal et al. (1998) compared results from 17 GCMs and found possibility of intense spells of summer monsoon rainfall over the Indian subcontinent due to enhanced convective activity in a warmer atmosphere.

3. HYDROLOGICAL CYCLE AND AVAILABILITY OF WATER UNDER CLIMATE CHANGE REGIME

The amount of runoff demonstrates high and low water availability in the monsoon and dry season, respectively. In Nepal, India and Bangladesh, about 70-80% of the annual runoff occurs in the monsoon months, which is virtually wasted into the seas. The remaining 20-30% occurs in the dry season (from November-May) when water demand is high for various sectoral uses. Excessive rainfall often causes moderate to severe floods in South Asia. Agriculture, infrastructure, industrial productivity, and housing and human settlements are seriously affected during a severe flood. Existing reservoirs in India can store some amount of monsoon water but not enough to meet the dry winter and summer demands. Bangladesh and Nepal each has only one reservoir. Drought is common in all countries in South Asia in the summer months. At times, these countries face very extreme episodes of drought, which are very destructive to agriculture and livestock population. In future, there is a high likelihood that the characteristics of the hydrological cycle will change with worsening flooding and drought. In an average year, more water will flow in the monsoon and less in the dry season resulting in an aggravating water availability situation when demand of water will increase by many folds due to population growth, changing living styles and economic activities.

Bangladesh

The tendency of the models to show increasing rainfall during monsoon means that the discharge volume will increase in the GBM rivers without increase in discharge capacity. Nevertheless, continued sedimentation will lead to decreased discharge capacity by gradually rising the river beds (Ahmed et al., 1998a). Consequently, there will be frequent overbank spillage and incidence of flood. Mirza and Dixit (1997) showed the relationship between increasing temperature and precipitation with flood intensity for Bangladesh and reported that, 2°C warming with a 10% increase in precipitation[1] would increase runoff in the Ganges, Brahmaputra and Meghna rivers by 19%, 13% and 11%, respectively. Mirza et al. (2003) estimated a 30% increase in flooded area in Bangladesh with a 2°C rise in global mean temperature. They also found a dramatic increase in flooded area between 0 and 2°C temperature increase compared to higher temperature ranges.

[1] This scenario closely matches with GCM projections for monsoon months June-July-August for 2100.

It is speculated that increased rainfall runoff in the vast GBM region, comprising a total catchment area of 1.75 million km^2, will also contribute to enhanced sediment flows along the GBM river systems. It is likely that the process of sedimentation in the lower parts of the delta (i.e., in Bangladesh) will be enhanced leading to increased rate of bed level rise in the channels and also in the floodplains. Increased availability of water in the rivers during monsoon in combination with reduced conveyance capacity of the rivers would increase potential for high intensity flooding in the country.

In recent years, there has been a discernible increase in frequency of high intensity floods: three of the five very high intensity floods of the past century have occurred in the last three decades. Simultaneously, continued heavy siltation in the channels, and subsequent rise in river bed levels, has significantly reduced the conveyance capacity of the rivers (Halcrow et al., 2001a). Poorly designed infrastructure without proper and/or inoperable drainage facilities has impeded drainage and increased both extent and duration of floods (WB, 2000). Assuming that the situation might not appreciably be changed in near future, climate change-induced peak-flood water volume is most likely to pass over the floodplains of the country. Consequently, flood vulnerability of the country will only increase in terms of both extent and frequency. The schematic pathways of increasing flood due to various climate-driven factors are provided in Fig. 2.

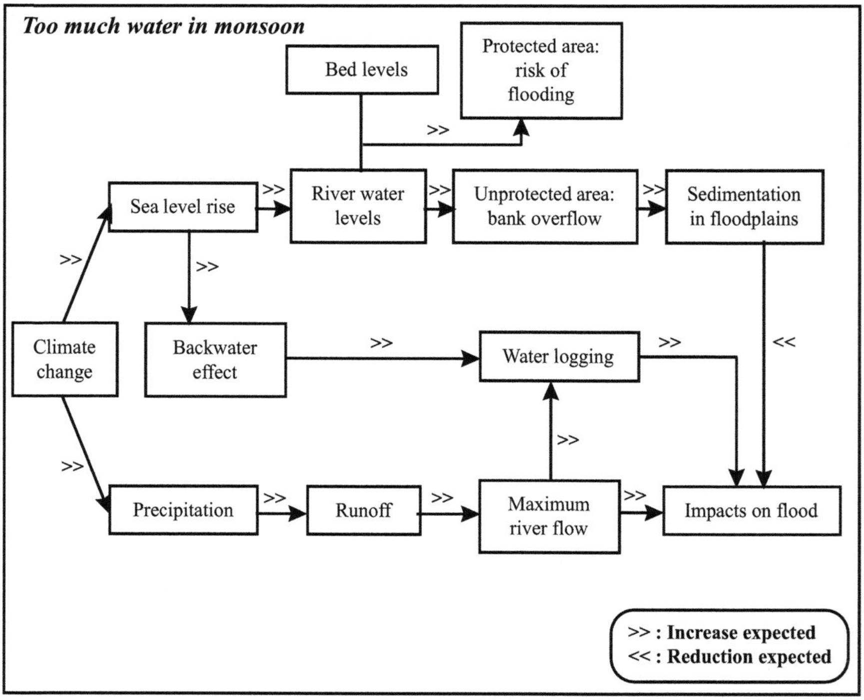

Fig. 2: Climate factors and flooding processes.

India

Impacts on water availability under climate change have been discussed in India's National Communications for the UNFCCC (Government of India, 2004). The results presented here are based on climate scenarios from one model and should be treated with cautions. The SWAT model has been used on each of the 12 major river basins[2] separately using daily weather generated by the HadRM2 control climate scenario (1981-2000). The model has been used with the assumption that every river basin is a virgin area without any man-made change incorporated, which is reasonable for making a preliminary assessment. However, a general country-wide framework has been created that can be used conveniently for adding the additional information at various scales. The model has been run using climate scenarios for the period 2041 to 2060, without changing the land-use pattern. The outputs of these two scenarios have been analyzed with respect to the possible impacts on the run-off, soil moisture and actual evapotranspiration. The model generates detailed outputs at daily interval on flow at sub-basin outflow points, actual evapotranspiration and soil moisture status. Further sub-divisions of the total flow, such as surface and sub-surface run-off are also available. It is also possible to evaluate the recharge to the ground water on a daily basis.

It may be observed that even though an increase in precipitation is projected for the Mahanadi, Brahmani, Ganga, Godavari, and Cauvery basins for the Climate Change Scenario, the corresponding total run-off for all these basins has not necessarily increased. For example, the Cauvery and Ganga show a decrease in the total run-off. This may be due to an increase in evapotranspiration on account of increased temperatures or variation in the distribution of rainfall.

In the remaining basins, a decrease in precipitation is projected. The resultant total run-off for the majority of the cases, except for the Narmada and Tapi, is projected to decline. As expected, the magnitude of such variations is not uniform, since they are governed by many factors such as land use, soil characteristics and the status of soil moisture. The Sabarmati and Luni basins are likely to experience a decrease in precipitation and consequent decrease of total runoff to the tune of two-thirds of the prevailing run-off. This may lead to severe drought conditions under a future Climate Change Scenario.

Nepal

Climate change can cause significant impact on water sector in Nepal. Based on the following criteria: certainty of impact, timing, severity of impact and importance of sector, Agrawala et al. (2003b) ranked sectoral climate change

[2] These river basins are: Cauvery, Brahmani, Godavari, Krishna, Luni, Mahanadi, Mahi, Narmada, Pennar, Tapi, Ganga and Sabarmati.

Table 2: Priority ranking of climate change impacts for Nepal

Resource/ranking	*Certainty of impact*	*Timing of impact (urgency)*	*Severity of impact*	*Importance of resource*
Water resources and hydropower	High	High	High	High
Agriculture	Medium-low	Medium-low	Medium	High
Human health	Low	Medium	Uncertain	High
Ecosystems/ Biodiversity	Low	Uncertain	Uncertain	Medium-high

Source: Agrawala et al., 2003.

impacts in Nepal and water resources was identified as the high impact sector (Table 2). Water resources and hydropower rank significantly higher than any other sector for several reasons which include: rising temperatures and glacier retreat, GLOFs and potential damage to hydropower stations.

The most critical impacts of climate change in Nepal are related to its water resources and hydropower generation, stemming from glacier retreat, expansion of glacial lakes, and changes in seasonality and intensity of precipitation. These impacts include:

- Increased risk of Glacial Lake Outburst Flooding (GLOF)
- Increased run-off variability (as a result of glacier retreat, more intense precipitation during monsoon, and potentially decreased rainfall in the dry season)
- Increased sediment loading (and landslides) as a result of GLOFs, as well as intense rainfall events
- Increased evaporation losses from reservoirs as a result of rising temperatures

Pakistan

The effect of climate change on water resources in Pakistan is expected to be significant (GoP, 2003). An analysis of changes in the hydrological regime can provide a basis for estimating the impacts of climate change on water resources, and can be used as a tool to recommend changes in the water management regimes. In general, increase in temperature would not only increase water demand because of higher evaporation rates, but may also increase rainfall due to additional moisture supplied to the clouds as a result of higher evaporation from the sea surface. Similarly, increased amount of rainfall may cause increase in magnitude and frequency of floods.

Impact of Water Resources on Electricity Generation

At present, 34% of total electricity generation in the country is based on hydel sources (HDIP, 1999). The share of hydropower to total generation is likely to increase in the future as government intends to increase it to reduce costs of power generation, which have risen significantly due to the increased share of fossil fuel-based generation.

The results of the analysis for hydropower generation are given in Table 3. In year 2020 under the no change in precipitation scenario, the increase in hydropower is 0.03% whereas under the increased precipitation scenario, the increase in hydropower generation is 2%. In the decreased precipitation scenario, there is a decrease of 1.5% in hydropower generation. Overall, impacts on hydropower generation, which are based on average base inflows, are expected to be insignificant. In drought scenarios, however, impacts on hydropower generation are likely to be significant (GoP, 2003).

Table 3: Changes in hydropower generation at main dams

Scenarios	*Changes in per cent*			
	2000	2010	2020	2050
0.3°C and +0% PPT	0	0.04	0.03	0.22
0.3°C and +1% PPT	0.02	0.86	1.98	4.32
0.3°C and –1% PPT	–0.01	–0.83	–1.46	–3.85

Source: Government of Pakistan, 2003.

Impacts on Groundwater

Changes in precipitation pattern and increases in temperature can impact groundwater recharge and uses. In Bangladesh, India and Pakistan, a sizeable portion of agriculture is highly dependent (>60%) on ground water. In western India and Pakistan, availability of ground water may be constrained due to reduced monsoon rainfall. In Bangladesh, ground water may be increased resulting from increased monsoon precipitation. However, in the coastal Bangladesh, groundwater aquifers will be at danger of pollution from saline water intrusion to be caused by sea-level rise.

4. CONCLUSIONS

Water resources availability in South Asia is highly sensitive to climate. Monsoon contributes maximum share to the annual runoff. However, high runoff availability is more or less useless as the water cannot be stored for dry season uses. There is simply not enough water in the dry season to meet present demands of various economic sectors. The paucity of water has led to inter-nation and intra-nation water conflicts. Although the climate and hydrologic models indicate overall increased availability of water in South Asia, in the dry season water availability will be smaller than now. Therefore, competition among the water user sectors will be fiercer. In addition to this, water related extreme events such as floods, droughts and salinity ingress may also increase. Increased frequency and magnitude of these events will pose additional threat to agriculture, infrastructure and livelihoods of hundreds of millions of people. Many of the South Asian nations have drafted their national water plans or water policies. Unfortunately, however, the very issue of climate change has not been featured in the plans/policies in an appreciable manner. It is important to take notice of the changes in climate system and find correlation between climate parameters and timely availability of water resources of a country in order to avoid future complications. It is expected that future revision of the plans/policies would consider climate change adaptation for the water sector and help people reduce climate related hazards towards maintaining livelihoods.

REFERENCES

Agrawala, S., Ota, T., Ahmed, A.U., Smith, J. and van Aalst, M., 2003. Development and Climate Change in Bangladesh: Focus on Coastal Flooding and The Sundarbans. Organization for Economic Cooperation and Development (OECD), Paris.

Agrawala, S., Raksakulthai, V., van Aalst, M., Larsen, P., Smith, J. and Reynolds, J., 2003. Development and Climate Change in Nepal: Focus on Water Resources and Hydropower. OECD, Paris.

Ahmad Q.K. and Ahmed, A.U. (Eds.), 2002. Bangladesh: Citizens' Perspectives on Sustainable Development. Bangladesh Unnayan Parishad (BUP), Dhaka.

Ahmad, Q.K., Ahmad, N. and Rasheed, K.B.S. (eds.), 1994. Resources, Environment and Development in Bangladesh with Particular Reference to the Ganges, Brahmaputra and Meghna Basins. Academic Publishers, Dhaka.

Ahmed, A.U. and Alam, M., 1998. Development of Climate Change Scenarios with General Circulation Models. *In:* Vulnerability and Adaptation to Climate Change for Bangladesh, S. Huq, Z. Karim, M. Asaduzzaman and F. Mahtab (eds.). Kluwer Academic Publishers, Dordrecht.

Ahmed, A.U. and Mirza, M.M.Q., 2000. Review of Causes and Dimensions of Floods with Particular Reference to Flood '98: National Perspectives. *In:* Perspectives on Flood 1998. Q.K. Ahmad, A.K.A. Chowdhury, S.H. Imam and M. Sarker (eds.). The University Press Limited.

Ahmed, A.U., 2000. Adaptability of Bangladesh's Crop Agriculture to Climate Change: Possibilities and Limitations. *Asia Pacific Journal on Environment and Development*, **7(1),** 71-93.

Ahmed, A.U., 2004. Adaptation Options for Managing Water-related Extreme Events under Climate Change Regime: Bangladesh Perspectives. *In:* Climate Change Vulnerability and Adaptation in South Asia. M.M.Q. Mirza and Q.K. Ahmad (eds). Francis and Taylor Publications Inc. (in press).

Ahmed, A.U., Alam, M. and Rahman, A.A., 1998. Adaptation to Climate Change in Bangladesh: Future Outlook. *In:* Vulnerability and Adaptation to Climate Change for Bangladesh. S. Huq, Z. Karim, M. Asaduzaman and F. Mahtab (eds.). Kluwer Academic Publishers, Dordrecht.

Bahadur, J., 1998. Himalayan Glaciers. Vigyan Prasar, New Delhi.

CICERO, 2000. *Developing Strategies for Climate Change: The UNEP Country Studies on Climate Change Impacts and Adaptations Assessment*, Report 2000:2. CICERO, Oslo, Norway.

Duan, K., Yao, T. and Thompson, L., 2004. Low-frequency of southern Asian monsoon variability using a 295-year record from the Dasuopu ice core in the central Himalayas. *Geophysical Research Letters*, **31,** pp?

Government of India, 2004. India's National Communications Report to the UNFCCC. Ministry of Environment, New Delhi.

Government of Pakistan, 2003. Pakistan's Initial National Communication on Climate Change to the UNFCCC. Ministry of Environment, Islamabad.

Halcrow and Associates, 2001a. National Water Management Plan. Ministry of Water Resources, People's Republic of Bangladesh, Dhaka.

Halcrow and Associates, 2001b. Options for Ganges Dependent Area. Several Volumes. Sir William Halcrow and Associates, Water Resources Planning Organization (WARPO), Ministry of Water Resources, Government of the People's Republic of Bangladesh, Dhaka.

HDIP, 1999. Pakistan Energy Yearbook, 1998. Hydrocarbon Development Institute of Pakistan. Ministry of Petroleum & Natural Resources, Government of Pakistan.

Kumar, K.R., Kumar, K.K., Ashrit, R.G., Patwardhan, S.K. and Pant, G.B., 2003. Climate Change in India: Observations and Model Projections. *In:* Climate Change and India: Issues, Concerns and Opportunities. P.R. Shukla et al. (eds.). Tata McGraw-Hill Publishing Company Limited, New Delhi.

Lal, M., Whetton, P.H., Chakraborty, B. and Pittock, A.B., 1998. The greenhouse gas-induced climate change over the Indian subcontinent as projected by general circulation model experiments. *Terrestrial, Atmospheric and Oceanic Sciences*, **9,** 673-690.

Li, C. and Yani, M., 1996. The Onset and Interannual Variability of the Asian Summer Monsoon in Relation to Land-Sea Thermal Contrast. *Journal of Climate*, **9(2),** 358-375.

Liu, X. and Chen, B., 2000. Climatic Warming in the Tibetan Plateau During Recent Decades. *International Journal of Climatology*, **20,** 1729-1742.

Mirza, M.M.Q. and Dixit, A., 1997. Climate Change and Water Resources in the GBM Basins. *Water Nepal*, **5(1),** 71-100.

Mirza, M.M.Q., 2002. Global Warming and Changes in the Probability of Occurrence of Floods in Bangladesh and Implications. *Global Environmental Change*, **12,** 127-138.

Mirza, M.M.Q., Warrick, R.A. and Ericksen, N.J., 2003. The Implications of Climate Change on Floods of the Ganges, Brahmaputra and Meghna Rivers in Bangladesh. *Climatic Change*, **57,** 287-318.

Munot, A. and Kothawale, D., 2000. Intra-seasonal, inter-annual and decadal scale variability in summer monsoon rainfall over India. *International Journal of Climatology*, **20(11)**, 1387-1400.

Pant, G.B., 2003. Long-term climate variability and Change Over Monsoon-Asia. *J. Ind. Gephysical Union*, **7(3),** 125-134.

Pradhan, P., Mool, P., Bajracharya, S. and Joshi, S., 2002. Experiences in Inventorying Glaciers and Glacial Lakes of Hindu Kush Himalayas. ICIMOD, Kathmandu.

Shakya, N.M., 2003. Hydrological Changes Assessment and Its Impact on Hydropower Projects of Nepal. *In:* Draft proceedings of the 'Consultative Workshop on Climate Change Impacts and Adaptation Options in Nepal's Hydropower Sector with a Focus on Hydrological Regime Changes Including GLOF', Department of Hydrology and Meteorology and Asian Disaster Preparedness Center, 5-6 March 2003, Kathmandu.

Subbiah, 2001. *Climate Variability and Drought in Southwest Asia.* Asian Disaster Preparedness Center, Bangkok.

Shrestha, A.B., Wake, C.P., Mayewski, P.A. and Dibb, J.E., 1999. Maximum Temperature Trends in the Himalaya and its Vicinity: An Analysis Based on Temperature Records from Nepal for the Period 1971-94. *Journal of Climate,* **12,** 2775-2789.

Watson et al. (Eds), 1998. IPCC Special Report on The Regional Impacts of Climate Change and Assessment of Vulnerability. Cambridge University Press, Camdridge, UK.

World Bank, 2000. Bangladesh: Climate Change and Sustainable Development. Rural Development Unit, South Asia Region, the World Bank, Dhaka.

Zeng-Zhen, H., Mojib, L., Erich, R. and Lennart, B., 2000. Intensified Asian summer monsoon and its variability in a coupled model forced by increasing greenhouse gas concentrations. *Geophysical Research Letters,* **27(17),** 2681-2684.

7 A Review on Current Status of Flood and Drought Forecasting in South Asia

M. Monirul Qader Mirza

1. INTRODUCTION

South Asia region is particularly vulnerable to natural hazards like floods and droughts. Other hazards, for example, cyclones, earthquakes, landslides and tsunamis are also common. However, floods and droughts occur more frequently than the other hazards. Circulation and variability of the monsoon, spatial and temporal variability of extreme events, large floodplain and coastal areas, high population density, economic situation, etc. contribute to the vulnerability to natural hazards. Every year, floods and drought cause enormous economic losses in the form of crop damage, destruction of infrastructure and human settlement, deaths of livestock population, diseases and nutritional problems, etc. Over the decades, huge efforts and investment in the form of structural measures were made to reduce vulnerability but there is virtually no sign of wane. In recent times, non-structural measures like forecasting and warning have been receiving more priority as the successes of structural measures proved to be inadequate. Advance warning of floods and droughts can provide adequate time for preparedness that may eventually reduce vulnerability and losses from hazards. Advance warning on medium (7-15 days) and long (1-3 months) range can provide adequate time for preparedness only. Short range (1-3 days) forecast is not really that helpful for preparedness related action programmes. Most of the South-Asian countries produce short-range forecasts. This is one of the major limitations towards preparedness in these countries. In order to make forecast usable, the existing lead time should be increased from days to month (and season). This paper reviews the status of flood and drought forecasting in South Asia.

2. FLOOD AND DROUGHT FORECASTING – PRESENT STATUS

2.1 Bangladesh

2.1.1 Flood Forecasting and Warning

Flood forecasting and warning system was established at the Bangladesh Water Development Board (BWDB) about three and a half decades ago. Since inception, its development can be divided into three distinct stages when flood forecasting went through significant developments. For the purpose of water level forecast, the FFWC collects measurements of water level and rainfall, satellite pictures and simulates the water level conditions. Every day during most of the monsoon season the model simulates the water level conditions during the previous seven days (hind-cast simulations) and during the coming three days (forecast simulation).

Evaluation of Performance

Chowdhury (2000) evaluated performance of flood forecasting and warning system in Bangladesh in the context of 1998 floods. The flood was very disastrous in nature and engulfed about 70% of Bangladesh. Hydro-meteorological analyses of the floods were documented which revealed that extreme rainfall, cross-border runoff, synchronization of flood peaks of the Ganges and Brahmaputra rivers, drainage congestion, etc. were primarily responsible for flooding. Long duration (67 days) was one of the important features of the flood. There were 46 monitoring stations on 26 rivers. In 1998, 41 of the monitoring stations were above danger level (Fig. 1). In Bangladesh danger level at a river location is defined as the level above which it is likely that the flood may start causing damages to nearby crops

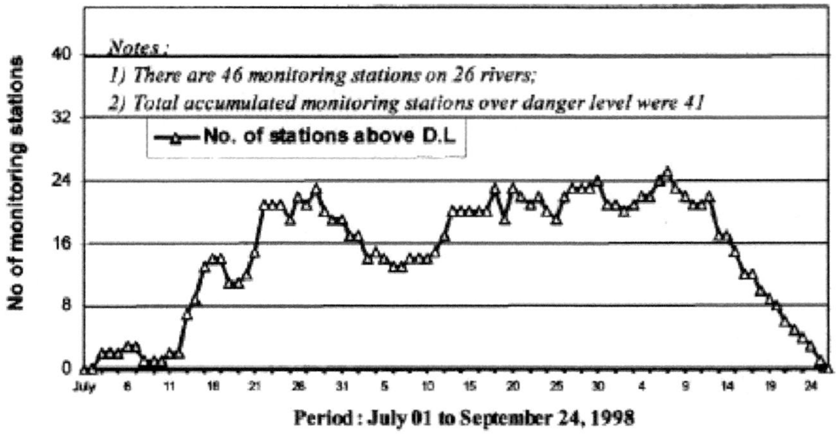

Fig. 1: Number of monitoring stations above danger level during 1998 floods in Bangladesh.
Source: Chowdhury, 2000.

and homesteads. In a river without any embankment, danger level is about the annual average flood level. In an embanked river, danger level is fixed slightly below the designed flood level of the embankment. The danger level at a given location needs continuous verification due to morphological changes in the river, breaching of embankments, etc. but it is not done continuously by the FFWC, whereby some danger levels may not be precise (www.ffwc.gov.bd).

Chowdhury (2000) applied correlation method (between the observed and forecasted values) in determining performance of forecasting and found it to be reasonable (statistically significant at 1% level) for 24-, 48-hour forecasts in most of the monitoring points. However, three monitoring points on the major rivers demonstrated poor quality forecasts. They were Chilmari and Bahadurabad (down of boundary Noonkhawa) on the Brahmaputra River in Northern part of Bangladesh, and Rajshahi (down of boundary Pankha) on the Ganges/Padma River in North-western part of the country. Forecasting quality improves gradually with progression towards the downstream areas from the points near the Indo-Bangladesh border. Difference between measured and observed values are higher in the bordering stations because of the uncertainty in the boundary conditions for the cross-boundary rivers set in the model. The magnitude of uncertainty decreases in the downstream areas and thus increases the 'forecasting accuracy'. Forecasting accuracy is higher for all the monitoring stations on the Ganges River. This is because flow of the river is regulated at Farakka (18 kilometres from the Bangladesh border) by barrage and day-to-day rise and fall of water levels are available. However, six-hourly information on water levels and rainfall (real and forecasted) would have produced better forecasting.

There are other factors, which influence accuracy. Measured, hind-casted, forecasted and transformed water levels are all determined with a given precision, but also influenced by errors. Measured water levels in Bangladesh have a precision of approximately ±10 cm (www.ffwc.gov.bd). The precision of a measured water level is the sum of the precision of the leveling of the station and the precision of the reading of the gauge (often a marked wooden stick). Error is also introduced by the human factors, which include misreading the gauge (rain/water level), or recording a wrong number in the computer or in the recording sheet. Other sources of errors are approximation of model parameters and their field level measurements.

Apart from progress in short-range forecasting, progress has been made regarding long-to seasonal forecasting. Chowdhury and Ward (2004) demonstrated that streamflows in Bangladesh could be reasonably estimated for 1 to 3 months in advance (especially for the Ganges and Brahmaputra rivers) by employing simple correlation subject to availability of rainfall data from upstream countries on a real-time and continuous basis. In the absence of rainfall data, streamflow forecasts are still possible from unusually warm or cold sea-surface temperatures in the tropics. Whitaker et al. (2001)

found a significant relationship between the natural variability of the Ganges annual flow and ENSO index. They further showed that the rate of change of ENSO index was also statistically related to the Ganges flow. A statistical model that combines all these indicators to forecast annual flow in the Ganges was proposed. This model uses current flow data, predicted ENSO data and its gradient to forecast flow in the Ganges with a forecasting lead-time of one year. However, these models are yet to be employed in flood preparedness and management in Bangladesh.

2.1.2 Drought Forecasting and Warning

Drought forecasting in Bangladesh is not as developed as floods despite its frequent occurrence in the dry months (November to May). During the last 55 years, Bangladesh suffered about 20 episodes of droughts. In recent times, the most important drought affected years were 1989 and 1994/1995. The persistent drought condition in North-Western (NW) Bangladesh in recent decades had led to shortfall of rice production of 3.5 million tonnes in the 1990s (Rahman and Biswas, 1995).

In Bangladesh, the Meteorological Department is responsible for weather forecasting and its dissemination. The BMD does not have any short, medium or long-range forecasting model like its Indian counterparts. Traditional methods of weather forecasting are in practice which include manual analyses and interpretation of synoptic or surface charts, isothermal analysis, upper air charts, constant pressure charts, isallobaric charts and prognostic charts. In addition to these, the BMD also uses SADIS data, GMS, INSAT, NOAA satellite imageries, etc. for forecasting purposes (Hossain, 2004).

2.2 India

2.2.1 Flood Forecasting

In India Central Water Commission (CWC) is responsible for analyzing and issuance of forecasts. The Central Water Commission is responsible for analyzing and issuance of flood forecasting and warming and currently the Commission forecasts at 159 stations. A basin-wide distribution of the stations is shown in Table 1.

Goswami (2001) detailed flood forecasting and warning in the Brahmaputra basin in India. Three major organizations, the CWC, India Meteorological Department (IMD) and Ministry of Water Resources, maintain their networks of hydrometeorological stations. In addition to them, many departments of state governments also maintain hydrometeorological stations. Network of rain gauges is sparse in the mountainous and upstream areas. Rainfall is measured by both ordinary and self-recording gauges. In most of the stations, water level is measured by conventional methods. Automatic water level recording is yet to be done on a routine basis. Water level data is transmitted via wireless and occasionally through telephone, telegraph or fax.

Table 1: Basin-wide flood forecasting stations in India

Basin	*Number of forecasting stations*
Ganges	80
Brahmaputra	27
Barak	2
Godavari	15
Krishna	8
Subarnarekha	1
Mahanadi	3
Brahmani-Baitarani	3
Burha Balang	1
Ruyshi Kulya	1
Vamsadhara	13
Narmada	4
Tapi	3
Sabarmati	2
Mahi	2
Banas	1
Damanganga	3
Total	159

Source: Government of India, Ministry of Water Resources

Modern satellites have immense potential for hydrometeorological applications. India's INSAT is a geostationary weather and communications satellite which has onboard data collection system (DCS) that transmits data received from ground-based Data Collection Platforms (DCP) to the computerised data processing facilities. Hydrological models are increasingly used for inflow and flood forecasting. The modified version of the SSARR model developed for the Yamuna River, the NAM model for the Damodar River and several other currently used models like TANK, NWSH, HBV, etc. indicate the country-wide endeavour being made in government agencies and educational institutions to develop suitable flood forecasting models (Goswami, 2001).

The accuracy level of flood forecasts is an important issue, however, not much written information and data are available. Available records show that in the early nineties, a forecast was considered to be reasonably accurate if the observed water level falls within ±15 cm of the forecast level (Rao, 1989). It was recently observed that the flood forecasts currently issued by CWC using conventional rainfall-runoff models had an accuracy of around 65% to 70% with a warning time of six to twelve hours (Rao, 2000).

2.2.2 Drought Forecasting

The India Meteorological Department (IMD) generates the short (1-2 days) and long range (monthly and seasonal scale) predictions whereas the National

Centre for Medium Range Weather Forecasting (NCMRWF) is responsible for the medium range predictions. The short and medium range forecasts are for weather (i.e. temperature, rainfall) over meteorological subdivisions of India. Long range forecasts are made for larger regions such as the all-India scale or two or three sub-regions of the country based on a 16-parameter statistical model (Table 2).

The IMD's monsoon forecast was found to be reasonably accurate. Since 1988, all but one monsoon rainfall forecasts have proved to be fairly accurate. The IMD's long range forecast was inaccurate in 2002 (FrontLine, 2003; Gadgil et al., 2003). Gadgil et al. (2003) also analyzed the possible causes of prediction failure by the empirical models. They made two important conclusions: *First*, that the failure of monsoon was a part of natural variability and *second*, even with overall decrease in errors of predictions, there are possibilities that the models, whether physical or empirical, will fail once in

Table 2: The 16 climatological parameters of the IMD model

Temperature	El Nino (same year)[3]
	El Nino (previous year)[3]
	South Indian Ocean Sea Surface Temperature (February+ March)*[1,2,3]
	Arabian Sea SST (November to January)*[1,2]
	East Coast of India minimum temperature (May)[2,3]
	Northern Hemisphere Surface Temperature (January +February)[1]
Wind	Northern hemispheric wind pattern at 20 km height (50 milibar); that is, E-W extent of trough and ridge (January +February) [2,3]
Pressure	Southern Oscillation Index (SOI) (March to May)
	Argentina pressure (April) [1,2]
	Darwin Pressure Tendency (April to January)*[2,3]
	Indian Ocean Equatorial Pressure (January to May)
	European Pressure Gradient (January)*[2]
Snow	Himalayan Snow Cover (January to March) [3]
	Eurasian Snow Cover in December [1,2]

Source: Rajeevan, 2001

Note: *Since 2000, these four parameters have replaced the following old ones: Northern India Minimum Temperature (March), Wind pattern at 6 km height (milibar) and the location of the high pressure ridge along the 75°E longitude in April, Zonal wind at 30 km height (10 milibar) in January and Sea surface pressure at Darwin, except for Darwin Pressure Tendency, which is related to the old parameter Sea Surface Pressure at Darwin, there is no one-to-one correspondence between the other replaced ones and the new ones.

1. Parameter subset northwestern Indian homogenous region
2. Parameter subset for Peninsular India
3. Parameter subset for northeastern India

a while. When dealing with complex and chaotic systems one must be ready for surprises.

2.3 Nepal

2.3.1 Flood Forecasting

In Nepal floods are often triggered by: (1) torrential rainfall/cloudbursts in the middle mountains and foothills; (2) glacier lake outburst in high Himalayas; (3) failure of landslides/debris flow dams in the rugged mountain regions, and (4) infrastructures and their failures. Therefore, flood forecasting in Nepal has three major challenges: GLOF (Glacier Lake Outburst Flood), rainfall generated floods and failure of landslides/debris flow.

In the monsoon, up to 93% of total annual precipitation occurs during the main season (June to September), and up to 37% of the mean annual precipitation occur in a single event within 24 hours. Diurnal rainfall exceeding 200 mm within 24 hours disrupts both the slopes and channel equilibrium at local as well as at regional scales. More than 24 hour precipitation events exceeding 375 mm have been recorded in different regions of Nepal between 1959 and 1993. Nepal has very limited capacity in forecasting high intensity rainfall and subsequent flooding.

Forecasting of GLOF is another challenging task. Many big glaciers have melted rapidly, forming numerous glacial lakes due to global warming. Accumulation of water in these lakes has increased rapidly due to the increased rate of snow and ice melt. Glacial lake outburst floods (GLOFs) result from a sudden discharge of large volumes of water and debris from these lakes caused by some triggers, such as snow or ice avalanche or quakes or a storm event. Nepal undertook several steps in forecasting GLOFs which include: inventories of glacier lakes and assess the hazards to existing communities and infrastructures and undertaking of the Tsho-Rolpa GLOF Reduction Project jointly implemented by the HMG Nepal and the government of the Netherlands.

2.3.2 Drought Forecasting in Nepal

Although western Nepal is vulnerable to droughts, there is no operational drought forecasting system in Nepal.

2.4 Pakistan

2.4.1 Flood Forecasting and Warning

The necessity of flood forecasting in Pakistan received attention after the devastating flood of 1973. The flood unfolded many shortcomings of the system in place during the flood. The subsequent floods of 1988 and 1992 again revealed many serious system discrepancies pertaining to areas of flood forecasting and dissemination and it was badly felt that the system still needed to be further enhanced (Sheikh, 2001).

Flood Forecasting Division (FFD) stationed in Lahore is responsible at the national level for the issuance of flood forecasts and their warnings. The FFD is a branch organization of Pakistan Meteorological Department, government of Pakistan. Since floods are basically caused due to rainfall which requires the identification of weather systems, conventional weather charts, satellite clouds images from NOAA and geostationary satellites, data from a network of Quantitative Precipitation Radars are used for forecasting. For quantitative flood forecasting, the hydrological data is obtained from the Provincial Irrigation Departments, which maintain a big network of flow measurement stations at the specific sites on the rivers and *nallahs*. The Water and Power Development Authority (WAPDA) also has an active involvement in the process by providing the river and rainfall data from its telemetry river and rain sensors located within the upper catchments of the rivers. Sheikh (2001) summarized existing flood forecasting and warning facilities in Pakistan which include uses of satellite imageries, weather radars, 69 HF radio stations, and hydrologic models.

2.4.2 Drought Forecasting

In Pakistan drought is a hazard that inflicts heavy damage to its economy and livelihoods. Most of the geographical area is vulnerable to droughts. Drought prone areas are classified based on *aridity index* and *% households having access to piped water supply*. Based on the aridity four classes of drought prone areas – severe, high, moderate and low – were identified.

In Pakistan, several organizations are involved in monitoring drought related hydro-meteorological parameters. They include: Pakistan Meteorological Department (PMD), WAPDA, Provincial Drainage and Irrigation Authorities and district governments. The PMD under the Ministry of Defence is mandated to monitor meteorological parameters including droughts. It maintains a network of about 200 stations across various regions of the country. Ahmad et al. (2004) reported that the PMD was planning to install another 350 meteorological stations to strengthen existing drought monitoring system. A drought and environmental monitoring centre (DEMC) was established within the PMD. For the characterization of drought events and drought prone areas, the PMD uses *% normal method*, *aridity index* and *standard precipitation index* as drought indicators (PMD, 2003). Details of these methods are also described in Ahmad et al. (2004).

Ahmad et al. (2004) argued that introduction of advance technologies can lessen vulnerability. They include: (1) Improvement and lining of 36,000 water courses; however, farmers have to bear 55% of the estimated cost. (2) Under the On-Farm Water Management Programmes, construction of water tanks with a capacity of storing 180 m^3 water is planned. Each of these tanks will supply water to 5-10 farms. (3) Use of potential of *rod-kohli* system or diversion of water from the mountain torrents. Currently only 0.7 million ha land is irrigated using the system but can be expanded to two million ha with an estimated cost of $85 million. (4) Introduction of drip and sprinkler irrigation system and can learn from India and Iran as they are much advanced in these systems.

3.0 SUMMARY AND CONCLUSIONS

Floods and drought are frequent visitors to South Asia and cause substantial loss of lives and property. South Asian nations – Bangladesh, India, Nepal and Pakistan – have differential capacity in terms of drought and flood forecasting. Comparatively, flood forecasting systems are strong in Bangladesh and Pakistan compared to droughts. In Bangladesh, so far, drought forecasting has not received priority as compared to floods although a large part of the country is highly vulnerable to it. Drought forecasting in Pakistan is weaker than the floods. Nepal's flood problem due to topography, sparse meteorological stations in the mountain areas, heavy rainfall potential and climate change is unique in nature so as the forecasting systems. Economic reasons are also a barrier in developing advanced forecasting and warning systems. India is in an advantageous position in terms of technology and resources. In India drought forecasting is at advanced state than floods. Advanced hydrodynamic models are not in use in India for flood forecasting. Floods are particularly a problem for the northeastern rivers, Brahmaputra and Barak, and use of advanced model can benefit both India and Bangladesh.

For floods, South Asian countries can forecast short range (3-day) advance forecasts. However, because of weak outdated dissemination network, the short-range forecasts have severe limitations to meet users need and, therefore, any viable real-time response plan is practically difficult. So, the best way to make these products more usable is by enhancing the lead-time from three days to one month or one season. For these purposes, oceanic and atmospheric parameters like the ENSO cycle have to be added to the model. For example, Indian drought forecasting is useful because they are taking ENSO climate cycle as one of the important parameters to develop their drought model. Other countries (Bangladesh, Pakistan, and Nepal) may be recommended to follow similar approach.

REFERENCES

Ahmad, S., Hussain, Z., Qureshi, A.S., Majeed, R. and Saleem, M., 2004. Drought Mitigation in Pakistan:Current Status and Options for Future Strategies. International Water Management Institute (IWMI), Working Paper 85, Colombo, Sri Lanka.

Bhuiyan, S., 2004. Flood Forecasting, Warning and Response System. *In:* Proceedings of the National Workshop on "Options for Flood Risk and Damage Reduction in Bangladesh", 7-9 September 2004, 1-10 pp. Dhaka.

Chalise, S.R., 1994. Mountain Environments and Climate Change in the Hindu Kush-Himalayas. *In:* Mountain Environments in Changing Climate. M. Beniston (ed.), London, Routledge, 382-404.

Chowdhury, M.R., 2000. An assessment of flood forecasting in Bangladesh: The experience of the 1998 flood. *Natural Hazards*, **22,** 139-163.

Down to Earth, January 15, 2003. Centre for Science and Environment, New Delhi.

Gadgil, S., Srinivasan, J., Nanjundiah, R.S., Kumar, K.K., Munot, A.A. and Kumar, K.R., 2002. On Forecasting the Indian Summer Monsoon :The Intriguing Season of 2002. *Current Science*, **83(4),** 394-403.

Gyawali, D. and Dixit, A., 2001. Water and Science: Hydrological Uncertainties, Development Aspirations, and Uningrained Scientific Culture. *Futures*, **33,** 689-708.

Government of Assam, 2004. Economic Survey of Assam 2003-2004 (http://www.assamgov.org/ecosurvey/Flood.htm).

Government of India, 2005. *Annual Report 2004-2005*. Ministry of Water Resources, New Delhi.

Goswami, D., 2001. Paper presented in the conference on "Regional Cooperation for Flood Disaster Mitigation in the Hindu Kush-Himalayas", Kathmandu, 15-18 May 2001, organized by the International Centre for Integrated Mountain Development (ICIMOD) and World Meteorological Organization (www.southasianfloods.org, accessed in June, 2005).

Hossain, A.N.H.A., 2004. An Overview on Impacts of Flood in Bangladesh and Options for Mitigation. *In:* Proceedings of the National Workshop on "Options for Flood Risk and Damage Reduction in Bangladesh", 7-9 September 2004, Dhaka.

Hossain, M.A., 2004. Weather Forecasting for Flood Disaster and Mitigation. *In:* Proceedings of the National Workshop on "Options for Flood Risk and Damage Reduction in Bangladesh", 7-9 September 2004.

Kattelmann, R., 2003. Glacial Lake Outburst Floods in the Nepal Himalaya: A Manageable Hazard? *In:* Flood Problem and Management in South Asia (M.M.Q. Mirza, A. Dixit and A. Nishat Eds.), Kluwer Academic Publishers, Dordrecht, the Netherlands, 145-154.

Khanal, N.R. 1999. Study of Landslide and Flood Affected Area in Syangja and Rupandehi Districts of Nepal. A Report submitted to Mountain Natural Resources Division, ICIMOD, Kathmandu.

NASA, 2005 (accessed on May 21, 2005). Floods in Pakistan and Iran (http://visibleearth.nasa.gov/view_rec.php?id=6771).

Nayak, P.C., Sudheer, K.P. and Ramasastri, K.S., 2005. Fuzzy computing based rainfall-runoff model for real time flood forecasting. *Hydrological Processes*, **19,** 955-968.

Pakistan Meteorological Department (PMD), 2003. Spatial and temporal analysis of 1999–2002 drought over Pakistan, Islamabad.

Rahman, A. and Biswas, P.R., 1995. Drought Devours Resources. *Dhaka Courier*, Bangladesh, **11(42).**

Rajeevan, M., 2001. Prediction of Indian summer monsoon: Status, problems and prospects. *Current Science*, **81(11),** 1451-1457.

Rao, D.P., 2000. Disaster Management. Proceedings of Map India 2000. April 10-11, 2000, New Delhi.

Rao, M.S., 1989. Keynote Address on Floods, Symposium on Preparedness, Mitigation and Management of Natural Disasters, New Delhi, August 2-4.

Sarma, J.S., 2004. *Review and Analysis of Drought Monitoring, Declaration and Management in India.* International Water Management Institute (IWMI) Working Paper, Drought Series No. 2, IWMI, Colombo. Sri Lanka.

Sheikh, M.M., 2001. Flood Forecasting System in vogue in Pakistan with a Case Study on the 1992 Flood in Jhelum River.

Singh, R., 2004. Practices and Current Status of Flood Forecasting and Management in India. Unpublsihed Paper, Water Resources Development and Training Centre (WRDTC), Indian Institute of Technology, Roorkee, India.

8 Hydrometeorology of Floods and Droughts in South Asia – A Brief Appraisal

S. Nandargi, O.N. Dhar, M.M. Sheikh, Brenna Enright and M. Monirul Qader Mirza

PART 1: HYDROMETEOROLOGY OF FLOODS

1. INTRODUCTION

Flooding is a common event in South Asia. In fact, it has been said that after Bangladesh, India is the worst flood affected country in the world (Agarwal and Narayan, 1991). It is possible for extreme floods to inundate up to approximately 10 million hectares or nearly 70 percent of Bangladesh (Mirza, 2003a). By contrast, about 40 million hectares or nearly 12.5 percent of India is flood prone. Different types and extents of flooding occur in different regions of the subcontinent. The definition of flooding with respect to each area where it occurs is given in Box 1.

Region specific flood scenarios arise from different rainfall patterns. For example, more than 75 to 80 percent of India's annual rainfall occurs during the monsoon (June to September) or even up to middle of October in some

Box One: Country Specific Flood Definitions

India: A flood is defined when the flood waters in a river cross the danger level (D.L.) at a particular site, called gauge/discharge site (G/D site) (Dhar and Nandargi, 1998). The D.L. for a G/D site is usually 1 m above the warning level.

Bangladesh: The danger level at a specific G/D site is the level above which flood water is likely to cause damages to nearby crops and homesteads (FFWC, 2005).

Nepal: For flood plains of the Inner and Outer Terai, the flood is defined as the condition when the channel becomes incapable of carrying the discharge and the water rises above the channel bank.

parts of the country. However, India's rainfall is not uniformly distributed all over the country and there are large inter-seasonal, seasonal and yearly variations of rainfall in both space and time. Bangladesh experiences similar rainfall patterns as India. Rainfall is highest in the North- and South-eastern parts of the country and progressively reduces towards north-west. In fact, nearly 85 percent of the country's rainfall occurs during June to October (Ahmad and Ahmed, 2003). Similarly, annual precipitation in Nepal is highly influenced by the monsoon. The precipitation patterns demonstrate a very high variance which is attributed to high topographical variability. The precipitation pattern in Pakistan is also influenced by the topography.

The erratic distribution of rainfall during the course of the year in South Asia is one of the main causes of floods (and droughts) in various parts of the region almost every year. Despite these large variations, the rainy season has been coined as the *"presiding deity"* in South Asia because the agricultural prosperity and well being depends to a large extent on the timely arrival and withdrawal of monsoon and its associated rainfall. As such, this paper will carry out a temporal and spatial hydrometeorological synthesis for the region so as to get an understanding of the climate related hazards that are prevalent now and in the future.

2. HYDRO-METEOROLOGICAL FACTORS RESPONSIBLE FOR FLOODS

In general, heavy rainfall of fairly long duration is the main cause of floods in South Asia during monsoon months. The monsoon system is maintained by the strong differential heating of the Indian Ocean in the south and the mountain ranges in the north. The two main events that are responsible for generating floods during the monsoon season include cyclonic disturbances and break monsoon situations. Besides, glacial lake outburst is a new issue of concern in the Himalayas, causing severe flooding and flash floods.

2.1 Cyclonic Disturbances

Cyclonic disturbances include low pressure areas, depressions, deep depressions and severe cyclonic storms. Table 1 gives the WMO Tropical Cyclone Operational Plan (TCP-21) definition for each disturbance type for South Asia. Most of these disturbances originate from the Bay of Bengal and very few from the Arabian Sea and during their passage over the region, the cyclones strengthen considerably and cause fairly widespread heavy to very heavy rainfall. Consequently, rivers originating or flowing through cyclone affected zones are liable to experience frequent and severe flooding during each monsoon season. Figure 1 shows cyclone affected zones within Bangladesh. Pakistan is also affected by cyclones originating in the Arabian Sea, which experiences at least one cyclone per year.

Table 1: Cyclonic disturbance definitions

Types of Disturbances	*Wind Speed*	
	Knots	*Kph*
Low	< 17	< 31
Depression	17-33	31-61
Cyclonic storm	34-47	62-88
Severe cyclonic storm	48-63	89-117
Severe cyclonic storm of hurricane intensity	64	>118

Source: WMO

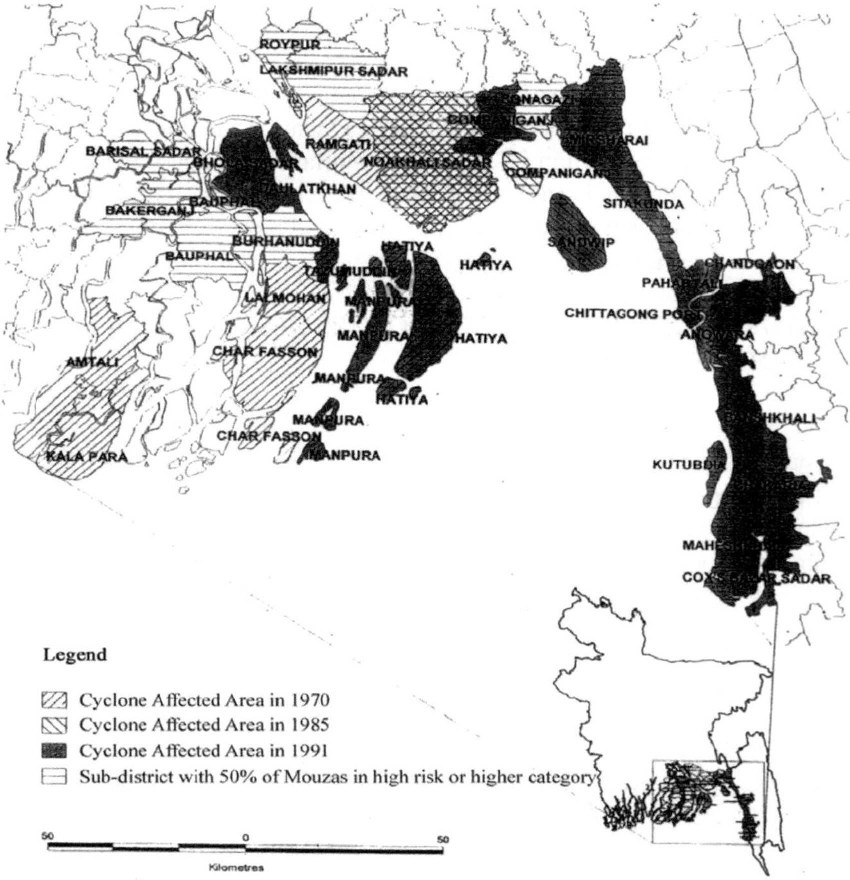

Fig. 1: Areas affected by the 1970, 1985 and 1990 cyclones
Source: Cyclone Shelter Preparatory Study 1996

2.2 Break Monsoons

This peculiar weather situation is specific to the South Asian monsoon system. Break monsoons normally occur when the axis of the seasonal monsoon trough over the Indo-Gangetic plains moves from its normal position to northwards and remains close to the foot-hills of the Himalayas on the sea level and 850 hpa synoptic charts. When this happens, heavy rainfall occurs mostly in the foot-hills of the Himalayas (mostly central and eastern Himalayas) while the remaining regions, except the southeastern part of the Indian peninsula, experiences severe drought conditions. The duration of a break monsoon is generally 2 to 3 days but can sometimes last up to 10 days or more (De et al., 1998).

2.3 Glacial Melting

Due to global warming, many of the big glaciers are rapidly melting to form glacier lakes (UNEP, 2005). This will consequently instigate the sudden discharge of large volumes of water and debris and subsequent flooding in downstream river catchment areas. This event has been coined as a Glacial Lake Outburst Flood (GLOF) (UNEP, 2005; Dahkal, 2003).

Based on the study of satellite images, the WWF-Nepal Programme has reported, five GLOF events in Nepal from 1977 to 1998 and the UNEP has issued a warning that 20 big glacial lakes in Nepal are at the risk of GLOF, which could trigger huge loss of lives and properties (UNEP, 2005; Dahkal, 2003). A description of the most significant GLOF in Nepal is outlined in Box 2.

Box Two: Recent notable GLOF event in Nepal

The United Nations Environment Program (UNEP) in cooperation with the International Center for Integrated Mountain Development (ICIMOD) performed an inventory of glaciers and glacial lakes in Nepal in 2001. The most significant GLOF event occurred in 1985. The event produced surges of water 10 to 15 metres high and the associated debris flooded the Bhote Koshi and Dudh Koshi Rivers for 90 kilometres. The peak flow rate was 2000 m^3/s which destroyed the Namche Small Hydel Project.

3. FLOOD PRONE REGIONS OF SOUTH ASIA

India: There are two main regions in India which experience floods on widespread scale during the monsoon season. These are the northeast region comprising the Brahmaputra basin and its tributaries and the north Indian region of the Ganga basin and its tributaries (Fig. 2). The main rivers and their tributaries in these two regions experience floods during every monsoon season. The central Indian region and the northern peninsular region also experience occasional floods but their frequency is far less (Dhar and Nandargi, 2003).

Bangladesh: Bangladesh is a flood prone country with approximately 20 percent of the land being routinely inundated every year and another 40 percent is at regular risk from flooding (Choudhury, 2003; Mirza, 2003a).

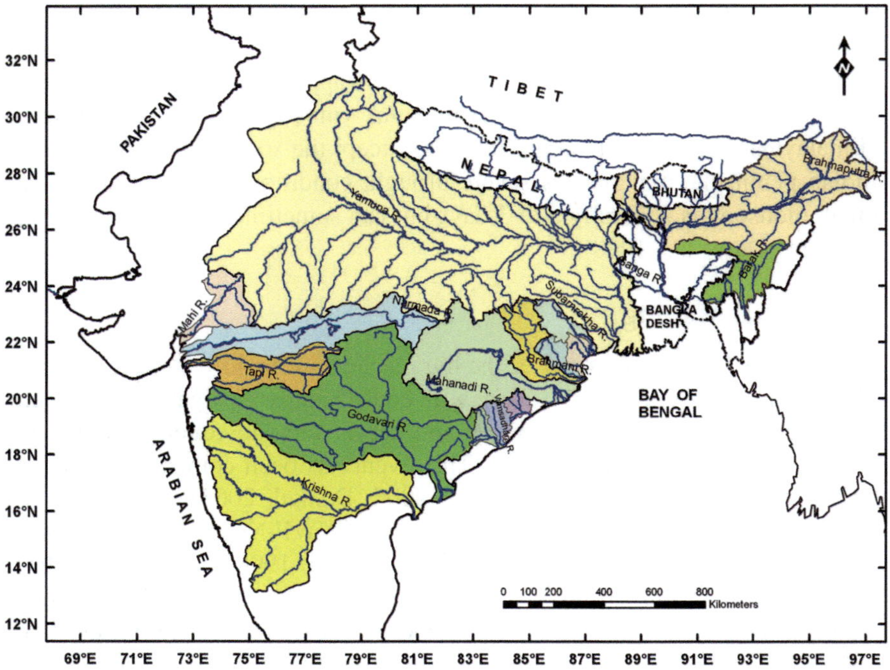

Fig. 2: Flood prone rivers of India.

The northern, central and southwestern districts are mainly inundated due to the onrush of flood waters primarily from West Bengal and Bihar in India and the coastal districts are flooded from storm surges and sea level rise (Choudhury et al., 2003). A map outlining the flooded areas during the 1998 flood is found in Fig. 3 and gives a good indication of the flood prone areas in Bangladesh.

Nepal: The most susceptible districts include Sarlahi, Rautahat, Chitawan, Makwanpur and Sindhuli, which border the Marin Khola, Bagmati, East Rapti, Manahari and Kamala rivers. There are also many districts that are prone to landslide and debris flows. They include Makwanpur, Dhading and Kavre Palanchok in Central Nepal as well as Taplejung, Panchthar and Okhaldhunga in Eastern Nepal (Shrestha et al., 2003).

Pakistan: Floods in Pakistan are mainly caused by heavy rainfall over the upper catchments of the main rivers of Indus Basin namely Indus, Jhelum, Chenab, Ravi and Sutlej whose upper catchments lie in India.

4. MAGNITUDE AND FREQUENCY OF FLOODS IN SOUTH ASIA

4.1 India

Out of about 120 major and medium rivers of the Indian sub-continent, every year some 50 or more rivers experience floods during the monsoon

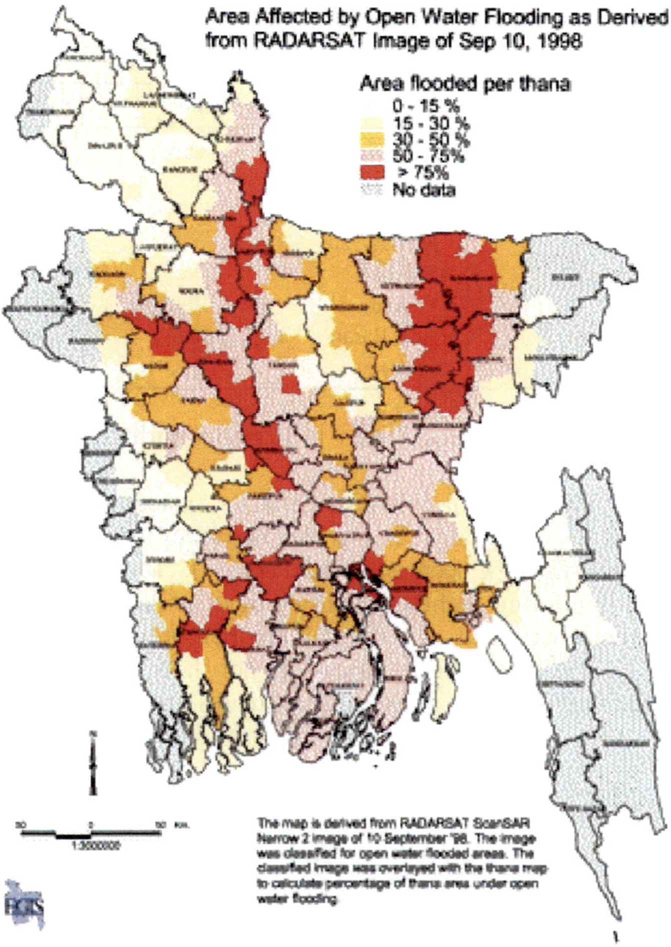

Fig. 3: Flood affected areas during the 1998 flood in Bangladesh
Source: Mirza, 2003a

season. On each of these rivers G/D sites record flood deviations from their respective D.Ls. In the river systems of Brahmaputra and Ganga, there are 18 G/D sites which have experienced over 50 floods during 1986-2003 (Nandargi and Dhar, 2003). Flood magnitudes and frequencies are higher in the Brahmaputra and the Ganga rivers and their tributaries having their origin in the Himalayan region. On the Brahmaputra and the Ganga river systems, eight G/D stations experienced over 100 floods in 18 years (1986-2003) (Table 2).

Table 2 shows that Road Bridge site on the Beki River and Dibrugarh site on the Brahmaputra River are the worst flood affected sites of India. Just as Cherrapunji and Mawsynram are the highest rainfall receiving stations in India (and the world), similarly, Road Bridge site on the Beki River and Dibrugarh site on the Brahmaputra River are India's worst flood affected

Table 2: G/D sites on the Brahmaputra and the Ganga river systems which experienced > 100 floods during 1986-2003

G/D site	*River*	*State*	*No. of floods*
Dibrugarh	Brahmaputra	Assam	157
Nematighat	”	”	126
Road Bridge	Beki	”	177
N.H.Crossing	Manas	”	101
Numaligarh	Dhansiri	”	138
Farakka	Ganga	West Bengal	108
Jhanjharpur	Kamala Balan	Bihar	104
Baltara	Kosi	”	150

sites. On an average they experience 9 to 10 floods during a monsoon season every year.

However, Himalayan tributaries of the Brahmaputra and Ganga river systems show an increasing trend in the frequency of floods while their southern tributaries show decreasing trend (Fig. 4). The entire Brahmaputra river system is showing slightly decreasing trend while the Ganga system is showing increasing trend so far total number of monsoonal floods is concerned. Central Indian river systems do not show any noteworthy trend as the number of floods recorded is very small in that region (Fig. 4).

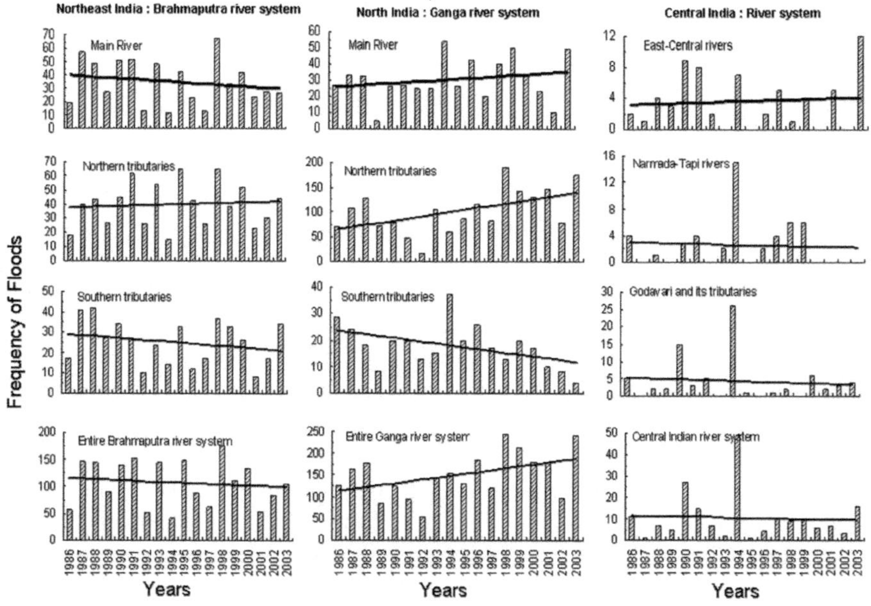

Fig. 4: Yearly frequency of floods in three major rivers.

4.2 Bangladesh

Since Bangladesh is located within the flat, low lying delta of the Ganges, Brahmaputra, and Meghna (GBM) river basins, it experiences annual flood of various degrees. Table 3 gives an overview of water levels for the years 1987, 1988 and 1998 for the four main rivers viz. the Brahmaputra, the Ganges, the Padma and the Meghna that flow through Bangladesh. From past events it is seen that each river has the capacity to make the water levels to rise approximately by 1 m or more above the danger level for periods as long as two months.

Table 3: Water level of major rivers in Bangladesh

River	*DL (m)*	*Annual Peak*			*Days Above DL*		
		1987	*1988*	*1998*	*1987*	*1988*	*1998*
Brahmaputra	19.50	19.68	20.62	20.37	13	27	66
Ganges	14.25	14.80	14.87	15.19	55	23	27
Padma*	6.00	6.99	7.43	7.50	56	47	72
Meghna	6.25	6.91	7.66	7.33	30	68	68

*Combination of flows of the Brahmaputra and Ganges rivers.
Source: FFWC, 1998

4.3 Nepal

A study on the Bagmati river basin which is located in the middle of the country and represents east to west spatial variation (Shrestha et al., 2003) reveals that its plains are highly prone to flooding and associated damage within and outside of Nepal's boundaries. It is seen that the Bagmati Basin has experienced a total of 10 floods in a single year and a maximum of five flood days for a single flood event. The study concluded that duration of flood events in the Bagmati Basin are increasing, which is verified by hydrological records of Nepal.

4.4 Pakistan

It is seen from the past 30 years (1971-2000) of flood data that once in every four to five years the country experienced catastrophic flood. Some recent floods with their damages are shown in Table 4.

PART 2: HYDROMETEOROLOGY OF DROUGHTS

1. INTRODUCTION

Drought (defined in Box 3) is one of the worst natural disasters (third in terms of severity) (Fig. 5) because its onset is slow, affected area is widespread

Table 4: History of some exceptionally high floods in Pakistan (1957 to 1997)

Year	*River*	*Discharge (m^3/s)*	*Flooded Area (km^2)[1]*
1957	Chenab (Marala)*	31148.53**	N/A
	Ravi (Jassar)	9287.93	
1959	Jhelum (Mangla Dam)	29591.10	N/A
	Chenab (Marala)	24658.17	
	Ravi (Jassar)	8410.10	
1973	Indus (Guddu)	30688.16	41957.81
	Chenab (Marala)	21803.97	
	Ravi (Jassar)	6427.92	
1975	Indus (Guddu)	28387.53	35340.39
	Chenab (Marala)	16494.56	
1976	Indus (Guddu)	33970.93	82879.62
	Jhelum (Mangla)	13593.79	
	Ravi (Balloki)	7191.74	
1988	Indus (Guddu)	31682.47	11395.95
	Chenab (Marala)	21265.24	
	Ravi (Shahdara)	16310.50	
1992	Indus (Guddu)	30778.12	39212.42
	Jhelum (Mangla Dam)	30865.36	
	Chenab (Marala)	23930.28	
1995	Jhelum (Mangla Dam)	8560.81	16881.54
	Chenab (Khanki)	17854.25	
1996	Chenab (Marala)	21715.06	N/A
	Ravi (Balloki)	6654.46	
20th July, 1996	Lahore (Punjab Province)	500 mm rainfall in 24 hrs caused heavy urban storm flooding in the city	N/A
1997	Jhelum (Mangla Dam)	17068.83	N/A
	Chenab (Marala)	21960.42	

* Discharges only at the sites mentioned in the parentheses are given.

** Exceptionally high flood limits for different rivers and reaches have the following threshold (i.e. normal values) values:

Indus (Guddu)	22653.5 m^3/s
Jhelum (Mangla Dam)	8495.05 m^3/s
Chenab (Marala)	16990.1 m^3/s
Ravi (Jassar)	5663.4 m^3/s
Ravi (Shahdara)	5097.03 m^3/s
Ravi (Balloki)	5097.03 m^3/s

[1]*Source:* Pakistan Water Gateway (PWG), 2002

and causes adverse impacts like starvation/malnutrition of human beings and cattle, soil degradation, loss of crops, loss of other economic activities, spread of diseases, migration of people and livestock. In order to understand the severity of a drought, it is necessary to differentiate between drought, aridity and desertification.

Droughts are influenced by climatic fluctuations over an extended period. As with floods, the erratic distribution of rainfall over the course of the year in South Asia is one of the main causes of droughts in various parts of the region almost every year. Droughts can occur virtually in all climates. Severe droughts in South Asia mostly occur in the pre-monsoon periods and post-monsoon period. The pre-monsoon droughts occasionally extend throughout the monsoon period due to late onset of monsoon and weak monsoon activities (Ahmad et al., 2003; Choudhury et al., 2003; Mirza, 2003a).

Box Three: Scientific Definition of Drought

Scientifically, drought is defined on the basis of non-availability of rainfall, decrease or non-availability of the surface water and depletion in the soil moisture and goundwater level. Droughts can, therefore, be categorized in the following three types:

(a) Meteorological drought – Occurs when actual rainfall over an area is less than three-fourth of the normal value i.e. the long-term climatological mean. Here drought persists throughout the season which leads to substantial deficiency of seasonal rainfall over the specified region. Thus both spatial and temporal scales are important in meteorological drought.

(b) Hydrological drought – When there is marked depletion of surface water causing very low stream flow and drying of lakes, reservoirs and rivers, it is called a hydrological drought. This occurs on the local, regional or sub-continental scales which spread in horizontal direction. In this spatial distribution, drought can last for a few weeks during a season with intermediate breaks of spells of good rains.

(c) Agricultural drought – When soil moisture is inadequate to support healthy growth of crops resulting in very low yield, is usually termed as agricultural drought. As water level goes deeper, even groundwater is unable to meet the needs.

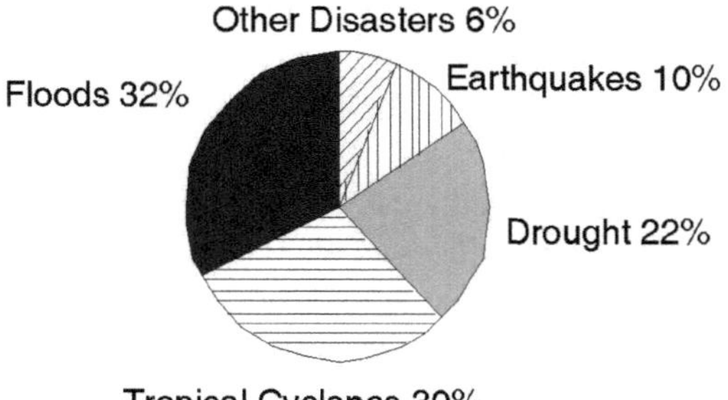

Fig. 5: World's significant damages caused by natural disasters. (Courtesy: WMO, 1994)

2. CAUSES OF DROUGHT

The occurrence of drought is closely related with the precipitation on point or regional scale since droughts are caused from a lack of precipitation over an extended length of time. During long durations of drought, evapotranspiration causes the loss of surface water and soil moisture. Consequently, groundwater recharge is slow and inadequate and rivers experience lower discharges. The situation is exacerbated if flow interventions take place in the upstream parts of the rivers.

It is seen from past data that the major causes of drought (or deficiency of rainfall) were:

(a) Late arrival and early departure of monsoon causing decrease in total duration of monsoon period and therefore deficient rainfall activity.
(b) Prolonged 'break' monsoon conditions.
(c) Formation of less number of or no cyclonic disturbances during the monsoon season and restriction of their movement to the east of 80°E over the country.
(d) Strong ENSO warming, especially during the monsoon period along with negative SOI (Southern Oscillation Index) resulting in deficient rainfall over the country.

3. DROUGHT PRONE REGIONS OF SOUTH ASIA

India: Despite an estimated national average freshwater potential of 2464 cubic metres per capita per year, several parts of India face acute water scarcity, droughts and famines. The Indian states of Gujarat, Haryana and Punjab are highly susceptible to drought, particularly within the agricultural sector due to reduced rainfall in post-monsoon months and inadequate availability of ground water.

Bangladesh: Since the western and northwestern regions of Bangladesh generally experience high temperatures and have characteristically lower soil moisture holding capacities, they suffer from drought during the pre-monsoon season (Choudhury et al., 2003).

Nepal: Drought conditions are a permanent characteristic in the semi-arid Trans Himalayan region of Nepal where the annual rainfall is less than 500 mm. This area is made up of the Manang, Mustang and Dolpa districts (Shrestha et al., 2003).

Pakistan: The most affected areas due to droughts since 1998 are outside the Indus Basin with the provinces of Punjab, Sindh and Balochistan being the most susceptible. Out of sixteen districts in the Sindh Province, eight were affected by drought during the period from 1998 to 2002 with four districts being severely affected. The Asia-Pacific Network (APN) Project focuses on drought in Bahawalpur, Punjab and Mirpurkhas, Sindh. Special attention is given to the Cholistan desert in Bahawalpur and the Thar Desert in Mirpurkhas (Ahmad et al., 2003).

4. CONCLUSIONS

To summarize, it can be said that in India, droughts in the past affected mostly the crop producing regions of the country which in turn badly affected the economy of the country. On the contrary there are monsoon flood years during which country experiences large number of floods in the Ganga and Brahmaputra river systems of the country as described in detail in Part I of this article. There is no doubt that floods cannot be totally eliminated, but their impact on the society and country at large can be considerably minimized by taking proper remedial measures well in advance and in a planned manner. One such measure is forecasting of floods which is now being adopted by most of the South Asian countries. Some efforts are being made to inter link rivers to divert waters from a flooded river to a region whose rivers are almost dry. Proper storage and utilization of this water which otherwise goes waste will be useful to overcome moderate to severe drought conditions to a large extent.

REFERENCES

Agrawala, S., Raksakulthai, V., van Aalst, M., Larsen, P., Smith, J. and Reynolds, J., 2003. Development and Climate Change in Nepal: Focus on Water Resources and Hydropower. Organisation for Economic Co-operation and Development, Paris.

Agarwal, A. and Narayan, S., 1991. State of India's Environment: Floods, Centre for Science and Environment, N. Delhi.

Ahmad, S., Bari, A. and Muhammed, A., 2003. Climate Change and Water Resources of Pakistan: Impacts, Vulnerabilities, Coping Mechanisms, Climate Change and Water Resources in South Asia, Proceedings of Year End Workshop, January, 2003, pp. 138-201.

Ahmad, Q.K. and Ahmed, A.U., 2003. Regional Cooperation in Flood Management in the Ganges-Brahmaputra-Meghna Region: Bangladesh Perspective. *In:* Flood Problem and Management in South Asia. Kluwer Academic Publishers, Netherlands, pp. 181-198.

Awan, S.A., 2002. The Climate and Flood Risk Potential of Northern Areas of Pakistan. Commission on Science and Technology for Sustainable Development in the South. http://www.comsats.org.pk/main_introduction.htm.

Choudhury, A.M., Quadir, D.A., Neelarmi, S. and Ahmed, A.U., 2003. Climate Change and its Impacts on Water Resources of Bangladesh. Climate Change and Water Resources in South Asia. Proceedings of Year End Workshop, Kathmandu, January, 2003, pp. 21-60.

De, U.S., Lele, R.R. and Natu, J.C., 1998. Breaks in southwest monsoon. Pre-published Scientific Report No. 1998/3, IMD, Pune, pp. 1-25.

Dhar, O.N., Rakhecha, P.R. and Mandal, B.N., 1981. Some facts about Indian rainfall - A brief appraisal from hydrological considerations. *Ind. J. of Power & River Valley Devp.*, **31,** 117-125.

Dhar, O.N., Mandal, B.N. and Rakhecha, P.R., 1984a. Rainfall distribution over India during the monsoon months in the absence of depressions and cyclonic storms. *Mausam*, **35(3),** 309-314.

Dhar, O.N., Soman, M.K. and Mulye, S.S., 1984b. Rainfall over the southern slopes of the Himalayas and the adjoining plains during 'breaks' in the monsoon. *J. of Climatol., UK,* **4(6)**.

Dhar, O.N. and Nandargi, S., 1995, Zones of severe rainstorm activity over India. *Int. J. Climatol.*, **15(7),** 811-819.

Dhar, O.N. and Nandargi, S., 1998. Floods in Indian rivers and their meteorological aspects, *Memoir Geolog. Soc. of India, Bangalore*, **41,** 1-25.

Dhar, O.N. and Nandargi, S., 2003. Hydrometeorological aspects of floods in India. *In:* Natural Hazards. Kluwer Acad. Publ., Netherlands, Vol. 28, pp. 1-33.

Dixit, A., 2003. Floods and Vulnerability: Need to Rethink Flood Management. *In:* Flood Problem and Management in South Asia, Kluwer Academic Publishers, Netherlands, pp. 155-179.

FAO, 1999. Cyclones, Storm and Floods Cause Devastation and Serious Food Shortages in Parts of Asia. FAO Global Information and Early Warning System on Food and Agriculture, http://www.fao.org/documents/.

Government of India, 1972. Report of the Irrigation Commission, Vol. I. Min. of Irrigation and Power, Govt. of India, N. Delhi.

India Meteorological Department (IMD), 1971. Rainfall and droughts in India. Report of the Drought Research Unit, IMD, Pune.

India Meteorological Department (IMD), 2003. Weekly Weather Bulletin and IMSP News Letter of 2003.

Integrated Regional Information Networks (IRIN), 2005. Pakistan: Winter rains alleviate drought conditions,http://www.irinnews.org.

Kulshrestha, S.M., 1997. Drought management in India and potential contribution of climate prediction. Joint COLA/CARE Technical Report No.1, Centre for the Application of Research on the Environment.

Kulshrestha, S.M., 2002. Flood Management in India. Joint COLA/CARE Technical Report No. 3. Centre for Ocean-Land-Atmosphere Studies, Calverton, USA.

Mirza, M., 2003a. Extreme Hydro Meteorological Events in South Asia: Adaptation and Coping Mechanisms. *In:* Climate Change and Water Resources in South Asia. Proceedings of Year End Workshop, Kathmandu, January, 2003, pp. 236-244.

Mirza, M., 2003b. Three Recent Extreme Floods in Bangladesh: A Hydro-Meteorological Analysis, Flood Problem and Management in South Asia. Kluwer Academic Publishers, Netherlands, pp. 35-64.

Nandargi, S. and Dhar, O.N., 2003. High frequency floods and their magnitudes in Indian rivers. *Geo. Soc. of India*, **61(1),** 90-96.

Nyrop, R.F. (ed.), 1983. Pakistan: A Country Study. US Library of Congress, Washington.

Office for the Coordination of Humanitarian Affairs (OCHA), 2000. OCHA Situation Report No. 3: Pakistan - Drought, http://www.un.org.pk/drought/ocha-rpt3.htm

Pakistan Water Gateway (PWG), 2002. Floods in Pakistan, www.waterinfo.net.pk/pdf/fs2fp.pdf.

Shrestha, K.L., Shrestha, M.L., Shakva, N.M., Ghimire, M.L. and Sapkota, B.K., 2003. Climate Change and Water Resources of Nepal. *In:* Climate Change and Water Resources in South Asia. Proceedings of Year End Workshop, Kathmandu, January, 2003, pp. 91-137.

Sikka, D.R., 2000. Monsoon floods in India. Joint COLA/CARE Technical Report No. 4, Centre for Ocean-Land-Atmosphere Studies, Calverton, USA, pp. 1-54.

Sikka, D.R., 2003. Evaluation of monitoring and forecasting of summer monsoon over India and a review of monsoon drought of 2002. *Proc. Ind. Nat. Sci. Acad.*, **69(5),** 479-504.

Umrani, A.P., 2001. Living with Drought, http://www.dawn.com/2001/04/09/ebr9.htm.

UNEP, 1999 and UNEP, 2005.

UN, 2001. Drought – Pakistan Update No. 13, Resident Coordinator of the United Nations System's Operational Activities for Development in Pakistan, www.un.org.pk/drought/rcreport13.htm.

UN, 2002. Drought – Pakistan Update No. 14, Resident Coordinator of the United Nations System's Operational Activities for Development in Pakistan, www.un.org.pk/drought/rcreport4.htm.

World Meteorological Organization (WMO), 1994. On the Frontline – Public Weather Services, WMO No. 816, Geneva.

World Meteorological Organization (WMO), 2004. *WMO Bulletin*, **53(1),** 1-92.

9 The El Niño-Southern Oscillation (ENSO) and Stream-flows in the Greater Ganges-Brahmaputra-Meghna (GBM) Basins – A Climate Outlook

Md. Rashed Chowdhury

1. INTRODUCTION

The Ganges-Brahmaputra-Meghna (GBM) River system is the third largest freshwater outlet to the world's oceans; only the Amazon and the Congo River exceed it. This watershed encompasses a number of countries in the South Asian region, including China, India, Nepal, and Bangladesh (Fig. 1). Of these, China contributes solely to the flow of the Brahmaputra, and Nepal to the flow of the Ganges (Nishat and Faisal, 2000). The flows in the Ganges and Brahmaputra are highly seasonal, heavily influenced by the monsoon

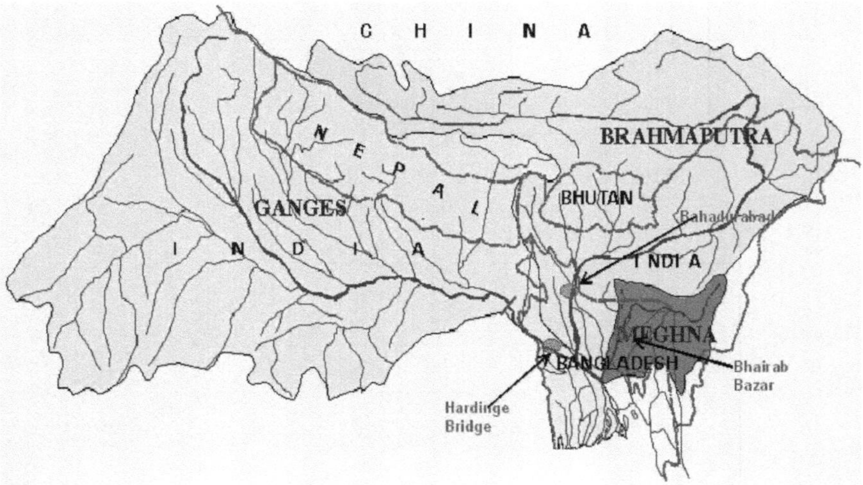

Fig. 1: Bangladesh, India, China, and Nepal along the principal channels of Ganges, Brahmaputra, and Meghna (Inset – Bangladesh occupies between lat. 20-30°N -26-45°N, lon 88-0°E - 92-45°E)

rainfall, and both the rivers have their headways outside the country (see Brammer, 1990; Haque, 1997; Rasid and Paul, 1987 and references therein for details on environmental settings in Bangladesh).

By and large, the seasonal rainfall in upstream India is a major factor that contributes significantly to the stream-flows in the major rivers of Bangladesh. Therefore, inside Bangladesh, the stream-flows in the major rivers are the results of monsoon rainfall in the upstream India (Chowdhury and Ward, 2004). While there are some initiatives in conducting research between the links of Bangladesh climate and the El Niño-Southern Oscillation (ENSO) climate cycle (see for example, Hossain et al., 2001; Chowdhury, 1994; Douglas et al., 2001, Chowdhury, 2003; Chowdhury and Ward, 2004) – any particular study relating to ENSO and basin-wide stream-flow variability is virtually non-existent. This study is primarily intended to focus on this issue. Therefore, the objectives here are as follows: (i) to quantify the variability of key stream-flow features of the seasonal hydrographs for the GBM River system at the lower riparian part of Bangladesh, and (ii) to understand the impact of ENSO cycle on hydrologic variability and changes in the greater GBM basins. As the stream-flows in Bangladesh are the results of rainfall in India, a brief brainstorm discussion on the basin-wide rainfall-runoff relationship is necessary to visualize the picture of how the stream-flows in Bangladesh are highly correlated to the rainfall of upper catchments in India. This relationship is reported in Chowdhury and Ward, 2004.

2. DATA, BASIC INDICES AND METHOD

The following section briefly describes the data used in this study. A more detailed description of these hydro climatic and oceanic data is available in Chowdhury (2003) and Chowdhury and Ward (2004, 2005).

Rainfall data: Due to lack of accessibility to the upstream rainfall data, an attempt was made here to explore the globally available precipitation data (popularly known as *Hulme* data) to gather rainfall information from the upstream GBM basins (Hulme, 1994). Rainfall data (outside of Bangladesh: 1950-1998) for the Brahmaputra, Ganges, and Meghna basins were collected from the available *Hulme* Global Precipitation data. A standardized anomaly of this area-average rainfall index is presented in Fig. 2.

Stream-flow data: Stream-flow data were collected directly from three major river points i.e., Ganges at Hardinge Bridge (Hbr), Brahmaputra at Bahadurabad (Bbd), and Meghna at Bhairab Bazar (Bbz) (see Fig. 1 for location of these stations). Standardized anomalies of maximum annual flow in these three monitoring points are computed (Fig. 3).

ENSO data: Commonly used indices like the Southern Oscillation Index (SOI)[1] and temperature anomalies show that El Niño events (the year of

[1] The Southern Oscillation (or atmospheric) component of ENSO is measured by the SOI usually defined as the anomalous Sea Level Pressure (SLP) in the eastern Pacific at Tahiti (17.6°S, 149.6°W) minus SLP in the western Pacific at Darwin, Australia (12.4°S, 130.9°E).

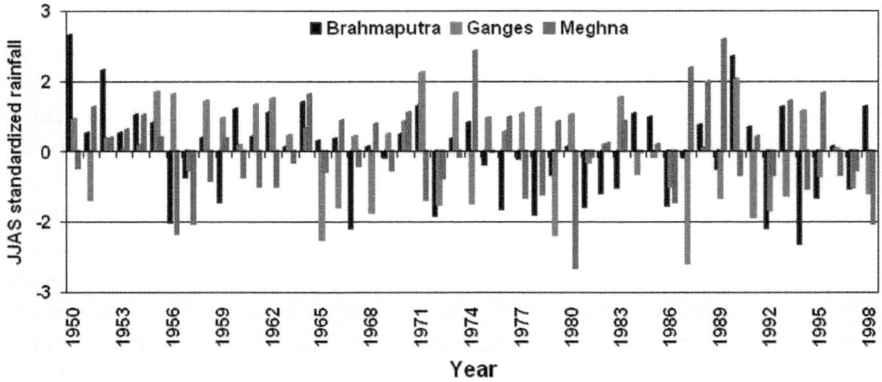

Fig. 2: Year-to-year standardized deviation of seasonal (JJAS) rainfall in upstream India – Basin-wide: 1950-1998 (Source: UEA Climate Research Unit (CRU) *Hulme* (1994) Global Precipitation data, Grid: 2.5 × 2.5 deg (1° = 111 km). (See Table 1 for El Niño and La Niña years).

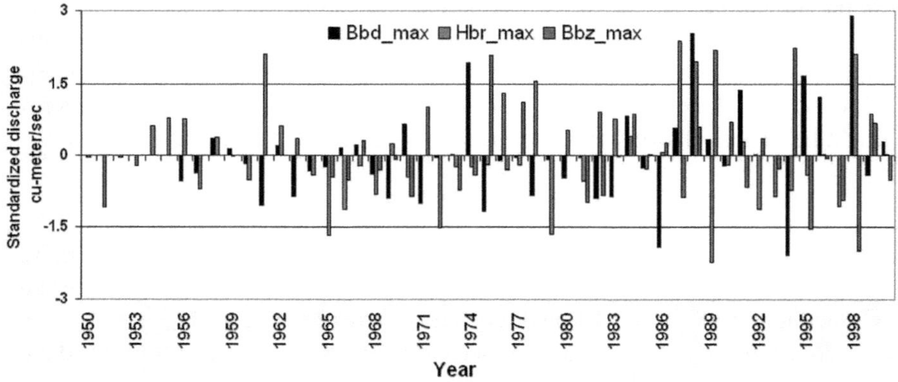

Fig. 3: Year-to-year standardized deviation of annual maximum flows in the lower riparian part of Bangladesh. (Note: Bbd_max: Standardized deviations (SD) of maximum flows at Bahadurabad point (Brahmaputra), Hbr_max: SD of maximum flows at Hardinge Bridge point (Ganges), and Bbz_max: SD of maximum flows at Bhairab Bazar point (Meghna). (See Table 1 for El Niño and La Niña years)

occurrence of ENSO) over the past 50 years occurred in 1951, 1953, 1957-58, 1963, 1965-66, 1969, 1972-73, 1976-77, 1982-83, 1986-87, 1991-95 and 1997-98 while La Niña events occurred in 1949-50, 1955-56, 1964, 1970-71, 1973-75, 1988-89, 1995-96, and 1998-2000 (McPhaden, 2001). It has been observed that some El Niño and La Niña events persist for as many as two full years. Observations also revealed that neither of these El Niño or La Niña events is of equal duration nor do they all evolve in the same way. To complicate the matter, there is hardly any definitive objective procedure available now for classifying the intensity of these El Niño and La Niña events. Therefore, based on a subjective ranking process using Niño 3.4[1] sea surface temperatures (SST), average SOI for six months, and season-by-season cold and warm episodes, five extreme El Niño/La Niña years are identified (Table 1) (see Chowdhury, 2003 for more details about this subjective procedure; also see Hossain et al., 2001).

SST and atmospheric circulation data: The National Center for Environmental Prediction's (NCEP) historical monthly fields of the global SST have been used in these analyses. The Extended Reconstructed SST (ERSST: version 1) data are taken from the National Oceanic and Atmospheric Administration (NOAA) National Climate Data Center (NCDC) (Smith and Reynolds, 2002). For atmospheric circulation, the NCEP/National Center for Atmospheric Research (NCAR) reanalysis has been used (Kalnay et al., 1996).

SST Indices for five wettest (highest flow) and driest (lowest flow) years: Based on average seasonal flows in excess of $\tilde{e} + s$ ($\tilde{e}$: mean) as higher (increased flows or wettest) and less $\tilde{e} - s$ as lower (decreased flows or driest), five extreme wettest and driest years have been identified (Table 1). The indices of SST and circulations are constructed (Figs 5 and 6).

Data limitations: Because data accumulation is a major task in this study – assuming that the globally available data are accurate – concerted efforts were made to collect data from the reliable sources in Bangladesh. Despite all possible efforts, a complete access to the discharge data was extremely difficult and, therefore, some discharge estimates are uncertain.

Table 1: Wettest and driest years in Bangladesh (1950-2000)

Criteria	*Wettest (highest flow) years*	*Driest (lowest flow) years*
Stream flow		
Ganges (at Hardinge bridge)	1961, 1978, 1987, 1988, 1998	1965, 1972, 1979, 1989, 1992
Brahmaputra (at Bahadurabad)	1974, 1988, 1991, 1995, 1998	1961, 1975, 1982, 1986, 1994
ENSO years		
Strong El Niño*		1951, 1958, 1972, 1982, 1997
Moderate El Niño	1963, 1965, 1969, 1974, 1987	
Strong La Niña*	1964, 1973, 1975, 1988, 1998	
Moderate La Niña	1956, 1970, 1971, 1984, 1999	

* El Niño reflect periods of exceptionally warm SST across the eastern tropical Pacific while La Niña represent periods of below average SST across the eastern tropical Pacific.

[1] The Niño 3.4 region is bounded by 120 W – 170 E and 5S-5N.

3. FINDINGS

Results of hydro-meteorological variability – correlations of basin-wide rainfall (in India) and stream-flows (in Bangladesh) – in the greater GBM basin system are not reported here; it is available in Chowdhury and Ward (2004).

3.1 ENSO and Features of River Hydrograph

Features of the river hydrograph are studied here because of their anticipated significance for human activity, such as agriculture (because the flows of Brahmaputra and Meghna are highly correlated, the following discussions are limited to the Ganges and the Brahmaputra only). The mean daily water-level data (1950-2000) of the Ganges at Hardinge Bridge (Hbr) and Brahmaputra at Bahadurabad (Bbd) have been analyzed. From the variation of onset and recession of water-level, the river situations are explored and related to the strong (or major) El Niño/La Niña years. The 2nd decade of May (May 11-20) and the 2nd decade of September (September 11-20) are taken as the separation points for onset and recession respectively (Environment and Geographical Information Services (EGIS), 2002) (Fig. 4).

Observation revealed that there is a trend of decreased flows in the Ganges and Brahmaputra during the strong El Niño years; the reverse is true in the case of La Niña years (see Fig. 3). While considering the water-level based onset/recession, observations clearly reveal that the dry years in the Ganges are abundant in the strong El Niño years indicating that water-level is lower than the average (or lower, henceforth) on the date of onset (May 20). The dryness is quite severe in the Ganges, where a shortage of water of about 160 cm is observed on 10 July (this is an important day in the seasonal crop plantation cycle) (Fig. 4: top panel). However, during the strong La Niña years, the average water-level, instead of showing any considerable rise, is found to be very close to other normal years. This is an indication that the Ganges is gradually drying and displaying a trend to even lower water-level during the normal years.

The average monthly water-level in the Brahmaputra also provided correlation with the ENSO cycle indicating El Niño to dry (lower water-level) and La Niña to wet (higher water-level) (Fig. 4: bottom panel). However, unlike the Ganges, the Brahmaputra displays a lower water-level during the El Niño years and considerably higher than average (or higher, henceforth) water-level in the La Niña years.

Further, the interannual variability and trends in the hydro-meteorological characteristic, in terms of water-level (in cm) and duration (in days), are studied and related to the ENSO cycle (Table 2). When compared to the water-level on May 20 and September 20 (separation points for onset and recession), the Ganges displays a trend of *'late onset – late recession'* in the strong El Niño years. A trend of 'early onset – normal recession' is visible

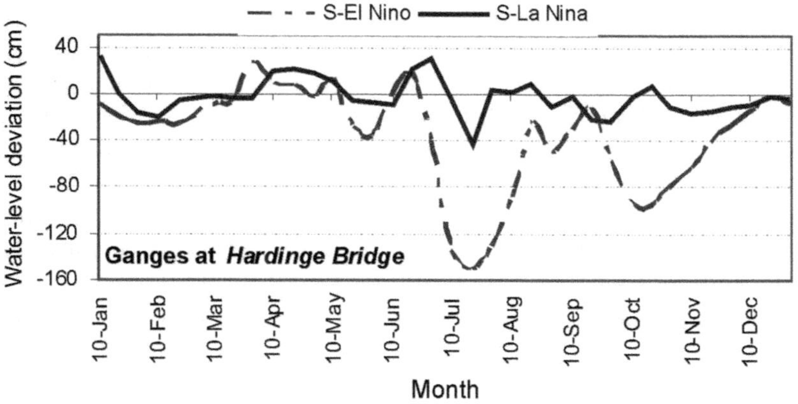

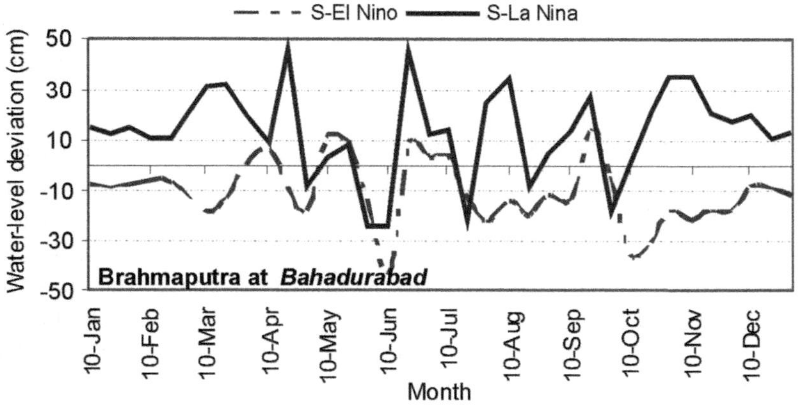

Fig. 4: Deviations of monthly average water-level (cm) during the ENSO years (Top: Ganges at Hardinge Bridge point, and bottom: Brahmaputra at Bahadurabad point) (Note: See Table 1 for El Niño/La Niña years, S stands for strong).

Table 2: ENSO years and deviations of onset and recession (water level)

		Ganges at Hardinge Bridge				*Brahmaputra at Bahadurabad*			
		Onset (May 20)		Recession (Sep 20)		Onset (May 20)		Recession (Sep 20)	
		(+/-) cm	(+/-) days	(+/-) cm	(+/-) days	(+/-) cm	(+/-) days	(+/-) cm	(+/-) days
Strong El Niño years	AV	-22	11	26	3	-25	9	27	0
	SD	27	9	35	9	72	10	62	14
Strong La Niña years	AV	22	-1	0	4	7	-2	29	4
	SD	53	16	90	10	82	11	42	3

Note: AV: Average of five extreme years, SD: Standard deviations, See Table 1 for El Niño/La Niña years.

in the strong La Niña years. Like the Ganges, the Brahmaputra displayed a similar response during the strong El Niño years. However, unlike the Ganges, the Brahmaputra displayed an 'early onset – late recession' during the strong La Niña years. From Table 2, relation between strong El Niño years and onset/recession is interpreted as follows: during the strong El Niño years, the water level (WL) was 22 cm lower than the average WL on May 20, and it took 11 days more to reach the average WL on May 20 (late onset). Similarly, the WL was 26 cm higher than the average WL on Sep 20, and it receded 3 days more to reach the average WL on Sep 20 (late recession).

From the long-term (1950-2000) water-level data record, a trend of delayed onset (1-2 weeks) and a late recession (1-2 weeks) have been observed in the recent decades (1980-2000) (details are not reported here). Added to this, either shortage or excess of water due to climate extremes (El Niño or La Niña) in the recent decades further aggravates this problem. This is very prominent in the Ganges basins. Therefore, an adverse impact during the harvesting period of *Aus* (paddy) and *Jute* has been observed (the harvesting time of *Aus* and *Jute* is during the end of May and beginning of June – and any variation in flooding onset/recession causes serious impact on the cropping seasons in Bangladesh). If the period of onset of flooding is earlier than the normal then the harvesting of *Aus/Jute* is delayed, which causes huge damage to crops. On the other hand, if the onset of shallow flooding is late then *Broadcast Aman* (paddy) cannot thrive well in May/June because of a lack of water. Similarly, the recession cycle in the recent decades are found to be interrupted – mostly delayed. If the flood recession is late, and the duration of flood peak is above the normal, then it floods the highland *Transplanted Aman* (paddy) crops causing severe damage to the vegetation and flowering stage of *Transplanted Aman* crops (Rasid and Paul, 1987).

3.2 ENSO and Seasonal Stream-flow

The links between ENSO and stream-flows in the GBM basin show that strong El Niño years are associated with decreased flows (dry), while both the strong/moderate La Niña years are associated with increased flows (wet). This is because, during the strong El Niño years, the whole basin experiences less rainfall (Figs 2 and 3), and the basin-wide deficiency of rainfall cause rivers in Bangladesh to become drier because of decreased flows. On the other hand, during the La Niña years (strong/moderate), there is a significant increase of rainfall along the greater GBM basins causing increased flows along the whole catchment area, which, in turn, severely floods Bangladesh. It is important to mention here that, in the case of moderate El Niño years, the basin-wide rainfall/runoff picture in downstream GBM basins (Bangladesh) is relatively different from upstream India. While severe scarcity of rainfall exists in India, Bangladesh experiences high rainfall along the Ganges and Brahmaputra basin areas. This exceptionally high and prolonged local rainfall contributes increased flows and flooding in Bangladesh (Chowdhury, 2003). Therefore, it is very difficult to quantify any basin-wide correlation between the strength of El Niño and the variability of Bangladesh climate. For example, from the recorded wettest and driest years for the Ganges and Brahmaputra rivers, it is clear that neither all the wettest years are necessarily La Niña years nor all the driest years are El Niño years. Again, because of wet climate during the moderate El Niño years, the ENSO-rainfall relationship is different in Bangladesh from the one in the Indian monsoon regime. So, understanding the broad-scale features of the GBM monsoon with particular emphasis to the stream-flows in Bangladesh is essential.

ENSO is the largest source of year-to-year climate variability. Therefore, knowing the ENSO condition ahead of time would provide substantial opportunities to provide early warnings of climate extremes in the greater GBM basins. However, because of weak teleconnections to ENSO, the inclusion of studies of atmospheric circulation and variation of SST in the tropical Pacific would be a viable choice to enhance the seasonal stream-flow forecast in Bangladesh (also see Chowdhury, 2003; Chowdhury and Ward, 2004). This will provide information related to mean atmospheric response to different large-scale modes of oceanic variability.

4. SST AND STREAM-FLOW LINKS – PHYSICAL MECHANISM

Basin–wide monsoon rainfall patterns control flood peaks of the Ganges, Brahmaputra, and Meghna rivers (Mirza, 2003). Zahn (2003) described that the monsoon subsystems – the Asian, Indian and African monsoons – are connected to variations in SST in the Pacific, which, in turn, are linked to the El Niño – Southern Oscillation (ENSO). During El Niño years, the SST anomalous behaviour in the Pacific-warming from the long-term average produces a negative ENSO index. As compared to La Niña, the average warming in the SST of the south-central Pacific is much higher during the El Niño years. The SSTs in the Indian Ocean during La Niña are warmer than in El Niño years, which tend to affect the atmospheric circulation pattern in this region (Chowdhury, 2003; also see Chowdhury, 1994).

Composite averages of 4-month SSTs and surface circulation anomalies for the five wettest years, the five driest years, and the five wettest *minus* five driest years are presented in Figs 5 and 6 (SST-based indices are presented in Figs. a, c, d and circulation-based indices is presented in Fig. b). It has been revealed that the development pattern of SST in the ENSO years influences the seasonal flows in the Ganges (Fig. 5). The indices show that a increased flow (at Ganges) is expected when the average SST anomalies in the month of October-November-December-January (ONDJ, lag –2) and February-March-April-May (FMAM, lag –1) (top panel of Figs d, c) are positive in the domain of Niño 3.4 region, South central to Western Pacific Ocean (180E-140E, 50N–30N; and 120E-180E, 0-30E), and in the Indian

Ocean (40E-100E, 20N-10S). A gradual change of SST pattern of the Niño 3.4 region to negative anomalies in the month of June-July-August-September (JJAS, lag 0) (top panel of Fig. a) will further enhance the possibility of increased flows in the Ganges River. The gradual change from positive to negative anomalies in the Niño 3.4 region indicates a clear *biennial signal* – a signal consistent with the maturity of ENSO in its typical life cycle – which tends to influence the flows in the Ganges. The atmospheric circulation indicated that zonal wind – (easterly) is to be associated with increased flows and (westerly) to be associated with decreased flows (Fig. 5: b). The development pattern of this atmospheric component is also a signal of classic ENSO. Therefore, the stream-flow forecasting in the Ganges is possible from the year-to-year climate variability of the ENSO cycle (see Douglas et al., 2001).

From Fig. 6 (d, c), it is clear that the increased flows in the Brahmaputra River are partly influenced by the positive SST anomalies in the domain of Western Pacific (90E-140E, and 0-30N) and Indian Ocean (50E-90E, 0-20N) during the month of ONDJ (lag –2) and FMAM (lag –1). An ENSO-type anomalous SST has been observed in the concurrent season JJAS (lag 0) (Fig. 6a). However, monitoring from the previous ONDJ (lag –2) and FMAM (lag –1), any signal consistent with the maturity of ENSO life cycle is not distinctly visible (Figs 6 d, c). Moreover, the *'biennial signal'* in the Niño 3.4 region is clearly absent here. So, the quantitative correlation of the ENSO cycle and seasonal flows in the Brahmaputra has been found to be relatively weak. Any definitive reason for this behaviour is still not clear; further researches are needed. Note that the circulation-based indices for FMAM and ONDJ are neither reported (in Figs 5 and 6) nor discussed here because of lack of any physically justifiable relation in the context of our understanding of ENSO and stream-flows in the GBM basin system.

Based on ENSO climate cycle and the variation of tropical SST, the outlook for stream-flow forecasting in the Ganges is good. However, this is not the case for the Brahmaputra where the outlook for ENSO-based seasonal prediction is difficult, and the outlook for SST-based prediction is also low. More studies would be essential to explore other potential sources of predictability, like the accessible Bangladesh hydrological indicators and large-scale oceanic /atmospheric patterns as predicted by the General Circulation Models (GCM) (see Chowdhury and Ward, 2005 for more details).

5. CONCLUSION

Significant variations in the timing of onset, peak, and recession of flooding have been observed to be changing dramatically in the lower riparian part of the Ganges. As a result, the Ganges tends to be drying in the south-western part of Bangladesh. A wider range of regional cooperation and water sharing can save this river from further human sufferings.

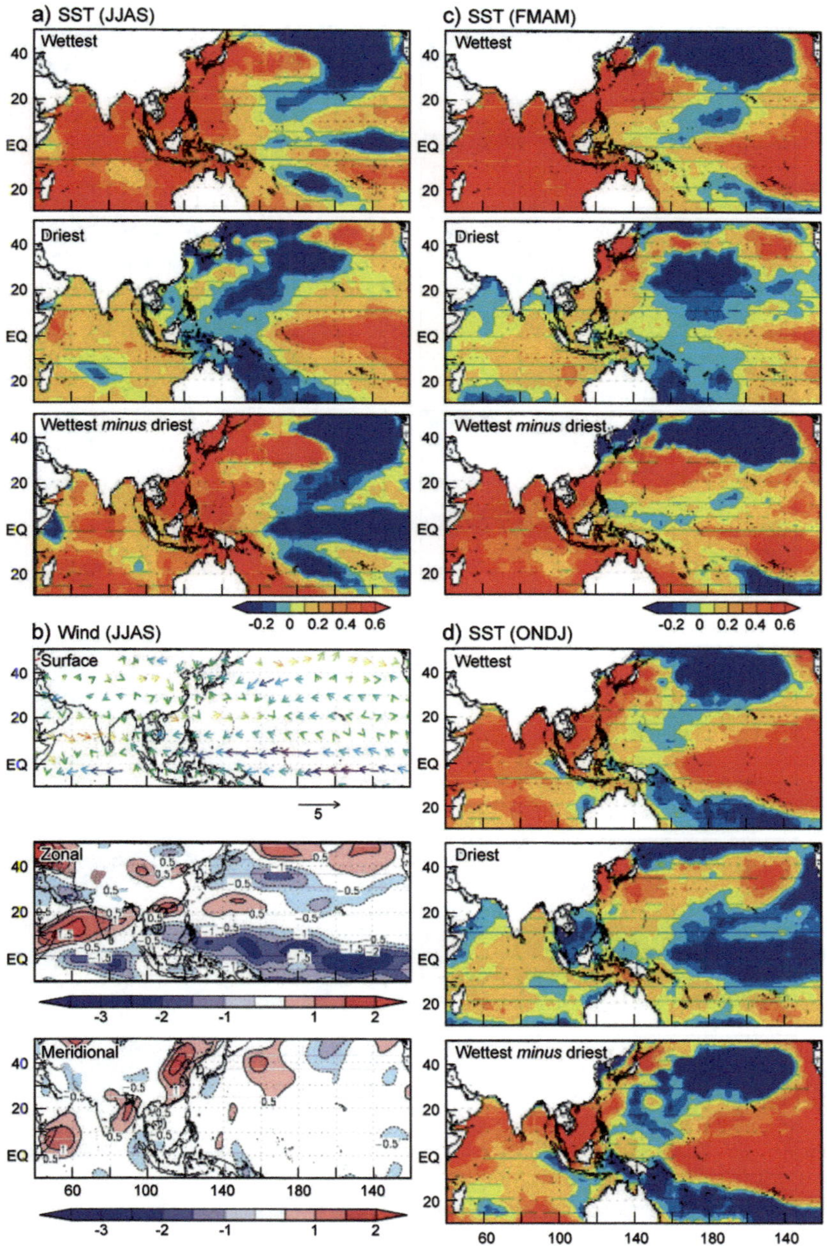

Fig. 5: Ganges River – SST based indices for JJAS, FMAM, and ONDJ are presented (Figs. a, c, and d). Composites of average SST anomalies (°C) for the 5-wettest (W), 5-driest (D), and wettest *minus* driest years are shown in the top, middle, and bottom panels respectively (see Table 1 for wettest and driest years). Circulation based indices for JJAS is presented in Fig. b. Composites of average wind anomalies (m s^{-1}) for the wettest (W) *minus* driest (D) years are shown (surface, zonal, and meridional wind anomalies are shown in top, middle, and bottom panels respectively) [wind anomalies for FMAM and ONDJ are not shown in this figure]. Note that JJAS, FMAM and ONDJ stand for Jun-Jul-Aug-Sep, Feb-Mar Apr-May and Oct-Nov-Dec-Jan respectively.

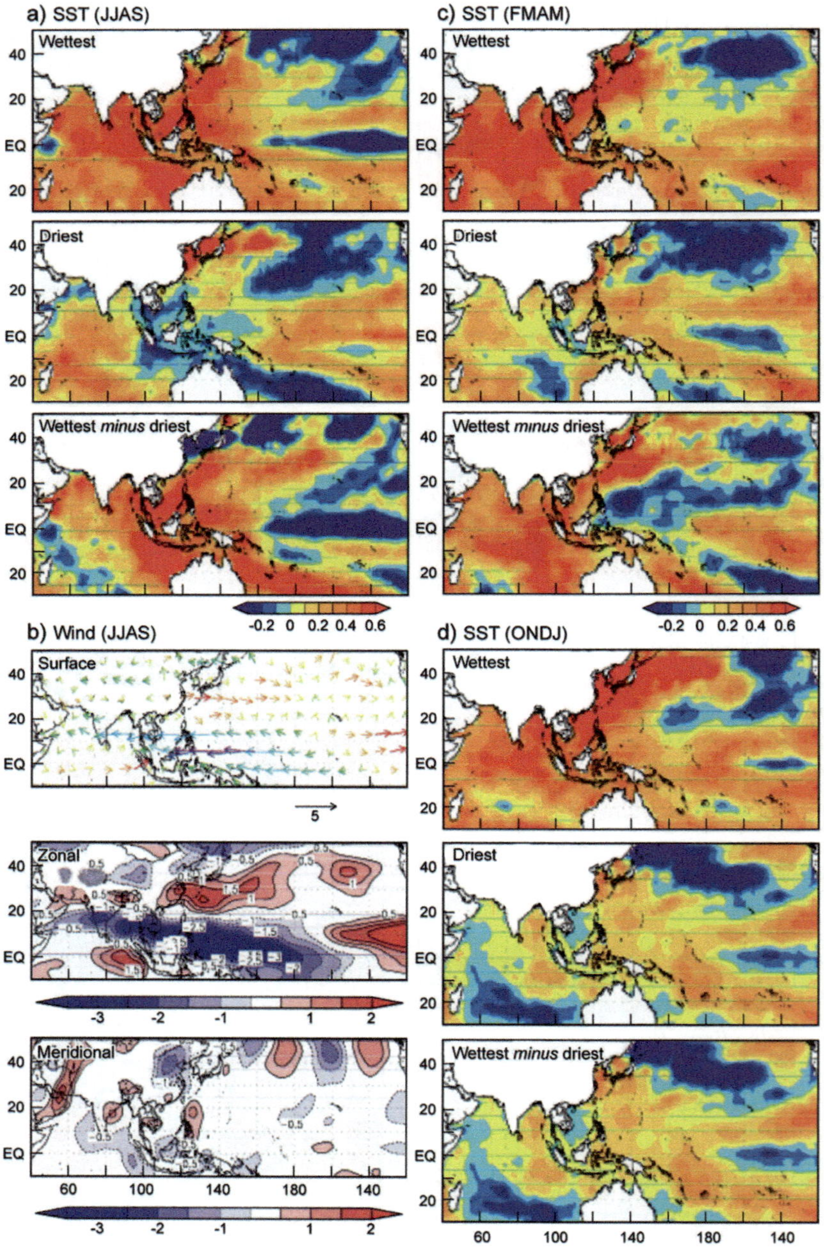

Fig. 6: Brahmaputra River – SST based indices for JJAS, FMAM, and ONDJ are presented (Figs. a, c, and d). Composites of average SST anomalies (°C) for the 5-wettest (W), 5-driest (D), and wettest *minus* driest years are shown in the top, middle, and bottom panels respectively (see Table 1 for wettest and driest years). Circulation based indices for JJAS is presented in Fig. b. Composites of average wind anomalies (m s^{-1}) for the wettest (W) *minus* driest (D) years are shown (surface, zonal, and meridional wind anomalies are shown in top, middle, and bottom panels respectively) [wind anomalies for FMAM and ONDJ are not shown in this figure]. Note that JJAS, FMAM and ONDJ stand for Jun-Jul-Aug-Sep, Feb-Mar Apr-May and Oct-Nov-Dec-Jan respectively.

Climate and ocean-driven factors such as the SST and circulation process are significantly affecting the climate variability in the greater GBM basins. Therefore, in addition to regular exchange of data and knowing the ENSO condition ahead of time, seasonal stream-flow forecasts can be enhanced from the variation of SST in the tropics. However, at present, several countries in the GBM basin do not have the resources to identify the impacts of an unusually warm/cool SST or ENSO episode. These countries, therefore, have to rely on the latest ideas and technology by sharing climate information, and regular exchange of data between the local institutions and those regional and international research institutes would be essential.

Acknowledgements

The Postdoctoral Program at the International Research Institute for Climate Prediction (IRI) (the Earth Institute of Columbia University), administered by the University Corporation for Atmospheric Research (UCAR), initially supported this research. Grateful acknowledgements are, therefore, extended to Drs Reid Basher and Neil Ward for their supervision during the entire period of this research. Special thanks are due to Mr. Ousmane Ndiaye for providing help on GrAds. Acknowledgements are also due to the officials of the Flood Forecasting and Warning Center (FFWC) of Bangladesh for their cooperation; in particular, cooperation from M/s. Salim Bhuyian, Golam Mostafa, and Sudhir Goswami is highly appreciated.

REFERENCES

Brammer, H., 1990. Floods in Bangladesh. I. Geographical Background to the 1997 and 1998 Floods. *The Geographical Journal*, **156(1),** 12-22.

Chowdhury, M.R (2003): *The El Niño - Southern Oscillation (ENSO) and Seasonal Flooding: Bangladesh, Theoretical and Applied Climatology*, 76: 105-124, Springer-Verlag, Vienna.

Chowdhury, M.R. and Ward, M.N., 2004. Hydro-Meteorological Variability in the Greater Ganges-Brahmaputra-Meghna (GBM) Basin. *International Journal of Climatology*, **24,** 1495-1508.

Chowdhury, M.R. and Ward, N., 2007. Seasonal Flooding in Bangladesh – Variability and Predictability. *Hydrological Processes*, **21:3**, 335-347, Wiley Inter Science.

Chowdhury, A.M. (1994). Bangladesh Floods, Cyclones and ENSO, Paper presented at the International Conference on Monsoon Variability and Prediction, International Center for Theoretical Physics (ICTP), Italy, 9-13 May. 1994.

Douglas, W.W., Wasimi, S.A. and Islam, S., 2001. The El Niño Southern Oscillation and Long-Range Forecasting of Flows in the Ganges. *International Journal of Climatology*, **21,** 77-87.

EGIS, 2002. Technical Note on Flood and Erosion Monitoring. EGIS Support Project for Water Sector Planning, 30, M/O Water Resources, Government of Bangladesh, 67 pp.

Haque, C.E., 1997. Hazards in a Fickle Environment: Bangladesh. Kluwer Academic Publishers, Dordrecht, 380 pp.

Hossain, E., Alam, S.S., Imam, K.H. and Hoque, M.M., 2001. Bangladesh country case study: Impacts and response to the 1997-98 El Niño event. *In:* Once Burned Twice Shy? Michael H. Glantz (ed.), 44-50, United Nations University Press.

Hulme, M., 1994. Validataion of large-scale precipitation fields in general circulation models. *In:* Global precipitation and climate change, M. Desbois and F. Desalmand (eds.), Springer-Verlag, 387-406.

Kalnay, E., Kanamitsu, M., Kistler, R., Collins, W., Deaven, D., Gandin, L., Iredell, M., Saha, S., White, G, Woollen, J., Zhu, Y., Leetmaa, A., Reynolds, B., Chelliah, M., Ebisuzaki, W., Higgins, W., Janowiak, J., Mo, K.C., Ropelewski, C., Wang, J., Jenne, Roy, Joseph and Dennis, 1996. The NCEP/NCAR 40-Year Reanalysis Project. *Bulletin of the American Meteorological Society*, **77(3),** 437–472.

McPhaden, M.J., 2001. El Niño and La Niña: Causes and Global Consequences. Encyclopedia of Global Environmental Change, R.E. Munn, Michael C. MacCracken and John Perry (eds.), John Wiley and Sons, Volume 1, 356-370.

Mirza, M.M.Q., 2003. Three Recent Extreme Floods in Bangladesh: A Hydro-Meteorological Analysis. *Natural Hazards*, **28,** 35-64.

Nishat, A. and Faisal, I.M., 2000. An assessment of the Institutional Mechanism for Water Negotiations in the Ganges-Brahmaputra-Meghna System. *International Negotiations*, **5,** 289-310.

Rasid, H. and Paul, B.K., 1987. Flood Problems in Bangladesh: Is there any indigenous solution? *Environmental Management*, **11(2),** 155-173.

Smith, T.M. and Reynolds, R.W., 2002. Extended Reconstruction of Global Sea Surface Temperatures Based on COADS Data (1854-1997), *Journal of Climate*, **7,** 1495-1510.

Zahn, R., 2003. Global change. Monsoon Linkages. *Nature*, **421,** 324-325.

10 Changes in the Coastal and Marine Environments

S.N. de Sousa, Ahsan Uddin Ahmed, M.D. Kumar, T.G. Jagtap, S. Sardessai and A. Hassan

1. BIOGEOCHEMISTRY OF THE NORTH INDIAN OCEAN

The Indian Ocean is bounded in the north by the Asian landmass at much lower latitudes. Presence of this massive landmass in the north deprives the Indian Ocean of the deep convection centres of the northern hemisphere, thus affecting the renewal and circulation of its deep waters. Consequently, the deep circulation in the northern Indian Ocean depends largely on advection of waters from the south.

The surface circulation in the northern Indian Ocean undergoes seasonal reversal in response to the reversal of monsoon winds caused by the differential heating and cooling of the northern Indian Ocean, the northwest Pacific and the Asian landmass (Prasanna Kumar et al., 2004a; Zahn, 2003). At intermediate levels, the two high salinity water masses that are formed in the northwestern region – The Persian Gulf Water, characterized by σ_θ of 26.6, and the Red Sea Water, characterized by σ_θ of 27.2 – enter the Arabian Sea through the Gulf of Oman and the Gulf of Aden, maintaining their density levels at 300 m and 800 m, respectively. Similarly, a high salinity and highly oxygenated water, variously named – 'The Indian Ocean Central Water', 'Subtropical Sub-surface Water' or 'Sub-Antarctic Mode Water' – forms and sinks in the southern ocean at latitudes 30-40 °S (Warren, 1981). As this water from the southern source moves northwards, it mixes with the Banda Sea throughflow water. The mixture is then carried into the Arabian Sea through the northern branch of the South Equatorial Current.

The two northern arms of the Indian Ocean, the Arabian Sea and the Bay of Bengal, though situated at similar latitudes and subjected to similar atmospheric forcing, exhibit widely differing hydrographical and hydrochemical characteristics. While the Arabian Sea is an area of negative water balance due to the large excess of evaporation over precipitation and runoff, the Bay of Bengal, on the other hand, is an area of positive water balance owing to the large excess of precipitation and runoff over evaporation. The contrasting hydrographic features of the two basins create a hydrological imbalance in the region which, coupled with the seasonally reversing surface circulation, plays an important role in exchanging the water masses between them and maintaining their salt balance (Prasanna Kumar et al., 2004a).

1.1 The Arabian Sea

Compared to other oceanic areas, the Arabian Sea is fairly strongly stratified near the surface with a steep temperature gradient within the upper 200 m. Nutrient levels are high just below, or very close to the euphotic layer. These factors make the Arabian Sea a potentially fertile area, since any vertical disturbance that can disrupt or shift the thermocline barrier can lead to injection of nutrients into the euphotic layer and fuel increased biological production (Ryther and Menzel, 1965).

1.1.1 Upwelling

The onset of the SW monsoon gives rise to a vigorous atmospheric and oceanic circulation in the northern Indian Ocean, causing intense upwelling at several locations (Wyrtki, 1973) – off Somali coast, off Oman coast, off southwest and southeast coasts of India and open ocean upwelling in the northwestern Arabian Sea – resulting in nutrient enrichment of surface waters.

A western boundary current, the Somali Current, developing along the Somali coast, causes Ekman transport of surface waters resulting in upwelling along the coast. This brings to the sea surface cold (17-20 °C) and nutrient-rich (nitrate: 13.1-19.9 μM, phosphate: 0.92-1.53 μM and silicate: 10.8-15.7 μM) deep waters (Smith and Codispoti, 1980). Waters with similar properties are normally found at 100-200 m depths in the equatorial region of the Indian Ocean (Smith and Codispoti, 1980; Swallow, 1984; Morrison et al., 1998). Similarly, strong southwesterly winds blowing parallel to the southeast Arabian coast cause Ekman transport of surface waters offshore giving rise to coastal upwelling (Elliot and Savidge, 1990; Savidge et al., 1990) bringing cold and nutrient-rich deep waters to the surface. A temperature difference of at least 5 °C between the upwelled water and the ambient water was measured by Elliot and Savidge (1990), while the study of Banse (1987) showed nitrate concentrations in the upwelled water, varying between 0.1 and 8.0 μM. Similarly, a shore-base study along the Dofar coast, southern Oman, by Savidge et al. (1990) showed nitrate concentration in excess of 20 μM.

Coastal upwelling associated with the SW-monsoon has also been reported along the west coast of India (de Sousa et al., 1996a&b; Naqvi et al., 2000). Water from depths >50 m having higher salinity, higher nutrient concentrations and lower dissolved oxygen content appeared to rise to about 5 m towards the coast. The upwelled water, however, did not reach the surface as a 5 m thick freshwater layer capped the water column, thus suppressing the effects of upwelling (de Sousa et al., 1996a&b). At some locations, however, the upwelled water did reach the surface, thus giving it a patchy distribution of properties.

While coastal upwelling along the coasts of Somalia, Oman and SW-coast of India lead to enrichment of surface water with nutrients, making them highly productive, advection of these upwelled waters offshore in the form of filaments help to spread these nutrient-rich waters to the offshore regions (Prasanna Kumar et al., 2001). This offshore advection, however, is limited to about 100-200 km range. It was thus believed that the open ocean waters of the Arabian Sea were, in general, oligotrophic (nutrient depleted). Measurements made by de Sousa et al. (1996b), on a meridional section along 64 °E, indicated presence of high concentrations of nutrients (nitrate >4 μM) in surface waters north of 15 °N. These waters were found to be relatively colder by 0.5 °C and fresher by 0.1 psu (Prasanna Kumar et al., 2001). Presence of such waters at surface in the open ocean was attributed to upwelling caused by the wind regime of Findlater jet (Muraleedharan and Prasanna Kumar, 1996; Bauer et al., 1991; Burkill et al., 1993).

1.1.2 Winter Convection

Bauer et al. (1991) while studying the oceanic mixed layer response to large-scale wind forcing in the Central Arabian Sea, observed that, during the NE-monsoon, the mixed layer depth, as derived from the σ_t profiles, showed deepening from 50 m at 12 °N to 70-80 m in the region north of 14 °N, indicating convergence associated with weak Ekman forcing. They presented a simple two-dimensional model based on which they concluded that the mixed layer depth is influenced by Ekman layer dynamics, i.e., cooling of sea surface coupled with weak Ekman downwelling results in deepening of the mixed layer to the top of the permanent thermocline. Such deepening of the mixed layer in the Central Arabian Sea, associated with NE-monsoon, had also been observed by Morrison et al. (1998). The study of Prasanna Kumar and Prasad (1996) indicated that the observed deepening of the mixed layer was caused by the formation of the Arabian Sea High Salinity Water (ASHSW) at the surface due to the lowering of SST, which, in turn is forced by a combination of enhanced evaporation in the presence of cool dry air and reduced solar radiation, and its subsequent sinking. This sinking of surface water gives rise to convective mixing (Morrison et al., 1998) injecting nutrients from the thermocline region into the surface layer, as observed by de Sousa et al. (1996b).

1.1.3 Biological Response

It was seen in sections 1.1.1-1.1.2 that processes of upwelling (both coastal and open ocean) during summer monsoon, and convective mixing during winter monsoon bring nutrients from the deeper layer to the surface euphotic layer in the eastern, western and northern Arabian Sea, while horizontal advection of these waters makes the nutrients available in the offshore regions (Fig. 1). Quite often, the amount of nutrients supply is so large that it is not fully utilized by primary producers during the same season, the 'residual nutrients' being utilized during the subsequent season resulting in high productivity rates in October-November, long after the end of the monsoon (Olson et al., 1993).

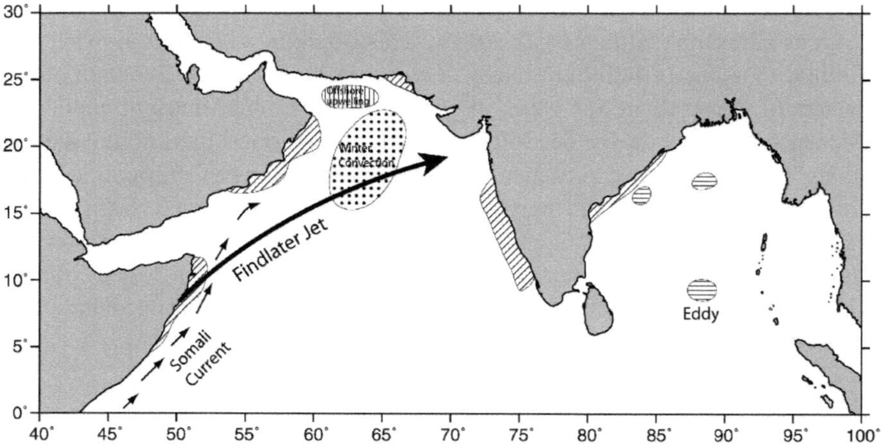

Fig. 1: Locations and processes injecting subsurface nutrients into the euphotic layer in the Northern Indian Ocean.

Presence of high concentrations of nutrients in the euphotic layer drives high rates of primary and new production, thus making the Arabian Sea one of the most productive regions of the world's oceans (Burkill et al., 1993). High biological production with a maximum of 5.7-6.4 $gC.m^{-2}.d^{-1}$ (mean of 1.8 $gC.m^{-2}.d^{-1}$), representing bloom conditions, was measured in the northern and western Arabian Sea during the IIOE (Ryther and Menzel, 1965; Ryther et al., 1966). Similar concentrations of primary productivity were measured by Smith and Codispoti (1980) in the upwelling waters of Somalia, while the study of Curie et al. (1973) showed high concentrations of chlorophyll-*a* and zooplankton biomass in the upwelling waters off the coast of southeast Arabia, associated with high concentrations of phosphate (~2.0 μM). The synthesis of Brock et al. (1991) showed peak pigment (chlorophyll-*a* and pheophytin-*a*) concentrations during August-September in the northwestern Arabian Sea, extending from Oman coast to about 65 °E. Similarly, the recent studies of Smith et al. (1998) showed high concentrations (1.5 $gC.m^{-2}.d^{-1}$) of primary production in the northwestern Arabian Sea, during the SW-monsoon of 1995, while the production during the NE-monsoon of 1996 was still higher (1-2 $gC.m^{-2}.d^{-1}$). Measurements made by the Indian scientists (Prasanna Kumar et al., 2001) around the same time indicated equally high concentrations (0.77 $gC.m^{-2}.d^{-1}$) of primary productivity in the Central Arabian Sea (64 °E).

Biological production shows high variability (latitudinal, seasonal and inter-annual) in the Arabian Sea. For example, Bauer et al. (1991) had reported productivity rates in the range 0.2-0.7 $mgC.m^{-3}.h^{-1}$ between the equator and 10 °N increasing northwards to a maximum of 3 $mgC.m^{-3}.h^{-1}$, while the simulated model of Brock et al. (1994) showed high seasonal variability in the mixed layer PP with major peak during the summer monsoon and a minor peak during the winter. Similarly, the productivity shows inter-annual variability associated with the variability in the intensity of coastal upwelling (Brock and McClain, 1992).

Early studies of Menon and George (1977) on the distribution and abundance of zooplankton along the SW-coast of India showed peak production during the period July-August, while it was the lowest during January-February. Similarly, the latest measurements (SW-monsoon of 1995) made by Indian scientists (Bhattathiri et al., 1996; Prasanna Kumar et al., 2001) showed peak concentrations of primary productivity in the range 440-1760 $mgC.m^{-2}.d^{-1}$ while the pigment concentration was 60 $mg.L^{-1}$.

Prasanna Kumar et al. (2004b) produced strong evidence to show the role of cold core eddies in driving high biological production, through nutrient enrichment, in the Bay of Bengal, based on the observations carried out during the SW-monsoon of 2001. Smith et al. (1998) suggested that a similar mechanism might be operating in the Arabian Sea; however, the study of Prasanna Kumar et al. (2001) has ruled out such a possibility.

1.1.4 Development of the Oxygen Minimum Layer (OML)

High biological production in surface waters sustained by nutrient enrichment through upwelling, convective mixing and advection, lead to the export of large quantities of organic matter to the sub-surface layers where it undergoes oxidation accompanied by consumption of dissolved oxygen. And, unless this loss of oxygen is quickly replenished, its continued utilization can lead to its severe depletion, affecting the transformations of several elements.

There are three main natural pathways through which the renewal can take place – advection, mixing and diffusion (both vertical and lateral), the last one being a very slow process. The balance between the consumption of oxygen due to oxidation of organic matter and its renewal through external sources determines its overall distribution in the natural water bodies, at any time. If the demand for oxygen is greater than the supply from these sources, then its concentration falls to suboxic levels (<0.1 $ml.L^{-1}$). It is at this stage that the process of denitrification starts taking place wherein the bacteria utilize the energy from nitrate to oxidize the organic matter, and in the process the nitrate is reduced first to nitrite (secondary nitrite) followed by the production of nitrous oxide and elemental nitrogen. When the entire nitrate is consumed, the environment becomes anoxic and further oxidation of organic matter proceeds through bacterial reduction of sulphate resulting in the production of hydrogen sulphide gas.

Figure 2 shows the vertical distribution of dissolved oxygen in the Arabian Sea along a N-S section between 58 °E and 64 °E longitudes and 2-24 °N latitude. A thick oxygen minimum layer (OML) with suboxic concentrations (<0.1 ml.L^{-1}) is seen in the depth range 150-1100 m, north of latitude 14 °N. This is one of the three extensive regions in the world oceans bearing oxygen minimum layers with concentration <0.1 ml.L^{-1}, the other two being the Bay of Bengal and the eastern tropical Pacific. However, the northern Arabian Sea OML is the most extensive (both depth-wise and area-wise) and most intense where the oxygen concentrations could reach very low levels of 0.02 ml.L^{-1} (Naqvi, 1987). As mentioned above, in extremely low oxygen environment (suboxic conditions), denitrification is the main respiratory process wherein nitrate is reduced with the production of secondary nitrite, and nitrous oxide and nitrogen gases. Large areas in the northern Arabian Sea come under intense denitrification zone, but there is no production of

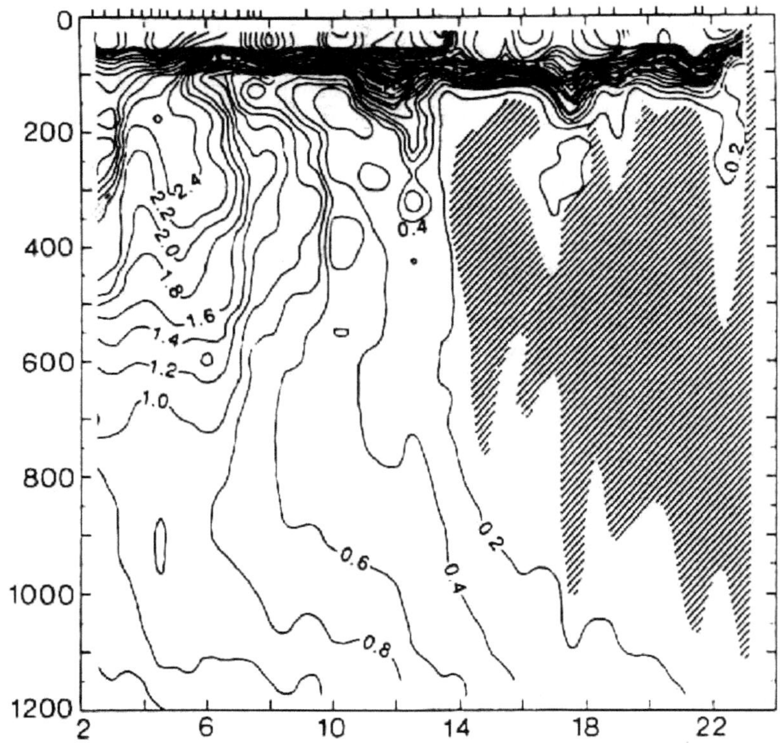

Fig. 2: Distribution of dissolved oxygen (ml.L^{-1}) along a N-S section through the Arabian Sea along 58-64 °E longitudes between latitudes 2-24 °N (December 1986). Vertical axis is pressure (db) and horizontal axis is latitude (°N); the shaded portion depicts the low-oxygen region where the oxygen concentration is less than 0.1 ml.L^{-1} (4 mmol.kg^{-1}). After Olson et al. (1993). Reproduced with permission from Elsevier.

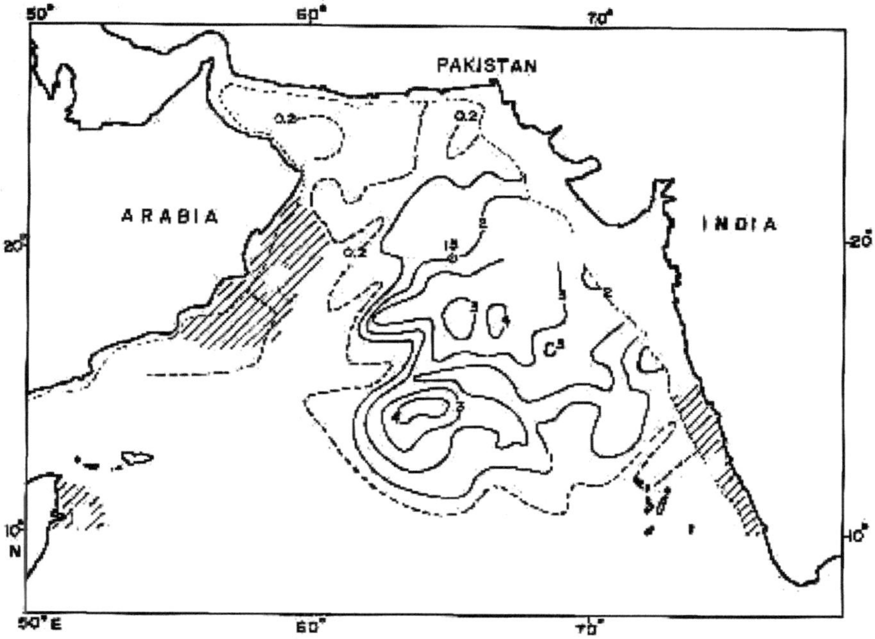

Fig. 3: The geographical extent of denitrification in the Arabian Sea, represented by secondary nitrite (NO_2^-) maximum (µM) which is an indicator of suboxic layer (O_2 <0.1 $ml.L^{-1}$). The hatched areas represent the upwelling zones. Vertical axis is latitude (°N), horizontal axis is longitude (°E). From Naqvi (1991), reproduced with permission from the author.

hydrogen sulphides. Figure 3 shows the geographical extent of the secondary nitrite maximum in the Arabian Sea, which can be identified with the denitrification zone representing the suboxic layer. As seen in the above figure, the secondary nitrite maximum layer with concentrations >2.0 µM extends from about 12 °N to about 22 °N hugging the Indian continental shelf.

There has been a considerable debate regarding the formation and maintenance of the OML in the Arabian Sea. Vertical mixing in the Arabian Sea being limited due to the presence of a strong tropical thermocline, the other hypotheses being debated were:

(a) Stagnation due to the sluggish water circulation in the Arabian Sea,
(b) High local respiration rates driven by the enhanced biological production in the euphotic layer, and
(c) Low oxygen concentration of the source waters.

It was earlier believed that the formation and maintenance of the Arabian Sea OML was due to stagnation caused by the sluggish nature of water circulation at intermediate levels. However, recent studies (Olson et al.,

1993; Naqvi and Shailaja, 1993) have shown that the circulation of intermediate waters in the Arabian Sea is not as slow as it was thought to be, and that the Arabian Sea intermediate waters may be renewed in time scales of 1-10 years. Regarding the *in-situ* respiration rates, Warren (1994) argues that, although the mean annual surface production in the Arabian Sea is high, only about 10% of that organic matter may escape the euphotic zone and be used for respiration in the oxygen minimum layer, thus leading to only a moderate oxygen demand in the subsurface water. This would imply that 'lavish' consumption of oxygen is not the principal cause for the oxygen depletion in the Arabian Sea. Supply by advection depends on the strength of the source waters. Two marginal seas (Red Sea and the Persian Gulf) supply water to the intermediate depths of the Arabian Sea; however, both these sources are weak in water supply, while their waters are highly saline (39.7 and 39.0 salinity, respectively) and contain moderately low oxygen of ~3.0 $ml.L^{-1}$ (Swallow 1984). Olson et al. (1993) suggested that the warm and saline waters from these marginal seas are only capable of carrying about 73% and 85%, respectively, of oxygen that waters of their density carry from the southern hemisphere. Swallow (1984) examined the various sources of water at intermediate depths in the Arabian Sea and their relative contributions to the oxygen pool in the oxygen minimum layer. Based on the earlier reports on the outflows from these marginal seas, and the salinity distribution pattern observed at intermediate levels in the Arabian Sea, he inferred that the waters from these marginal seas undergo rapid dilution on entering the Arabian Sea, thereby reducing their salinities to 35.9 and 35.4, respectively. To achieve such dilution, he suggested an alternate source of water with relatively lower salinity and lower oxygen content. Further, he suggested that the volume transport of water of this source must greatly exceed that coming from the northwestern sources.

A high salinity and highly oxygenated water, variously named – 'The Indian Ocean Central Water' or 'Subtropical Subsurface Water' or 'Sub-Antarctic Mode Water' – forms and sinks in the southern Indian Ocean at latitudes 30-40 °S (Warren, 1981). Warren (1994) suggests that this is the southern source supplying water to the Arabian Sea at intermediate depths and, although this water is highly oxygenated at the place of its formation, by the time it reaches the northern Arabian Sea it has already lost part of its oxygen in respiration due to its long travel time (~30 years). Swallow (1984) suggests that before entering the Arabian Sea, the Subtropical Subsurface Water mixes with the Banda Sea throughflow water having lower oxygen content (2-2.5 $ml.L^{-1}$) and lower salinity of 34.0-34.5 (Warren, 1981; Fieux et al., 1996). The mixture is then carried to the Arabian Sea through the northern branch of the South Equatorial Current, thus supplying the area with water, which is already depleted in oxygen.

From the above discussion it can be concluded that the development and maintenance of the Arabian Sea OML may not be due to a single process but rather due to the combined effect of supply through source waters with moderately low oxygen content aided by the moderately high respiration rates and moderately slow rate of renewal of the intermediate waters.

Morrison et al. (1999) have studied the effects of the OML on the biology of the area. Several distribution patterns were observed, like exclusion from the OML of most zooplankton biomass and many zooplankton and nekton species and groups; high abundance of a few species of vertical migrators during day-time within the OML; organism specific distribution boundaries at the upper and lower edges of OML, etc. Apart from these biological effects, marine denitrification affects the atmospheric greenhouse gas concentration directly through production of N_2O and indirectly through modification of the marine nitrogen inventory (Altabet et al., 2002).

1.1.5 Reducing Conditions along the Indian Coast

It was seen in Section 1.1.1 that coastal upwelling of moderate intensity occurring along the west coast of India, during the summer monsoon, brings to subsurface levels near the coast, cold, saline and nutrient-rich waters from deeper levels. The upwelled water, however, cannot in general, reach the surface as a 5-10 m thick cap of warm and relatively fresher water derived from the land runoff and local precipitation covers the surface (de Sousa et al., 1996a&b; Naqvi et al., 2000). High concentrations of nutrients in the euphotic layer, derived from upwelling as well as from land runoff, fuel high biological production in the area (Menon and George, 1977; Prasanna Kumar et al., 2001), while the presence of freshwater cap at the surface causes strong stratification leading to poor ventilation of sub-pycnocline water (Naqvi et al., 2000).

It was seen in Section 1.1.4 above (Fig. 3) that the oxygen minimum layer in the Arabian Sea, as identified with the secondary nitrite maximum, is nearly hugging the continental shelf along the west coast of India. As this water is pushed up over the shelf towards the coast, during the southwest monsoon upwelling, its oxygen content, which is already low (~0.5 $ml.L^{-1}$), is further reduced due to consumption fuelled by the high biological production and aided by the presence of the freshwater cap which inhibits vertical mixing, thus leading to reducing conditions.

In a series of measurements made during September-October 1999, along the western continental shelf of India, Naqvi et al. (2000) had observed severe hypoxic conditions prevailing over the entire shelf. They had further reported that the characteristics of the sub-pycnocline water changed progressively from hypoxic to suboxic to anoxic as the upwelled water moved up from the shelf break to the inner shelf, with the reduction product changing from secondary nitrite to hydrogen sulphide (Fig. 4). High concentrations of H_2S (~19 μM) were measured over the inner shelf north of 12 °N in areas where the oxygen concentration fell to 'zero' level (anoxic), whereas in waters where the oxygen concentration was above 'zero' ($0 < O_2$

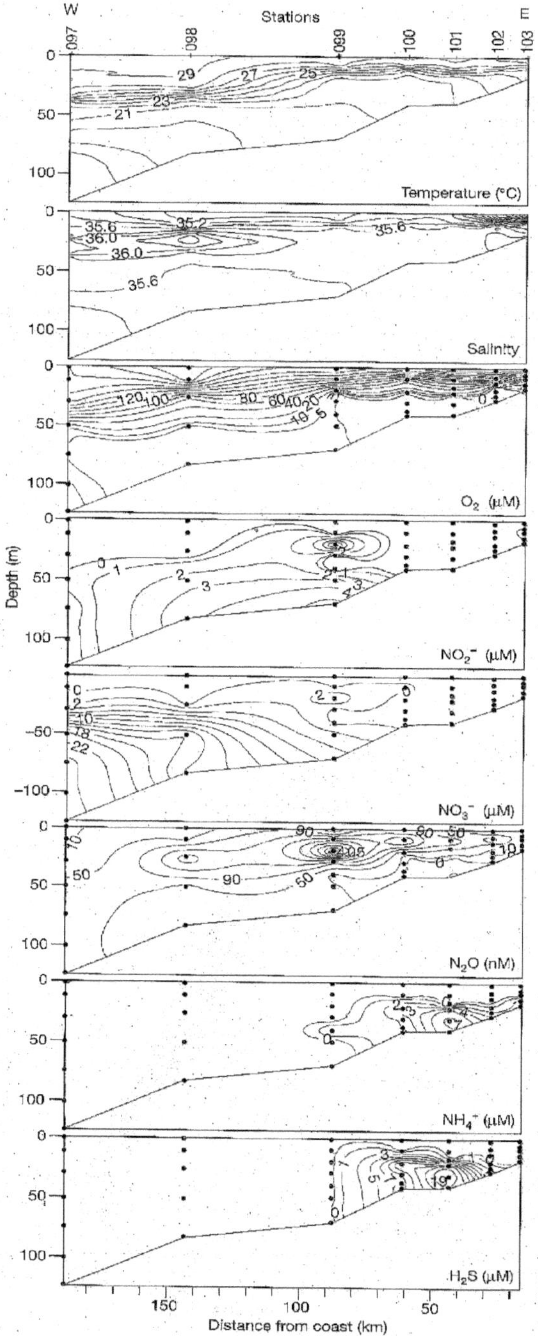

Fig. 4: Distribution of properties along the cross-shelf transect off Bombay during October 1999. Vertical axis is depth (m); horizontal axis is distance from the coast (km). From Naqvi et al. (2000), reproduced with permission from *Nature* (http:www.nature.com/) and author.

< 4.4 μM), i.e., denitrifying conditions, high concentrations of N_2O gas, with a maximum of 533 nM, were observed (Fig. 4). These high values of N_2O were also associated with high concentrations of secondary nitrite (NO_2^-). Comparing their data with the historical data collected under the UNDP/FAO (1973) project where neither the presence of H_2S nor 'zero' concentrations of oxygen had been reported, the authors suggest that the production of H_2S gas over the western Indian continental shelf is a recent phenomenon implying that the anoxia has intensified in recent years. However, the authors did not establish whether this intensification is brought about by an increase in nutrient loading or it is a natural process modified by physical forcing. From the coexistence of both NO_2^- and N_2O in some water samples, the authors suggest that denitrification is the major process of N_2O formation and accumulation in these waters.

Apart from the effects on cycling of gases, the development of the reducing conditions in the coastal waters is expected to have biological effects similar to those observed in the open ocean (Morrison et al., 1999). Naqvi et al. (1998) had reported severe mortality of threadfin bream (*Nemipterus Japonicus*), a demersal fish, associated with intense blooms of 'Noctiluca' dinoflagellate in the coastal waters off the west coast of India, during the southwest monsoon of 1998. The authors suggest that the death is caused due to severe oxygen depletion aided by choking of gills by the thick plankton biomass.

1.1.6 The Greenhouse and Radiative Gases

The issue of global warming due to increase in green house gases like carbon dioxide, nitrous oxide and methane has attracted the attention of ocean researchers to evaluate the contribution of oceans to these gases. The contribution of the Arabian Sea to the carbon dioxide pool of the atmosphere during different seasons, and the seasonal controls on the levels of pCO_2 in different sectors of the Arabian sea has been extensively studied by Sarma et al. (1998 & 2000), Sarma (2003) and George et al. (1994). Figure 5 shows a depth-wise distribution of pCO_2 on a N-S section along 64 °E in the Arabian Sea during different seasons. While the Arabian Sea acts as a perennial source of carbon dioxide contributing 80 $TgC.y^{-1}$ to the atmosphere (Sarma et al., 1998), the surface pCO_2 shows significant seasonal and spatial variability due to physical forcing like upwelling, water mass transport and winter convective mixing, and the associated biological processes. Levels of pCO_2 higher than the atmosphere have been reported in the open Arabian Sea throughout the year, whereas highest pCO_2 (>700 μatm.) has been encountered during the SW-monsoon along the west coast of India, which is attributed to the strong upwelling. Evaluation of surface water pCO_2 saturation in the Arabian Sea has been done by several researchers (Bange et al., 1998; Patra et al., 1998; Upstill-Goddard et al., 1999). The saturation levels during the southwest monsoon, reported by different studies range from 170% ± 55 % to 200% ± 74%.

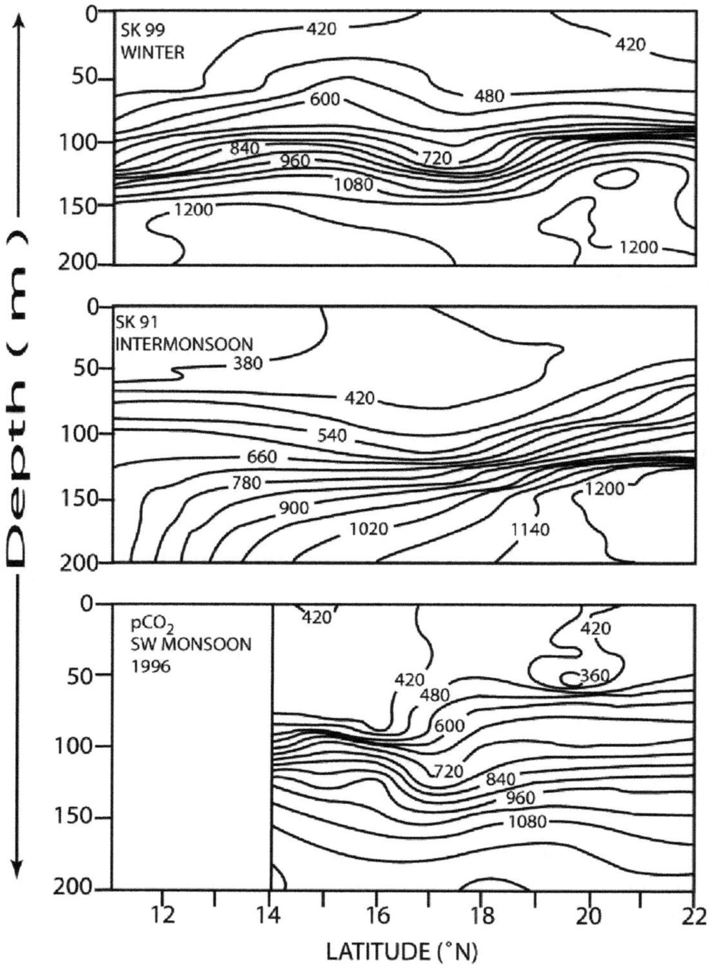

Fig. 5: Distribution of pCO_2 (μatm) along 64 °E in the Arabian Sea, during winter, intermonsoon and southwest monsoon. Reproduced from Sarma et al. (1998) with permission from the author.

Methane, another green house gas, is also produced in significant quantities in the Arabian Sea. It can be both produced and consumed in the oceans through microbial activity. The microorganisms responsible for its production (methanogens) are anaerobes active only in reducing environments. High spatial and temporal variability in surface saturation (110-2521%) of methane has been reported by Jayakumar et al. (2001) the maximum being observed during the southwest monsoon, resulting from coastal upwelling and fresh water runoff. For example, in the coastal upwelling waters off Oman the methane concentrations showed 156% saturation, with the upwelling filament areas showing 147% saturation (Bange et al., 1998). Similarly, measurements made by Jayakumar et al. (2001) have indicated presence of high concentrations 2.6-20.3 nM (140-1091% saturation) of this gas in the coastal surface waters off the west coast of India, with the concentrations showing a decreasing trend in the offshore direction. The above authors examined the possible pathways of formation and supply of this gas to the coastal waters – supply from underlying sediments, *in situ* formation, and inputs from the coastal wetlands.

Based on the studies of Karisiddaiah and Veerayya (1994, 1996) which suggested a diffusion flux rate of CH_4 from sediments to overlying waters in the inner continental shelf of western India as 0.039 $Tg.y^{-1}$ which is close to its net atmospheric flux rate from the Arabian Sea (Owens et al., 1991), Jayakumar et al. (2001) have concluded that the sediments cannot be considered to be a permanent source of CH_4 to the overlying water. The authors argue that the high primary production fuelled by coastal upwelling during the southwest monsoon would support rich zooplankton biomass in this area and provide conducive conditions for *in situ* CH_4 production in water, which is further aided by the presence of reducing conditions in subsurface waters. Coastal wetlands could also play a significant role in supplying CH_4 to the coastal waters. To investigate this aspect Jayakumar et al. (2001) measured the concentrations of CH_4 within an estuary (Mandovi) along this coast. Concentrations as high as 248.1 nM were measured at the freshwater-end of the estuary, decreasing to 57.9 ± 12.7 nM at the mouth, thus suggesting an additional riverine input. Wind speed dependent flux estimation reveals the coastal region of the Arabian Sea to be a stronger source of methane compared to the open ocean (Patra et al., 1998).

In the open Central Arabia Sea, a deeper methane maximum was observed. Whereas biological processes appear to control the production of methane in oxygenated waters, the maximum in deep anoxic waters is not related to any single chemical, biological or physical variable, but rather it suggests to be maintained by quasi-horizontal support.

Greater attention has been drawn in present days to the presence of nitrous oxide in the Arabian Sea due to its special biogeochemical significance in the processes like nitrification and denitrification, anoxia and hypoxia, and high biological production. Nitrous oxide is an intermediate product formed during nitrification as well as denitrification. It shows high degree of surface saturation in the Arabian Sea with large atmospheric fluxes. Patra et al. (2004) have indicated that severe weather conditions, like cyclones, contribute to high emissions of nitrous oxide from the Arabian Sea. Maxima in nitrous oxide have been reported at the periphery of the secondary nitrite maximum, with minima lying within this layer (Naqvi and Noronha, 1991). Surface water eutrophication and bottom water hypoxia in the coastal waters contribute to a large extent in the production and emission of N_2O.

In addition to producing greenhouse gases like CO_2, CH_4 and N_2O, the coastal waters of the eastern Arabian Sea act as a source of DMS (dimethyl sulphide) to the atmosphere. Concentrations of DMS as high as 525 nM were measured in the coastal waters off Goa during the hypoxic conditions in the later part and after the southwest monsoon (Shenoy, 2002).

1.1.7 Particle Fluxes

The highly productive areas in the northwestern region give rise to annual particle fluxes of over 50 $g.m^{-2}\ y^{-1}$ in the Western Arabian Sea as compared to the oxygen depleted Central Arabian Sea region where the annual particle fluxes of <27g $m^{-2}\ y^{-1}$ have been recorded through sediment traps deployed in these regions (Ramaswamy and Nair, 1994). The above authors had observed that biogenic material is dominant at sites located near the upwelling centres and is less degraded during peak flux periods. The interannual variability in the organic carbon flux in the Northern Indian Ocean is related to the strength and intensity of the southwest monsoon.

Water column ventilation probably exerts great influence on the abundance and type of organic matter preserved, and also on the redox state of many trace and minor elements, in the sediment. However, it is debatable whether it is the oxygen depletion or high primary production that controls the variability of organic carbon in sediments. While Demaison and Moore (1980) emphasized the anoxic character of bottom waters as the dominant factor of organic enrichment in sediments, others (Calvert, 1987; Pedersen and Calvert, 1990; Calvert et al., 1991) suggested that high primary productivity exerts first order control on accumulation of organic rich faecies in cretaceous, quarternary and modern oceans. Paropkari et al. (1992) have addressed this debate through the Arabian Sea properties, which exhibit high productivity sites as well as oxygen minimum layer at intermediate depths. Through their observations of organic rich depositional environments in the highly productive zones, as well as anoxic bottom waters, they have indicated that productivity is not the ultimate control for the organic enrichment of bottom sediments but the bottom water anoxia along with the various depositional parameters, like texture of sediments, rate of fine terrigenous sediments deposition and other geological features determine the degree of preservation.

1.2 The Bay of Bengal

Compared to the Arabian Sea, the Bay of Bengal is a less studied basin. As mentioned above, it is a region of positive water balance. The northern part of the Bay receives fresh water of about $1.625 \times 10^{12}\ m^3 y^{-1}$ (Subramanian, 1993) from the various large rivers draining into the Bay. This results in the formation of a fresh water lens about 20 to 30 m thick at the surface, during the summer monsoon. Physical processes, like seasonal reversal of surface currents under the influence of monsoonal wind reversal (Shetye et al., 1991), cold core eddies formation (Babu et al., 1991), coastal upwelling (de Sousa et al., 1981), and strongly stratified and turbid surface waters in the north caused by the immense discharge of fresh water laden with suspended load from numerous large rivers, and relatively weaker winds are the other dominant monsoonal features of the Bay of Bengal. The turbid conditions of surface water in association with the cloudy climatic conditions, especially during the monsoon, hinder the penetration of light to the subsurface layer.

1.2.1 Nutrient Supply

The formation of cold core eddies in the central and western Bay along with the coastal upwelling occurring along the southeast coast of India during the SW-monsoon help to inject nutrient-rich intermediate waters into the euphotic layer. Besides the above processes, the massive river system, together with the land runoff, draining the fertile agricultural lands of the bordering countries, is expected to discharge large quantities of nutrients into the Bay. However, Sen Gupta and Naqvi (1984) have observed that there is no significant addition to the inorganic nitrogen pool of the Bay, through the freshwater addition. Atmospheric precipitation is also expected to make a significant contribution to its nutrient pool in the surface water. On the other hand, the strong stratification caused by the presence of the freshwater cap at the surface prevents vertical mixing, thus depriving the surface water of the Bay from the supply of nutrient through this process.

1.2.2 Biological Production

Biological production shows spatial variability in the Bay. Prasanna Kumar et al. (2002) had reported integrated primary productivity of 89.4 $mgC.m^{-2}.d^{-1}$ at 20° N as compared to 220.7 $mgC.m^{-2}.d^{-1}$ at 9 °N, though similar amounts of nutrients were available in the euphotic zone in both the regions. The decrease towards the northern region may be the effect of light inhibition caused by the high turbidity. The suspended load, in the euphotic zone, ranged from 0.8 to 17.6 $mg.L^{-1}$ in the northern region (18-20 °N) whereas in the southern region (9°N) it varied from 0.2 to 1.1 $mg.L^{-1}$ (Bhosle, unpublished data). An examination of primary productivity against nitrate availability and suspended load indicates that though significant amounts of nutrients are available in the upper 20 m in the north associated with relatively higher chl-*a* content (0.28 $mg.m^{-3}$) in the upper 10 m, high suspended load appears to be the dominant factor in reducing the primary production with inhibitive effects on light penetration. Gomes et al. (2000), however, attribute the decrease in primary productivity to cloud cover.

1.2.3 The Chemical Response

The distribution of oxygen in the Bay follows a pattern similar to that of the Arabian Sea, however, with a less intense OML with concentrations <10.0 μM (<0.22 $ml.L^{-1}$) extending from 70 m to 700 m, mostly confined to the north of 11 °N. However, unlike the Arabian Sea, there are no reports indicating occurrence of denitrification in the Bay. Suppression of the regenerative processes through organic carbon sequestration, as explained below, could be one of the reasons for the oxygen not falling to sub-oxic levels. Towards the south the intermediate waters show relatively higher oxygen levels (15-35 μM) perhaps due to the penetration of a tongue of high salinity water into the central Bay. Sastry et al. (1985) and Murty et al. (1992) identified this water mass as the Arabian Sea high salinity water flowing into the Bay during the southwest monsoon.

1.2.4 Particle Fluxes

Although, the overall biological productivity in the Bay is lower as compared to that in the Arabian Sea, it is reported that the annual rates of particle fluxes are of similar magnitude in both the basins (Ittekkot et al., 1991). This indicates that the biogeochemical cycling in the Bay of Bengal is, perhaps, influenced by the addition of mineral particles of terrestrial origin. This occurs through suppression of regenerative processes by which the organic carbon binds itself to the mineral particles which then sink rapidly through the oxygen minimum layer (Ittekkot et al., 1991), thus sequestering the organic carbon to the sediment, and at the same time preventing the oxidation of the organic matter in the water column and effectively reducing the demand for oxygen. Thus this process could explain the relatively higher rates of particle fluxes as well as the lack of occurrence of denitrification in the Bay.

1.2.5 The Gases

The pCO_2 distribution in the Bay is seasonally variable and appears to be controlled by the prevailing surface circulation. The surface waters are reported to have low pCO_2 levels within the low salinity region with a large area in the Northwestern Bay acting as a sink for atmospheric CO_2 (Kumar et al., 1996). Studies carried out by few earlier researchers on the chemical characteristics of the eastern Bay of Bengal and the Andaman Sea have reported relatively higher levels of pCO_2 (by 20 μatm) upto a depth of 1200 m in the eastern Bay of Bengal as compared to the Andaman Sea, whereas the deeper waters of the Andaman Sea contained pCO_2 levels higher by 200 μatm than the eastern Bay (Sarma and Narvekar, 2001). This has been attributed to the combined effect of warmer waters (by 2 °C) in the deeper parts of the Andaman Sea and the effect of the Bay of Bengal intermediate waters, which have higher pCO_2 levels. However, all these studies are one-time measurements, and the seasonal variability in these properties will help in the quantification of fluxes of these chemical properties to the atmosphere and to the deep ocean.

The saturation levels of N_2O in both surface oxic and oxygen-depleted waters in the central Bay were significantly lower than in the western Bay and the Arabian Sea. However, methane showed elevated levels (upto 800 $nl.L^{-1}$) in the surface shelf waters close to the Ganges/Brahmaputra mouth in Bangladesh (Berner et al., 2003) while peak concentrations of 105 $nl.L^{-1}$ were observed below 45 m depth. The above authors related the methane levels in the surface waters to bacterial activity while the enrichment in deeper waters is apparently due to the gas charged sediments.

2. AIR-SEA FLUXES OF CO_2, N_2O, CH_4 AND $(CH_3)_2S$ IN THE NORTH INDIAN OCEAN

2.1 Atmospheric Gases: The Oceanic Relevance

Ocean-atmosphere interactions regulate global climate system. Presence of radiatively important gases in the atmosphere is mainly responsible for maintenance of the life sustaining temperatures at the earth surface. Although the prevalence of greenhouse gases in the early stages of the earth's atmosphere helped the origin of life, their recent anthropogenic emissions because of the increasing demands of the daily living, are posing a threat to future inhabitants of this planet. Approximately one-third (~2 Gt C y^{-1}) of the anthropogenic carbon dioxide (CO_2) released is expected to be absorbed by the oceans (Siegenthaler and Sarmiento, 1993) and thus reduce the extent of the global warming expected otherwise. Oceans have been playing the role of a buffer system in maintaining atmospheric composition since their formation. Therefore, it is important to understand the fluxes of radiatively important gases such as CO_2, N_2O, CH_4 and $(CH_3)_2S$ (dimethyl sulfide or DMS) between ocean and atmosphere and the processes that control the magnitude of their exchanges.

2.2 Air-sea Fluxes of Gases

The radiative gases of interest here are important components of the biogeochemical cycles of the earth system. Air and water media favour easy exchange of gases between them. The magnitude of flux of a gas between air and seawater is determined by its concentration gradient between the two media and exchange coefficient or piston velocity (primarily a physical function of surface turbulence) of the gas in question (Liss and Merlivat, 1986). Gas exchange coefficients are determined from wind speeds (a parameter normally used to characterize surface turbulence). Biogeochemical processes, on the other hand, control the abundance of these gases in seawater and determine the direction of the flux. Therefore, the extent of air-sea fluxes of these gases depend both on the strength of wind systems and nature of biogeochemical cycles in a specified region, which might vary significantly from place to place and from time to time.

2.3 South Asia Monsoon System and the North Indian Ocean Processes

The North Indian Ocean (north of equator) is bounded by landmass in the north. The presence of landmass intensifies ocean-atmosphere interactions through not only the wind driven circulation including upwelling and convections but also creates conditions conducive for the development of atmospheric turbulences. Summer monsoon winds are stronger in July-August favouring intense upwelling over large areas in the western (Somalia and Oman) and in a limited area in the southeastern Arabian Sea (Wyrtki, 1973). Though the winds are not as strong as in summer, convection is set up, between surface and subsurface waters, in winter monsoon over central and northern basins of the Arabian Sea as cool-dry winds blow from the north (Madhupratap et al., 1996). The influence of monsoon, the associated physical forcings and the consequent biogeochemical processes on the surface abundances of the gases considered here are schematically shown in Fig. 6. Atmospheric supply of nutrients through deposition of aerosols, particularly of land and anthropogenic sources, enhances the surface biological production in the North Indian Ocean. On the contrary, spreading of large volumes of freshwater discharged from the Indian subcontinent reduces the upward pumping of materials from subsurface layers in the Bay of Bengal (Naqvi et al., 1994). The biogeochemical responses in the regions affected by run-off or surface stratification caused by the Low Salinity Lens (LSL), with implications to development of atmospheric low pressure systems and air-sea exchange of gases, are depicted in Fig. 7.

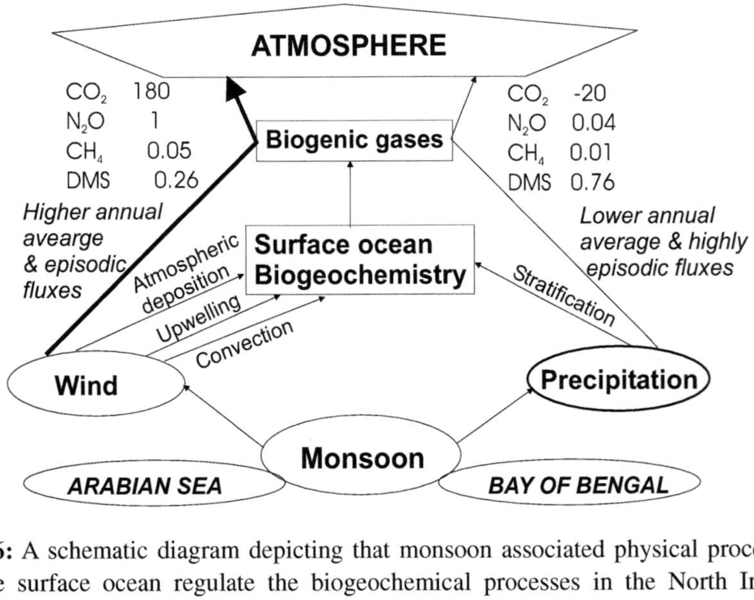

Fig. 6: A schematic diagram depicting that monsoon associated physical processes in the surface ocean regulate the biogeochemical processes in the North Indian Ocean. The two basins (the Arabian Sea and the Bay of Bengal) of the North Indian Ocean experience different rates of vertical mixing. While the entrainment is efficient in the Arabian Sea it is low in the Bay of Bengal because of stratification. Higher abundances of gases in the Arabian Sea, therefore, result in higher gas emissions than from the Bay of Bengal. The numbers shown against the considered gases indicate their fluxes in Tg y^{-1} from the Arabian Sea (left side) and the Bay of Bengal (right side).

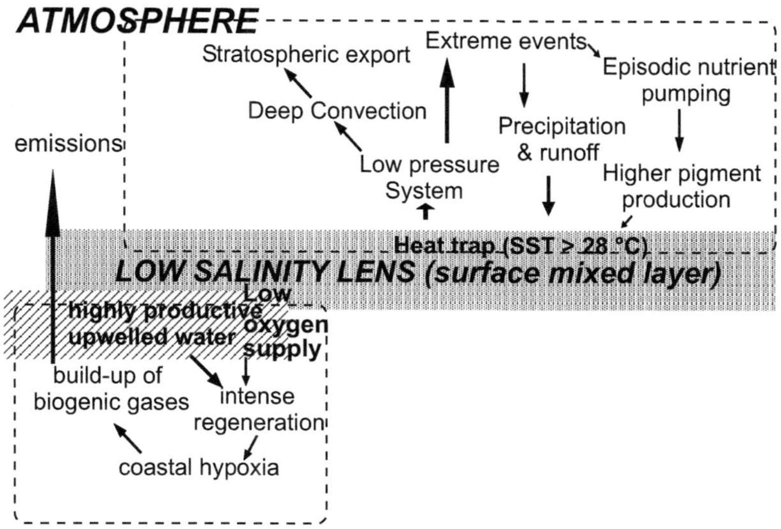

Fig. 7: Freshwater discharge leads to the formation of a low salinity lens (water layer) at the ocean surface. The LSL can trap excessive solar radiation due to limited entrainment and therefore conducive for region for the development of low pressure systems. The atmospheric disturbances result in episodic pumping materials from deep to surface ocean, with consequent influence on surface production and gas emissions. In near-shore regions (left) the LSL can lead to the formation of hypoxia by reducing the atmospheric oxygen supply required to meet the bacterial demands.

2.4 Abundance of Gases in Surface Waters of the North Indian Ocean

The surface circulation, the biogeochemical processes and the resultant abundances of gases are quite different between the two North Indian Ocean basins. In the Arabian Sea intense entrainment occurs in both monsoons leading to very high biological production in surface waters and enrichment of these gases, in general. The entrainment in the Arabian Sea occurs through upwelling in summer and convection in winter (Fig. 6). Therefore, the CO_2 concentrations are found to be nearly double to those in air in the core upwelling zones of the western (Kortzinger et al., 1997) and southeastern Arabian Sea (Sarma, 1998). Vertical exchange together with intense heterotrophic activities leads to CO_2 supersaturation in surface waters of the Arabian Sea (Sarma et al., 1998). In contrast, prevalence of low salinity water lens at the surface in the Bay of Bengal (Figs. 6 and 7) reduces upward pumping of CO_2 (George et al., 1994) and N_2O (Naqvi et al., 1994) and therefore their surface levels are lower than in the Arabian Sea.

The vertical diffusivity in the Bay of Bengal (0.16 $cm^2\ s^{-1}$) is found to be one-third of that in the Arabian Sea (0.55 $cm^2 s^{-1}$). Biological productivity triggered by river and atmospheric nutrient depositions, despite lower vertical exchange, make the Bay of Bengal a seasonal sink for atmospheric CO_2 (Kumar et al., 1996). While the open oceanic areas contained generally lower levels of gases, coastal and upwelling areas experience anomalously higher abundances due to intense biogeochemical processes facilitated by air-water and land systems. The highest concentrations of gases shown in Table 1 were found in coastal waters along the west coast of India during and immediately following the summer monsoon. Methane levels of about 13,000% (super)saturation are reported in estuarine waters of Goa (Jayakumar et al., 2001) whereas Berner et al. (2003) found very high levels of methane off Bangladesh coast. Surface waters in upwelling areas are 145-156% supersaturated with CH_4 (Bange et al., 1998). River and estuarine waters of Goa (Sarma et al., 2001), of Hoogly (Mukhopadhyay et al., 2002) and of Godavari (Bouillon et al., 2003) have been found to contain pCO_2 of 300-2500 µatm, 100-1500 µatm and 293-500 µatm, respectively. On the other hand, tidal mangroves of the Godavari delta region contained very high levels of pCO_2 of 1375-6437 µatm (Bouillon et al., 2003).

Table 1: Abundances, sea-air fluxes and contribution of North Indian Ocean to total gas emissions from South Asia

Gas	*Surface Abundance*	*Sea-to-air Flux*[a] *(Tg y^{-1})*	*Emission*[b] *from SA (Tg y^{-1})*	*Ocean contribution (%)*
pCO_2 (µatm)	200-700	32-204	715	4-22
N_2O (nM)	1-533	0.75-1.12	0.26	74-82
$(CH_3)_2S$ (nM)	0.1-525	0.13-1.3	4.3[c]	3-23
CH_4 (nM)	2-50	0.02-0.19	24	0.1-0.8

[a] pCO_2 range is the net after accounting for the Bay of Bengal as a sink.
[b] Data obtained from Mitra et al. (2002)
[c] Data given in Mitra et al. (2002) for CH_4 emissions from India are extrapolated to South Asia (to an approximation based on CO_2 data).

2.5 Extreme Events in the North Indian Ocean

The North Indian Ocean is an area of extreme or catastrophic events; seemingly regular hypoxia occur along the Indian coastal waters of the Arabian Sea following the southwest monsoon upwelling (Naqvi et al., 2000) and frequent atmospheric turbulences occur over the Bay of Bengal. Both the coastal hypoxia and atmospheric disturbances are associated with the strong surface stratification caused by the low salinity lens (Fig. 7). Anomalously higher concentrations of N_2O (Naqvi et al., 2000) and DMS (Shenoy, 2002) are found in oxygen deficient coastal waters of western India. Exceptionally higher levels of DMS (>500 nM) occurred along the west coast of India following upwelling driven plankton blooms. The emission of N_2O from this hypoxic region, only a fraction of the Arabian Sea area and restricted to a small period of the year, for instance, is equal to the annual average of the entire Arabian Sea, in general (Naqvi et al., 2000). On the other hand, episodic fluxes of DMS are estimated during highly turbulent conditions over the Bay of Bengal during a deep depression (the highest estimated emission rate of 41 μmol S m^{-2} d^{-1}, in the world oceans, Shenoy, 2002) and over the equatorial Indian Ocean during a turbulent event associated with the Inter-Tropical Convergence Zone (seven times more than in the previous year, Shenoy et al., 2002).

2.6 Contributions of North Indian Ocean to Regional Gas Emissions from South Asia

As the information on gas abundances during extreme events is sparse in the northern Indian Ocean, the errors associated with these flux estimates could be significant and the actual emissions might be substantially different. However, the values summarized here are qualitatively important as they reflect the different magnitudes of biogeochemical subsystems operating in the North Indian Ocean basins, driven primarily by monsoon associated processes. While the CO_2 supersaturated surface Arabian Sea perennially acts as a source (52-224 Tg C y^{-1}) of atmospheric gas, the Bay of Bengal absorbs ~20 Tg C y^{-1}. Therefore, net CO_2 flows in opposite directions across the air-sea interface between the eastern and western parts of the North Indian Ocean. On the other hand, lower vertical diffusion in the Bay of Bengal leads to lower emissions of other gases too by several times, under normal conditions, in comparison to those from the Arabian Sea. A comparison of these fluxes (listed in Fig. 6) against the emissions from South Asia reveals that the North Indian Ocean can contribute upto about 22% of CO_2, 23% of S gases and negligible (<1%) of CH_4 but a substantial 82% of N_2O to the total atmospheric loadings from this region (Table 1).

2.7 Flux Scenario under Global Warming Conditions

With the presently available information it is hard to predict the changes in air-sea fluxes in the North Indian Ocean with the changing climate although the biological productivity is shown to have increased in the last two decades by 13.8% (Gregg et al., 2003). However, the following changes can be predicted:

Increases in regional surface temperature and chlorophyll production due to anthropogenic forcing or global change will have a positive feed-back on: (a) formation of low pressure systems and (b) emissions of biogenic gases from the Indian Ocean. Warming the surface layers with simultaneous increase in biological productivity can enhance rates of metabolic/microbial processes in the surface mixed layer. Therefore, pCO_2 levels in the mixed layer may increase with concomitant emission to the atmosphere under turbulent conditions. The effect of temperature rise (0.6°C) on pCO_2 levels is through a reduction in pH by 0.01 units. At a total DIC of 1950 µmol kg^{-1} the pH drop results in a pCO_2 enhancement of 20 µatm; a rise that could prove important under turbulent conditions (a sea-to-air flux increase of 1.2 mmol m^{-2} d^{-1} at a normal wind speed of 5 m s^{-1} over the North Indian Ocean, but by 12 mmol m^{-2} d^{-1} with strong monsoon/cyclonic winds of 15 m s^{-1}). Anthropogenic carbon absorbed by the ocean constitutes only a minor fraction (<2.5%) of the total DIC in seawater. However, its influence on water column processes such as pH and solubility of carbonate minerals is important. Absorption of atmospheric CO_2 would lower pH of seawater, further enhancing pCO_2 and making the waters more corrosive to skeletal carbonates. This is the major reason for the upward migration of the aragonite saturation depth in the Arabian Sea between GEOSECS (1978) and WOCE (1995) expeditions (Sarma et al., 2002). Similarly, DMS concentrations will be modified. For instance, DMS flux is predicted to change differently in various regions of world oceans with an increase in the tropics north of the equator following Global Change when atmospheric CO_2 is doubled (Bopp et al., 2002), a behaviour similar to that of chlorophyll, which has increased in the North Indian Ocean (Gregg and Conkright, 2002).

Increase in chlorophyll (biological) production due to climate change can enhance sinking organic matter. Enhanced biological pump shall lead to higher oxidant demand and therefore to consumption of more N_2O at intermediate depths of the Arabian Sea. On the other hand, its production could be marginally higher in the subsurface waters of the Bay of Bengal. Nevertheless, with the warming of surface ocean enhanced emissions of N_2O and CH_4 could be expected, than at present, from the North Indian Ocean also because of the increased turbulence.

2.8 Present Status and Limitations

According to dynamic ocean processes, the southern boundary of the North Indian Ocean is about 10°S, as there exists a hydrographic front (Wyrtki, 1973). But hardly any data on gases in the surface ocean are available from the equatorial belt. Despite the fact that ocean observations in this region started several decades ago, the temporal and spatial coverage of the monsoon Indian Ocean remains poor. While the Arabian Sea received international attention, primarily due to high biological production and low oxygen containing subsurface waters, climatically important Bay of Bengal and equatorial Indian Ocean are receiving attention only recently, particularly for biogeochemical and gas exchange studies. Even the more accessible wetland regions are not covered except to some extent Sundarban and Andaman regions of India. Limited observations reveal that CO_2 is enriched in Indian estuarine systems but no major effort has been made to study estuarine systems of major rivers of the Indian subcontinent. The biological and biochemical processes responsible for very high levels of N_2O and DMS in surface and subsurface regimes of the North Indian Ocean are not well understood, let alone their spatial (horizontal and vertical) and temporal patchiness. Attribution of elevated N_2O emissions from hypoxic environments of central west coast of India to fertilizer run-off from rivers is premature as this region receives land run-off only during the southwest monsoon and from Goa, where the fertilizer consumption is scanty and probably the minimal in India. It appears that monsoon driven circulation, occurrence of low saline cap at surface and the associated, poorly understood, biogeochemical processes along the west coast of India are responsible for the seasonal occurrence of hypoxia and the region's importance as a chimney of greenhouse gases.

It is important that attempts to map abundances of these gases during extreme climatic events through automated systems in surface waters of the key biogeochemical and turbulent regimes be intensified to make realistic estimates of gas emissions during catastrophic events. This is because, the presently estimated fluxes will be underestimates as episodic events can emit the amounts of gases many folds higher than those computed using data measured on board ships under relatively calm conditions. During cyclones episodic upward pumping of subsurface waters have been documented through surface cooling, implying pumping of more materials into the surface layers and enhanced gas emissions to atmosphere. Churning of the surface ocean by tsunami waves will also be important although episodic emission periods might be short.

3. CORAL REEFS

Coral reefs are shallow water, tropical marine ecosystems that harbour the most diverse and productive communities on earth, unequalled by any other habitat. They are home for more than 25% of all known marine fish. Their species diversity, contain more phyla than the rainforests. Because these organisms can only thrive in warm, shallow and clear waters, they are mostly found in tropical oceans worldwide. These ecosystems perform important functions as atoll islands foundation, coastal protection structures and source for beach sand. They provide food and shelter for fish and invertebrates, and have high socio-economic value in the form of tourism promotion. Coastal areas with beaches and fringing reefs have become important locations for tourism. Activities like scuba diving and viewing corals through glass bottom boats, swimming and snorkeling have become very common in these areas. Biologically active compounds produced by reef organisms possess antiviral and antifungal activities and consequently, these compounds can be an important source for drugs and medicines and can support emerging opportunities in biotechnology. Many coastal communities depend on reef resources for food in the form of fish, and source of income through sale of ornamental fish, building material, and tourism related activities.

The total aerial extent of living coral reefs world over has been estimated at 284,803 km^2 (Goldberg and Wilkinson, 2004) out of which an area of 19,210 km^2 lies in South Asia. Among the South Asian countries, coral reefs are found mainly in India and Sri Lanka, and to some extent in Bangladesh.

Because of their delicate nature and their tolerance to a narrow range of environmental conditions, coral reefs are highly vulnerable to environmental stresses, both natural (ENSO events, storms, cyclones, tsunamis, diseases, etc.) and anthropogenic (industrial development, pollution, tourism, urbanization, agricultural runoff, sewage pollution, sedimentation, overfishing, coral mining, land reclamation, climate change, etc.).

3.1 Bangladesh

Corals are found only on Narikal Jingira Island (St. Martin's Island). There are about 66 species belonging to 22 genera and 10 families. *Porites*, *Favilites*, *Goniopora*, *Cyphastrea* and *Goniatrea*, are some of the most common genera; however, their abundance is low. The coral fauna consists of mollusks, echinoderms, zonathids and bryzoans. Colourful nudibranchs are also found. Other macroinvertebrates include sea urchins, sea stars, brittle stars and sea cucumber. There are 86 species of fish associated with corals, the most common being the damsel fish, parrot fish, surgeon fish, butter fly fish and angel fish (Mollah, 1997).

There are two types of threats to corals and coral reefs at Narikal Jinjira Island: natural and anthropogenic. The natural threats include cyclonic storms and tidal surges which cause shifting and overturning of boulders; turbidity caused by discharge of silt by Naaf River, and resuspension of sediment. The main anthropogenic threat to corals comes from direct extraction of coral colonies using boats, at an estimated rate of 40-100 $kg.day^{-1}.boat^{-1}$. About 30,000 coral colonies are thus collected annually (Mollah, 1997). Collection of rockfish is also an important occupation.

3.2 India

Coral reefs are found in four major areas of which two are adjacent to the mainland – Gulf of Kutchchh and Gulf of Mannar – and the other two are offshore island chains – Lakshadweep, and Andaman and Nicobar Islands. In addition patches of reefs are found in the intertidal areas along the west coast of India off Ratnagiri, Malvan, Redi and Mangalore coasts. Some patches of coral reefs are also found along the east coast of India, namely, Pondicherry, and on the coastal stretch between Parangipetai and Cuddalore.

The reefs of the Andamans and Nicobar Islands are of fringing type while those of the Lakshadweep group of islands are atolls. Barrier reefs are found along the east and west coasts. In Palk Bay, the corals occur in the form of fringing reefs along the mainland, about 25-30 km long and less than 200 m wide, while in the Gulf of Mannar, the reefs are developed around a chain of 21 islands. The total estimated coral reef area is about 2350 km^2 (Hoon, 1997).

The reefs in India house about 200 species of corals belonging to 37 genera. The most commonly occurring genera are *Acropora, Montipora, Porites, Pocillopora, Seriatopora, Stylopora, Montipora, Coscinaraares, Echinopora, Turbinria, Pseudosiderastrea* and *Millepora.* A profusion of blue coral *Heliopora coerulea* is found in the Lakshadweep group of islands.

Flora and Fauna

In Palk Bay, the reef contains seagrasses where turtles and dugongs are common. The area is also a breeding ground for squids. The extensive sea grasses present in the Gulf of Mannar offer shelter and food for green turtles, olive ridley turtles and dugongs. The reefs also harbour boring sponges, mollusks, worms, equinoderms, shrimps and variety of fish. Ornamental fish is an important source of income for the local population (Hoon, 1997).

The reefs of Andamans and Nicobar Islands offer important nesting places for leatherback, hawksbill olive ridley and green turtles and marine mammals, such as dugongs. Estuarine crocodiles are also found. Clams, gastropods and some species of crabs are found in plenty in the sand while seagrass beds harbour sea cucumber, starfish and brittle stars (Hoon, 1997).

The corals of Gulf of Kutchchh are mostly fringing type and show high state of degradation caused by mud deposits (Hoon, 1997). The flora and fauna consists of several species of algae, sponges, fishes, sharks, prawns, crabs, lobsters and mollusks. Marine mammals such as dugongs, dolphins and whales are also found.

In the Lakshadweep group of islands, the coral formation is built-up on the Lakshadweep-Chagos submarine ridge rising steeply from a depth of about 1500-3000 m (Hoon, 1997). These islands are flat, seldom rising more than 2 m. The reef biodiversity consists of crabs, bivalves, sea stars, brittle stars, sea cucumbers, sea urchins and a variety of fish. There are also green turtles and hawksbill turtles that feed on seagrasses.

3.3 Sri Lanka

Growth of coral reefs in Sri Lanka is influenced by monsoon through its impact on turbidity and freshwater input. About 2% of the 1585 km coastline of Sri Lanka contains nearshore fringing coral reefs (Rajasuria, 1997). Some offshore reefs are also found. However, the best reef systems, based on substrate cover, diversity and reef flora and fauna, are associated with offshore reefs in the northwest, southeast and eastern region, while the near-shore reefs are affected by human activities. The coral reefs contribute significantly to marine fish production. Export of ornamental fish is rated third both in volume and value, among the fish export products after prawns and lobsters (Rajasuria, 1997).

A total of 183 species belonging to 68 genera are found in Sri Lanka. Of these the dominant species belong to the families of *Acroporidae, Faviidae, Poritidae* and *Pocilloporidae.* A high diversity of butterfly fish has been recorded, while invertebrates like spiny lobsters, shrimps, and crabs are also found. Dolphins, whale sharks and sea turtles have also been sighted. The reef flora includes seagrasses and algae (Rajasuria, 1997).

Major causes for coral reef degradation include: sedimentation, use of explosives and bottom-set nets for fishing. Both industrial and sewage pollution have also contributed significantly to the reef degradation. Anchor damage caused during fishing operations when trawlers lay their anchors within the reef lagoon, is another cause for reef degradation. However, mining of corals by construction industry, to produce lime, has caused the worst destruction of most of the fringing reefs. A survey conducted by the Coast Conservation Department in 1984 has revealed that about 18,000 tonnes of coral is being supplied annually to the mining industry (Rajasuria, 1997).

3.4 Effect of ENSO Events

Coral reef organisms apparently live at or close to the temperature tolerance thresholds, and consequently, can tolerate changes in environmental conditions, especially temperature, only within a narrow range. When the sea surface temperature exceeds the seasonal maximum by about ≥1°C the corals start bleaching while extensive mortality occurs when the temperature anomalies are of the order of 3°C or more (IPCC, 2001). A major coral-bleaching event that occurred during the recent past was associated with the ENSO event of 1998. Approximately half of the reefs in the Indian Ocean and around South Asia were reported to have lost most of their living corals (Goldberg and Wilkinson, 2004).

In India, coral reefs were affected to varying extent: while there was no serious bleaching in the Andaman and Nicobar Islands, the coral reefs of Lakshadweep Islands suffered the maximum damage (Goldberg and Wilkinson, 2004). However, there are now clear signs of recovery. Surveys carried out subsequently (1999) indicated a recovery of 15-26% (Pet-Soede et al., 2000). In Sri Lanka, the recovery after the bleaching event was variable. While the shallow coral reef habitats (< 8 m) were severely damaged. The recovery was extremely low, hindered by the rapid deposition of sediment on dead coral, which prevented recolonization (Rajasuria and Karunarathna, 2000); in deeper waters (> 10 m), however, there was a complete recovery.

On the whole, while about 65% of coral reefs of South Asia were severely affected by the bleaching event of 1998, about 13% of them have shown complete recovery, while 10% are still in a critical stage (Goldberg and Wilkinson, 2004).

3.5 Effect of Tsunami of December 2004

The following information (Box 1) is based on the surveys carried out during January 2005, in the following coral reef areas in south western Srilanka: Kapparatota/Weligama, Polhena, Unawatuna, Hikkaduwa and Rumassala.

Box One: Summary of Post-tsunami Status

Damage caused to the inter-tidal and sub-tidal areas can be classified into five main categories: mechanical damage; smothering by sediment; litter; beach erosion and impact on fish community.

- Mechanical damage: Damage was moderate to high at all sites except at Rumassala, which sustained no damage from the tsunami. It was patchy with localized high impact. Some live coral was destroyed, and large quantities of coral rubble formed after the coral mass mortality of 1998 has shifted covering and killing reef biota as well as some seagrass.
- Smothering by sediment: Only low levels of smothering was observed, caused by terrigenous sediment washed into the sea, and resuspended sediment. Although not a major impact from the tsunami, this is a potential concern at all sites.
- Litter: Considerable amount of debris, both plant material and man-made objects, was carried with receding water and deposited both on the reefs and/or on beaches in all areas. A beach cleanup has been carried out in Hikkaduwa by SLSAC.
- Beach erosion: Some loss of beach width was observed, ranging between extreme at Unnawatuna and none at Rumassala.
- Impact on fish community: Where habitat has been destroyed, reduced fish abundance is observed, especially at Unawatuna and Polhena. It is noteworthy that Hikkaduwa has higher fish abundance than surrounding areas, a reflection of its management status.

Source: http://www.iucn.org/tsunami/docs/mpa-recommendations-ccd.pdf

3.6 Effect of Global Change

The concentrations of greenhouse gases, including CO_2, in the atmosphere are increasing rapidly. Presently, the concentration of CO_2 in the atmosphere is 367 ppm; however, at the present rate of emission, it is expected to go up to 463-623 ppm by the year 2050 (IPCC, 2001). An increase in the atmospheric concentration of the greenhouse gases will affect the corals and coral reef communities through rise in SST, rise in sea level, increase in UV-B radiation and increase in CO_2 concentration in surface seawater.

An increase in concentration of atmospheric CO_2 and hence oceanic CO_2, will result in the reduction of the concentration of $CaCO_3$ in water due to its decreased saturation. This, in turn, will affect the ability of reef plants and animals to make limestone skeletons (reef calcification), thus reducing the ability of coral to grow vertically and keep pace with sea level rise. The overall impact of rise in SST and elevated CO_2 levels could result in reduction in species diversity in coral reefs.

As was seen in case of ENSO event above, a rise in SST by ≥ 1 °C over the seasonal maximum can have a bleaching effect on coral reefs, and in serious cases it may lead to the death of the corals. Based on laboratory studies and field studies where thermal effluents from power plants and OTEC power cycles were discharged, Wafar (1990) summarizes that corals suffer sub-lethal effects such as bleaching and reduced growth rates at temperatures 3-4 °C, and near-total mortality at temperatures 4-6 °C above the ambient, over extended period of time. The projected rise in SST in tropical regions is expected to result in increased frequency and severity of bleaching and mortality events, although some changes to heat tolerance of corals and their symbiotic algae may occur through acclimation and adaptation (Ove Hoegh-Guldberg.oveh@uq.edu.au).

Laboratory experiments carried out by Jokiel (1980) demonstrated that a variety of reef organisms – sponges, bryozoans, tunicates, etc. – were killed after 1-2 days of exposure to UV radiation. However, these organisms could survive in intense sunlight if the UV portion was filtered out.

Global change will effect a rise in sea level due to two processes – water expansion due to rise in water temperature, and melting of sea ice and ice cap in the Arctic and Antarctica. While coral reefs in the atolls grow vertically at a rate of 1-3 $mm.y^{-1}$ those on fringing or barrier reefs grow comparatively faster at a rate of 10-12 $mm.y^{-1}$ (Wafar, 1990). Considering that the present analysis period is of 30 years, i.e., year 2030, this would mean that the atoll reefs will grow vertically by 30-90 mm, while the sea level will rise by 24-240 mm during the same period (applying the IPCC, 2001 projections). A comparison between these values indicates that in 30 years time, i.e., by the year 2030, the atoll coral reefs will be completely submerged, if the present rate of sea level rise continues. Healthy fringing and barrier reef corals, however, will be unaffected by the sea level rise as they would be able to cope up with the sea level rise due to their higher growth rates. The problem, however, will be with the unhealthy coral reefs or reefs that are bleached recently due to increase in SST, as this effect combined with the effect of decrease in $CaCO_3$ saturation will cause a serious threat to these reefs as these reefs will not be able to grow vertically faster enough to keep pace with the sea level rise.

4. THE BENGAL DELTA

In general terms, the Bengal delta is a monotonous flat land, consisting primarily of alluvial plains, which have emerged from the Bay of Bengal due to continued deposition of sediments carried out by the Eastern Himalayan River systems, and have been playing a major role towards draining huge perennial outflows of the Ganges-Brahmaputra-Meghna (GBM) River System of the region. The delta is predominantly formed by the sediments of the GBM system; however, the western banks of the Hooghly river is formed by the sediments carried by the rivers originating in the western hilly fringes of the delta (Morgan and McIntire, 1959). The Himalayan range demarcates the northern boundary of the delta, while the eastern parts of the delta are again demarcated by the Himalayan foothills and the Tipara hills. The southern boundary is provided by the Bay of Bengal where a large estuary is still nurturing the morphologically dynamic active delta. The delta is inhabited by Bangla speaking people and its major part falls in Bangladesh towards its eastern side.

4.1 Origin and Formation

The Bengal delta is formed on the Bengal Basin which is located in the northeastern part of Indian subcontinent and placed between the Indian Shield and Indo-Burman Ranges. In geological terms, the basin comprises three geo-tectonic provinces: (1) The Stable Shelf; (2) The Central Deep Basin (extending from the Sylhet Trough in the northeast towards the Hatia Trough in the south); and (3) The Chittagong–Tripura Fold Belt (Alam et al., 2003). It is located at the juncture of three interactive tectonic plates: the Indian, Burma and Tibetan (Eurasian) Plates. The Stable Shelf is formed with Precambrian metasediments and Permian–Carboniferous rocks. Sedimentation started on the stable shelf and deep basin in the Early Cretaceous (Desikacher, 1974). Sedimentation for most of the Bengal Basin and, therefore, formation of the Bengal delta has been continuous since Early Cretaceous period (Curray, 1991).

In addition to continuous sedimentation and gradual uplifting of the delta, there has also been continuous subsidence, which may be attributed to a number of factors including differential adjustments of the crust, collision with the various elements of South Asia, and uplift of the eastern Himalayas and the Indo-Burman Ranges (Alam, 1989; Alam, 1972). During the Cretaceous, a number of faults began tectonic movements (Desikacher, 1974). A major marine transgression occurred in Eocene period and the deep basin area observed deep-water sedimentation. A major switch in sedimentation pattern over the Bengal Basin occurred during the Middle Eocene to Early Miocene. Sedimentation in the Bengal Basin influenced occurrence of subsidence of the emerging delta (Alam et al., 2003; Alam, 1995). From Pliocene onwards, large amounts of sediment started filling the Bengal Basin from the west and northwest; and major delta building processes continued to develop the present-day delta morphology. The present basin configuration with the Bengal delta system on the north and the Bengal Deep Sea Fan on the south was established during the later part of Pliocene and Pleistocene (Alam et al., 2003).

4.2 Recent Changes in the Delta Environment

Besides undergoing constant morphological and environmental changes through natural processes, the delta has also been experiencing human induced changes in recent decades (Alam, 1996; Mirza, 2004). Consequently, there have been significant impacts, which will again be affected due to changes in monsoon regime (Mirza et al., 2006; this volume).

One of the most significant effects of anthropogenic activities that were felt in the Bangladesh part of the delta was when the Indian Government built a dam on the Ganges – The Farakka Dam – in 1975, to divert the water flow from the river for various activities. The construction of the dam, at a distance of about 18 km upstream from the Bangladesh border, has completely changed the micro-environment in the southwestern Bangladesh (Abbas, 1984; Khan, 1993). The main distributary, the Gorai river, has been choked and disconnected from the Ganges in the early 1990s; the salinity regime has penetrated so much upstream that livelihoods of people living in a large area just north of the Sundarbans had to be changed (Sarker and Koudstaal, 1999; EGIS, 2000). The changes induced in the hydrological pattern of the Gorai system has greatly reduced the potential of the affected areas; and raised the salinity of the shallow aquifer system thereby reducing availability of freshwater for drinking purposes, and threatening the well being of the Sundarbans forest itself (RVCC, 2003; Karim, 1995). There are reasons to believe that any further reduction in dry season flows in the Ganges and the Gorai rivers will be detrimental for the lives and livelihoods of people living downstream of the dam (Mirza, 2004).

Similarly, the coastal zone of Bangladesh has also experienced severe adverse impacts of structural interventions. Two examples of such structural intervention are:

- A large number of coastal embankments were built during 1968-85, with a vision to attain food grain self-sufficiency for Bangladesh. Although these structures initially helped to reduce tidal inundation within the poldered areas, they disturbed the natural aggradation processes of the tidal floodplains. Instead, sedimentation was guided by the poldered rivers within the river beds, thereby decreasing the discharge capacity of the rivers. The whole area is now subject to drainage congestion and saline water logging, thus reducing significantly the livelihood opportunities for poor farmers and fisher folks.
- The other engineering intervention that has also brought about dramatic results: In an attempt to raise coastal landmass, cross dams were built along the coastal District of Noakhali and a few others were planned during the 1970s up to 1990s (Eysink, 1983). The hasty interventions indeed helped to raise over 700,000 acres of land, but have gradually choked a number of small coastal rivers. There is now increasing understanding that the ecological balance in that part of the coastal area has thus been irreversibly destabilized and the people in the newly accreted land are increasingly exposed to inundation and water logging.

In the Bangladesh part of the delta, there have been continued encroachments into the wetland systems. For example, a large *beel* area, called the Chalan Beel, in the northwestern part of the delta has mostly been encroached and transformed into paddy fields. The intention was to bring the vast wetland under crop agriculture. However, such encroachments into wetlands have caused decreased sedimentation in the flood plains, while forcing increased sedimentation on the river beds, thereby contributing to higher levels of flood vulnerability due to increased drainage congestion. Meanwhile, increased occupation of the marginal and flood vulnerable low lands by the poor has increased flood vulnerability, despite building so many flood control and drainage structures throughout the delta.

5. THE SUNDARBANS: A UNIQUE MANGROVE ECOSYSTEM

The single largest patch of productive mangrove in the world is found in South Asia (between 88°85 and 89°55 E and 21°30 and 22°40 N), shared between Bangladesh and India. Mangroves of Bangladesh are primarily located in the Sundarbans, and in offshore islands of Bhola, Moheshkhali and in Chokoria of Cox's Bazaar district. The Sundarbans occupies an area of about one million hectares. About 62% of the forest covering an area of 577,000 hectares lies in Bangladesh, the rest lies within Indian state of West Bengal. Mangroves in Bangladesh other than the Sundarbans are severely degraded due to continued encroachment, whereas the Chokoria Sundarbans (about 7500 hectares) has been completely denuded during the past three decades (Hussain and Acharya, 1994).

The mangroves of the Bengal delta are ideally placed along the estuaries of the three mighty rivers: the Ganges, the Brahmaputra and the Meghna (GBM). However, the Sundarbans forest is now located in the fag end of the Ganges basin. The forest receives freshwater and sediment from a number of distributaries of the Ganges. About one third of the forest consists of water bodies in the form of rivers, channels and tidal creeks. The forestland is highly influenced by tidal interactions because of the presence of these water bodies.

The current landform of the forest exhibits a large number of fluvial and tidal geomorphological features that are formed over millennia due to continued deposition of weathered materials carried by the GBM river systems. Mudflats are observed along the estuaries or riverbanks; these are subject to direct wave action, flow and turbulence of the water currents in the river. During all high tides, the lower parts of mudflats remain submerged with brackish water. Backswamps or basins are also found to occur as low lying saucer shaped depressions where rainwater and sediment are collected. However, a part of the sediment is washed away during early monsoon season. Gradual sediment deposition on the edge of the riverbank gives rise to ridges or levees. Some levees have inclined slopes on their outer edges, with steep gradients towards the channel side. There are creeks and streamlets that are influenced by tides and maintain inter-connection between rivers and cross channels. The landform, therefore, is crisscrossed by a large number of rivulets and creeks.

The forest is influenced by semi-diurnal tides, with the tidal height varying between 3.5 and 5.0 m, while the mean tidal height is about 4.0

metres (Hussain and Acharya, 1994). In addition, lunar cycle significantly influences tidal heights. The mean water level of the rivers reaches its peak during the wet season (July-August) and is at its lowest level during the dry season (December-January). All areas below the lowest water level are regularly flooded by almost all high tides. Elevations above this level are subject to flooding during spring tide only while remaining exposed during neap tide. Besides tidal variations, the occurrence of monsoon floods and cyclonic surges raise the water levels of the rivers, causing submergence of elevated lands that normally do not undergo inundation. River flow and tidal currents play a vital role in creating the environmental conditions of the estuary.

The soil of the Sundarbans is saline due to tidal interactions, although the salinity is low compared to soil salinity in other mangrove forests of the world (Karim, 1988). Soil salinity, however, is regulated by a number of other factors including surface runoff and groundwater seepage from adjacent areas, amount and seasonality of rainfall, evaporation, groundwater recharge and depth of impervious subsoil, soil type and topography, etc. Soil salinity influences the floral distribution of the forest. Using the salinity scale established by Walter (1971), the forest areas have been divided into three zones (Karim, 1988).

Oligohaline (or miohaline) *Zone*: The zone is characterised by the soil containing less than 5 ppt of NaCl salt. This zone occupies a small area of the north-eastern part of the forest. It is occupied predominantly by freshwater type mangrove cover and dominated by high value 'sundri', the tall mangrove species.

Mesohaline Zone: The zone is characterized by NaCl content in soil in the range of 5 to 10 ppt. It covers the north-central to south-central part of the forest. The dominant species of this moderately-saline zone is gewa, intermixed with sundri and passur (Xylocarpus mekongensis), with a dense understorey of goran.

Polyhaline Zone: The NaCl content of the soil in this zone is higher than 10 ppt. This saline-zone covers the western portion of the forest. This zone supports increasingly saline tolerant species: primarily gewa with a dense understorey of goran, and dense patches of palm 'hantal' (Phoenix paludosa).

The Sundarbans delta has three zones, based on data of soil salinity in dry season. The boundaries of these zones, however, are not static and their precision remain tentative owing to high variability of salinity conditions within the zones. The general feature of soil salinity is that it increases from east to west. The increasing trend of soil salinity from north to south is not uniform throughout the forest.

The Sundarbans forest hosts one of the richest natural genepool for forest flora and fauna species in the world including the most notable one: the Bengal Tiger (*Panthera tigris*). The forest has been endowed with a number of commercially important mangrove species. A forest inventory completed by Chaffey et al. (1985) reported ten forest types consisting of eight dominant plants. The most common species are sundri (*Heritiera fomes*) from which the forest gets its name and gewa (*Excoecaria agallocha*). It is also reported that *H. fomes-E. agallocha* forest type covered the largest area (29.45%) followed by pure *H. fomes* (21%). *Heritiera fomes* constitutes about 65% of the total merchantable timber (Chaffey et al., 1985). The other important tree species include goran, golpatta (*Nypa fruticans*) and keora. In general, dicotyledonous tree species are represented in the forest by 22 families and 30 genuses, while Rhizophoraceae is represented by all the four known genera and at least six species. The shrubs are represented by 12 species belonging to 11 genus under seven families. Eleven different species of climbers belonging to six families have so far been identified in the Sundarbans. Orchids of 13 species and ferns of seven species are identified in the forest (Karim, 1994).

There is a wide variation of vegetation in the Sundarbans. It varies from multi-storied forest forming closed canopies to scrubby bushes with widely dispersed stunted trees. Moderate to relatively freshwater areas (oligohaline zone) support the best quality forest. While in the sea front areas (polyhaline zone) – exposed to tidal interactions and increased salinity – the forest consists of trees and shrubs of poor health. In the Bangladeshi part of the Sundarbans, only two thirds of the total forested area had canopy closure of 70% or more in 1984, while the rest of the forest had poor canopy density (Chaffey et al., 1985). The Indian part of the forest is reportedly degraded further, having very poor canopy density, especially along the south-western parts of the forest areas (FSI, 1999). The general vegetation types with the dominant species in respect to salinity zones are presented in Table 2.

River water is the primary source of nutrients and other compounds, supplying nutrient mineral, essential for meeting physiological demand of mangrove species, and toxicants. Phenological conditions of the mangroves are highly adapted to variable saline conditions: a few species shed leaves during periods of extreme dry conditions and high soil salinity, while most mangroves shed fruits during monsoon – taking advantage of the best natural condition to disperse the seeds by wave action, and when chances of survival of the propagules are higher due to low salinity of the river water. Historical records suggest that the mangroves of the Sundarbans have evolved under the influence of low salinity; *H. fomes* and other less saline water tolerant species occurred in great abundance in the western parts of the forest about 5000 years B.P. (Blasco, 1975).

In addition to the variety of mangrove species, the forest supports a variety of wild animals, birds, reptiles while the water bodies host a large range of sweet and brackish water fish, shrimps, crabs, mollusks, shellfish, turtles and snakes. According to Shah (1998), 53 species of pelagic (27 families), 124 species of demersal (49 families), 24 species of shrimp (five families), seven species of crab (three families), four species of lobster, two

Table 2: General vegetative types in respect to soil salinity zones

Salinity zone	*Land form types*	*Vegetation type/major species*
Oligohaline (i.e., low salinity)	Mudflat (outer zone)	Mixed. P. karka (nol khagra), N. fruticans, E. agallocha
	Mudflat (interior)	Pandanus foetidus (kewa katta)
	Mudflat-ridge (levees) Backswamps	Stratum A: H. fomes (D), S. apetala, A. officinalis Starum B: E. agallocha Stratum C: C. ramiflora (shingra)
Mesohaline (i.e., moderate salinity)	Mudflat (interior)	P. coarctata, N. fruticans, S. apetala, E. agallocha (D)
	Mudflat (inclined slope)	S. apetala, S. caseolaris, A. officinalis, N. fruticans, H. fomes (D), E. agallocha (D), C. decandra (D) etc.
	Backswamps	H. fomes (D) in association with A. cuculata and B. gymnorhiza as undergrowth
Polyhaline (i.e., high salinity)	Mudflats	Upper stratum: S. apetala (D), A. corniculatum with A. ilicifolius as undergrowth A. marina (D) in sea fronts with A. corniculatum and S. apetala
	Backswamps (at low elevations)	Mixed growth. E. agallocha (D) in upper story and C. decandra as undergrowth. Sparse canopy with degraded quality

Source: Ahmed, 1998. *Note:* D stands for dominant species.

species of snail, six species of mussel and three species of turtle currently inhabit the forest ecosystem. About 35 species of birds of prey are found in the forest, which include Accipitridae, Falconidae, and owls. About 84 species of migratory birds have also been found in surrounding areas of the forest (Hussain and Acharya, 1994). More than 50 species of reptiles and eight species of amphibians also occur in the Sundarbans. Snakes including king cobra and vipers are observed in the forest.

The creeks of the forest are nurturing grounds for the shrimp larvae and fries. Some fish and shrimp species caught from the forest provide protein to the diet of millions of people, while shrimp is exported to earn foreign currency. Important animals include the spotted deer (Axis axis), giant estuarine crocodile (Crocodylus porosus), gharial (Gavialis gangeticus), wild boar (Sus scrofa), Indian otter (Lutra perspicillata), and the most notable one

– the Bengal Tiger (Panthera tigris tigris). According to recent estimates, a total population of about 350 tigers, 40 to 70 thousand rhesus macaques, 50 to 80 thousand spotted deer, 20 thousand wild boars and 20 thousand smooth-coated otters live in the Bangladeshi parts of the forest. Apart from being a source of commercially important timber species and shrimp fries, the forest provides direct livelihoods for over a million households. Forestry Master Plan estimated that the Sundarbans forest provides about 0.3 million tons of fuelwood annually (MOEF, 1995).

5.1 Implications of Climate Change and Sea-level Rise on the Sundarbans

As indicated earlier in this book (referring to Chapter 6 on availability of water; Mirza et al.), global warming will cause: (a) a reduction of flows of the coastal rivers supplying freshwater to the forest during the low-flow season, leading to salinity intrusion; (b) an increase in susceptibility to saline inundation caused by sea-level rise; and (c) a change in morphological dynamics of the forest floor due to changes in tidal height as well as changes in inundation pattern. The impact on salinity intrusion might be compounded with climate change induced increased incidence of cyclonic storm surges in the Bay. Storm surges will inundate high levees and back swamps that normally do not get submerged with saline water, and thereby will be affected by salinity. In addition, increased rainfall runoff will provide increased freshwater discharge to all the major distributaries of the Ganges supplying freshwater to the Sundarbans. But increased sea level would cause backwater effect, thereby delaying the discharge process in the estuary. As a consequence, there would be relatively prolonged inundation in the Sundarbans areas in monsoon months (July-September) and also, increased rate of sedimentation/siltation in the back swamps and creeks inside the forest area (Ahmed et al., 1998). However, it may be expected that the impacts of dry season flow regime will be much higher compared to that for the monsoon season.

The mangrove species of the forest are highly influenced by salinity regimes of surface and groundwater systems and, more significantly, of soil. Not only soil salinity dictates vegetation types, systematic studies suggest that natural regeneration of vegetation and forest succession also depends on salinity regime (Karim, 1994; Siddiqi, 1994). Siddiqi (1989) noticed that germination of seedlings is directly linked to salinity, while a decrease in regeneration and growth is expected with increasing salinity. Seedling survival of mangrove species is also influenced by soil salinity. The seedling of Sundri (*H. fomes*) survives more in the less saline zone than in strongly saline zone (Siddiqi, 1989). Figure 8 provides ample evidence to highlight the relation between soil salinity and seedling half-life for the major mangrove species, *H. fomes*. Figure 9 represents effects of soil salinity on general health of major mangrove species of the forest.

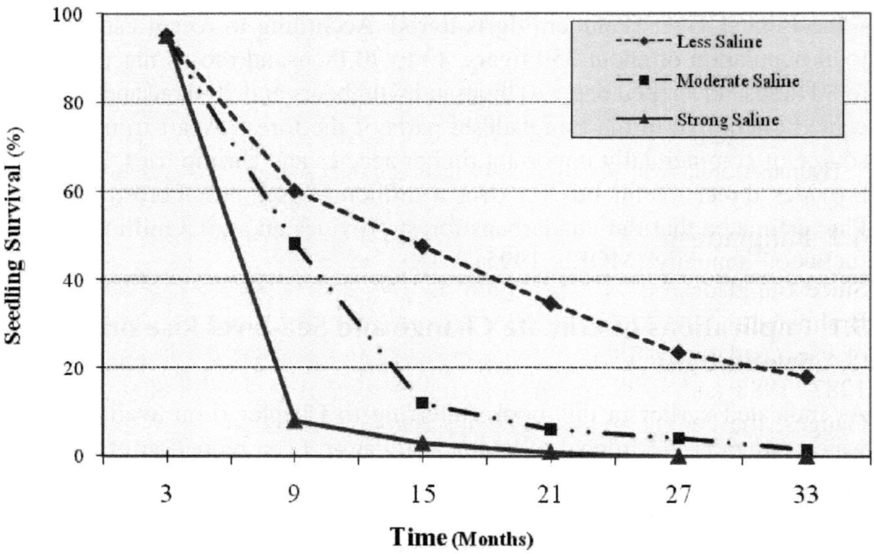

Fig. 8: Seedling half-life for *H. fomes* in different salinity zones.
Source: Siddiqi, 1989

Ahmed (1998) argues that, with increased soil salinity inside the forest under climate change compared to present salinity levels, it is highly likely that forest regeneration and succession will be adversely affected. This would, in turn, affect the long-term sustainability of the ecosystem. Furthermore, the highly dense human settlements just outside the forest area would restrict the species migration to less saline areas. A combination of the two effects would therefore threaten the very existence of the ecosystem.

Even if the forest manages to survive such threats due to increased salinity, it is highly likely that the present species composition will change (Ahmed et al., 1998). The Indian side of the forest bears witness to the increasing salinity and consequent gradual disappearance of freshwater thriving *H. fomes* trees (Chaudhuri and Choudhury, 1994). It is also reported that the dominant and economically important species is now replaced by low value shrubs and saline tolerant grasses. It is further reported that the remaining sundri trees, those that survived in the less saline areas, shrunk in size. Similar phenomenon is expected to occur due to increased salinity in the Bangladesh's side of the forest, as a result of climate change.

Based on the above ground observations, Ahmed (1998) reported that the existing oligohaline zone might be completely transformed into mesohaline zone, with a resulting overall degradation of the forest vegetation cover. It is likely that the areas with best quality standing timber predominated by long *Avicennia officinalis* and *H. fomes* will be replaced by inferior quality tree species, predominated by *H. fomes-E. agallocha* and *Ceriops-Excoercaria* forest types. In a similar fashion, a significant part of the existing mesohaline

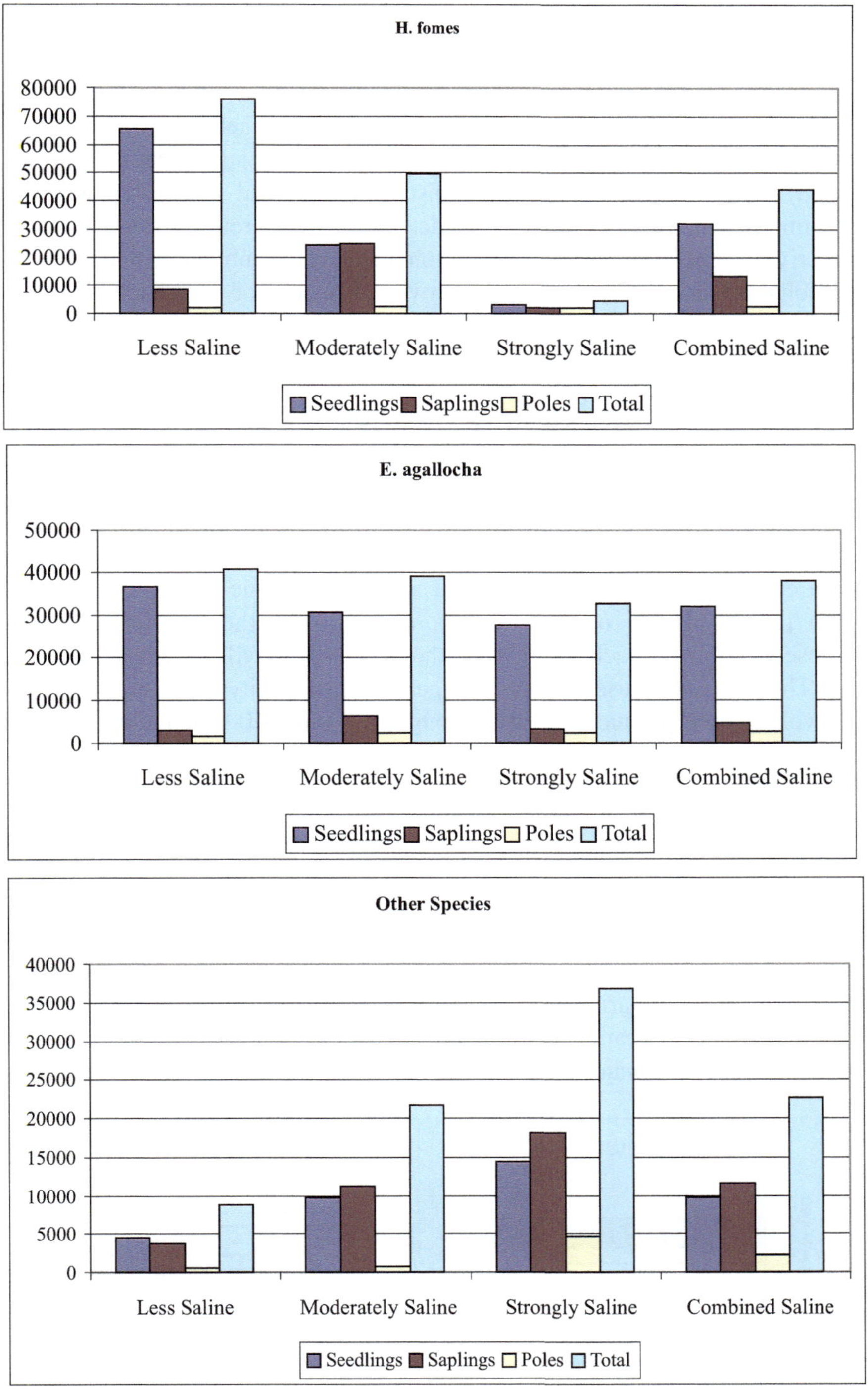

Fig. 9: Vegetative health of major mangrove species in relation to soil salinity.

zone, especially the south-western parts of the Sibsa river up to 22°00 N might be transformed into polyhaline zone. As a result, grasses, thorny herbs/ shrubs and trees of poor quality are likely to dominate, which will gradually replace woody tree species. Under such conditions vegetation canopy would become sparse and plant height would be reduced significantly (Ahmed, 1998). Chaffey et al. (1985) demonstrated that total merchantable wood volume per unit area of forest land declines with increasing salinity of soil and river water. Disappearance of oligohaline areas combined with decreasing mesohaline areas would result into over 50% loss of merchantable wood from the Sundarbans. Increase in salinity in the Indian side of the forest is likely to further degrade the forest vegetation, which will in turn negate the currently undertaken regeneration and conservation activities.

The issue of salinity will not be a factor in monsoon. As expected, increased flow volume due to increased rainfall runoff will push salinity regime southward. However, variations in water level along the estuary and changes in duration of inundation on the forest floor might cause adverse effects on forest vegetation. Increased drainage congestion in the estuaries, as a consequence of increased runoff and high oceanic stages, might result into increased rate of sedimentation/siltation in the submerged areas. Consequently, the back swamps of the Sundarbans will gradually be filled up. This will also cause changes in regeneration capacity of the major species as well as their productivity in the long run. Figure 10 schematically shows the detailed results of field-plot observations, as reported in EGIS (2005). Moreover, the low-lying mudflats would be more frequently submerged which could slow down forest succession process.

5.2 Concluding Remarks

The Sundarbans is a unique ecosystem located in South Asia. The balance between freshwater flows and saline inundation, as a function of climatic and hydrological variability, plays significant role towards defining the suitability of the mangrove vegetation cover of the forest. The potential reduction in freshwater flows in the dry season and subsequent salinity

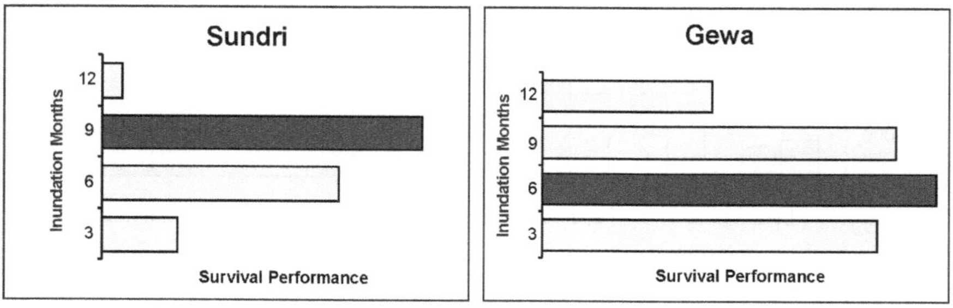

Fig. 10: Schematic representation of survival performance of dominant species with duration of inundation (From: EGIS, 2005).

intrusion in one hand, and an increase in duration of inundation in monsoon on the other hand can jeopardise the succession process of the biota, their regeneration, growth and survival under climate change conditions. This will affect the long-term health of the forest vegetative cover. Climate change, therefore, will have grave consequences on the unique mangrove ecosystem, which deserves priority actions now to safeguard the forest.

6. MANGROVES

Mangrove habitats from the tropics belt are of great ecological and socioeconomic importance. However, these resources are under constant threats from industrialization and the ever-increasing population, and their demands for shelter and food. Lately, the global climate variations have developed major concerns to these ecosystems because of their proximity to the sea (Jagtap et al., 2003).

6.1 Mangrove Status in the Indian Sub-continent

Total mangrove cover in the region, including Sri Lanka, has been estimated to be about 10,000 km^2 (Jagtap and Nagle, 2005). Major formations occur in the deltaic regions of Indus, Mahanadi and Sunderbans, and also in the Gulf of Kutchchh, Andaman and Nicobar group of islands, and along the west coast and Jafna peninsula of Sri Lanka.

The flora comprises more than 70 species of mangroves and a large number of associate macro and microphytes. Maximum species diversity has been recorded along the Indian coasts, while the Sunderbans delta of Bangladesh has the largest areal extent. However, various natural and anthropogenic processes have left these ecosystems, particularly, those of the Indian sub-continent, in a degraded form. Major natural forces causing destruction of mangroves include cyclones, storms, floods and occasional tsunamis, while among the anthropogenic pressures, urbanization, and agricultural and industrial activities form the major deteriorating elements. Other factors like cattle grazing, collection of firewood and fodder, dumping of wastes, etc. result in low productivity and formation of dwarf mangroves, while infestation by oysters, barnacles and different pastes on vegetative parts cause mortality of mangroves (Jagtap et al., 2002). The recent (December 2004) Asian tsunami has caused intensive damage to the mangroves from Andaman and Nicobar group of islands (Ramachandran et al., 2005).

Considering the socio-economic and ecological values, these habitats in the sub-continent have been protected vide the various Coastal Regulation Zone Acts. Conservation and restoration efforts are being made through the various government and private agencies, as well as by involving public sectors.

6.2 Impact of Sea Level Rise

As a result of global climate change, the sea level is expected to rise by 0.09 to 0.88 m by the year 2100 (IPCC, 2001). A relative annual sea level rise of 0.67 $mm.y^{-1}$ has been estimated by Gable and Aubrey (1990) for the South Asian region, while Das (1991), analyzing the tide gauge records and recent satellite altimeter data, has estimated that the sea level along the Indian coast will rise at the rate of 3-5 $mm.y^{-1}$.

The mangrove habitats, due to their proximity to the sea, will be the immediate targets of the projected sea level rise (SLR). These ecosystems can keep pace with a SLR of 0.08-0.09 m. $(100\ y)^{-1}$ but remain under stress at rates between 0.09 and 0.12 m. $(100\ y)^{-1}$ (Ellison and Stoddart, 1991). Presently, the mangroves occur between 0.4 and 0.9 m above the mean sea level (Ellison, 1989). Consequently, a sea level rise, at the present rate, will have a devastating effect on the low-lying mangrove areas (Fig. 11). Additionally, the individual components of the habitat may change their response pattern to the increased temperature and CO_2 levels, and alteration in the hydrological regime in the ambience (Jagtap and Nagle, 2005). The altered rates of precipitation and evaporation, as a consequence of increased temperature, will affect the salinity of the marine habitats, which, in turn will affect the mangroves ecosystems, as the individual mangrove species have particular salinity tolerance. Very low salinity is a prerequisite for the early

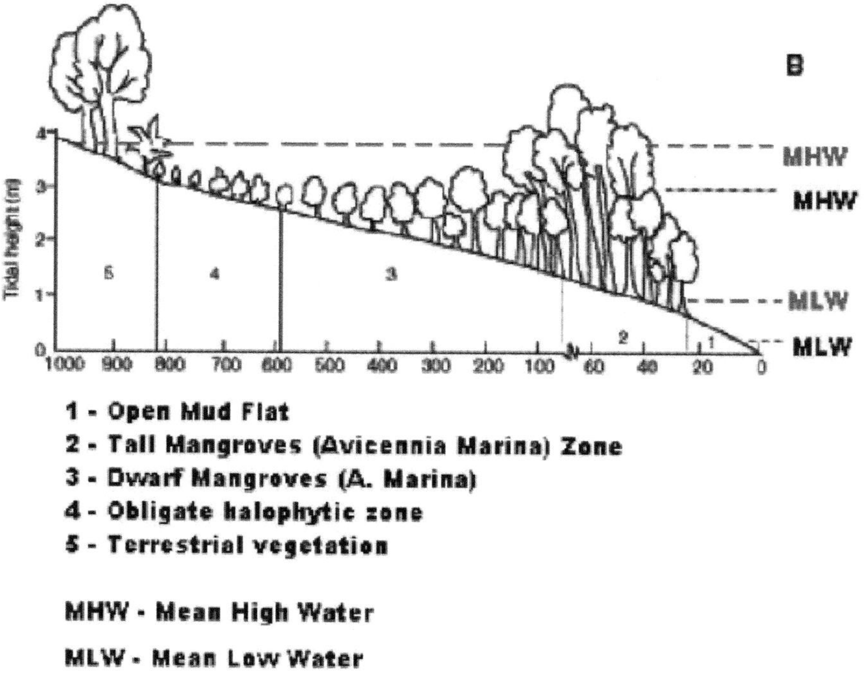

Fig. 11: Likely impact of SLR on mangroves from Gulf of Kutchchh, India. *Source:* Jagtap and Komarpant, 2003

growth of the seedlings of all species. Also, the change in temperature may lead to changes in the pattern of flowering and fruiting, and root respiration, growth and turn over rates.

Among the various kinds of the associated biota, the soft bodied animals and bivalves from the mangrove habitats are highly sensitive to changes in temperature. Consequently, mangrove biota, such as snails, crabs, shellfish, juvenile fish, etc., will be adversely affected by increase in temperature. While the biota with specific tolerances within the tidal spectrum will migrate farther landward, in regions with limited land margins there would be no scope for further expanse in response to sea level rise.

6.3 Recommendations

Being complex in nature, a multidimensional long-term approach is required for monitoring the impact of climate variation on this ecosystem. A reliable estimate of their area coverage should be made, using larger scale remotely sensed data. Coastal biodiversity regulations need to be strictly implemented for mitigating emissions of the green house gases. International programmes should be initiated to monitor the mangrove habitats in selected locations, to understand:

- Rate of atmospheric CO_2 conversion into biomass and impacts of temperature and saturated CO_2 levels on carbon fixation of mangrove species.
- Change in forest structure – density, biomass, annual growth, litter production, associated biota, etc.

7. IMPACTS OF CLIMATE CHANGE

As a consequence of man's activities, primarily the combustion of fossil fuels and change in land use and land cover, the emissions of green house gases (GHGs), like CO_2, CH_4 and N_2O, etc., have been increasing since pre-industrial era. However, the rates of emission of these gases have attained alarming proportions during the past 200 years, especially since 1750, leading to their accumulation in the atmosphere. For example, the atmospheric concentration of CO_2 has increased by 31.4 ± 4% over the period 1750-2000 while the CH_4 concentration has increased by 151 ± 25% during the same period, mainly as a result of emissions from the fossilized fuel, livestock, rice cultivation and landfills (IPCC, 2002). Increase in concentrations of green house gases in the atmosphere leads to warming of earth's temperature, e.g., global mean surface temperature has risen by 0.6 °C over the past 100 years, the year 1998 being the warmest year.

7.1 The Projections

The atmospheric concentration of CO_2 was 368 ppm in the year 2000, showing an increase by 31.4 ± 4% over the pre-industrial level, while the concentration of CH_4 had increased by 151 ± 25% over the same period (from 1750 to 2000). The projected atmospheric CO_2 increase over the next 100 years range from 540 to 970 ppm. Giving a margin of variation of –10 to +30% due to uncertainties, the range for the year 2100 is from 490 to 1200 ppm, which is 75-350% above the pre-industrial level. The globally averaged surface temperature is projected to increase by 1.4-5.8°C over the period 1990-2100, with the land areas warming more than the oceans (the projected temperature rise for the periods 1990-2025 and 1990-2050 being 0.4-1.1 °C and 0.8-2.6 °C, respectively). The most notable areas of warming are expected to be the landmasses of North America and northern and central Asia. While the warming will be less than the global mean change in South and Southeast Asia (2-3 °C over most of South Asia and the Northern Indian Ocean), in summer and in southern South America, in winter.

As a consequence of global warming, the sea surface temperature (SST) is expected to rise by 1-2 °C by the year 2100, while the globally averaged precipitation is expected to increase over the high latitude regions in both summer and winter. Of special interest is the projected increase in precipitation in southern and eastern Asia during summer. Models also suggest an increase in the wind intensity of tropical cyclones by 5-10% leading to increase in the precipitation rates by 20-30%. The rise in surface temperature is also expected to result in further reduction of the extent of snow cover, permafrost and sea ice cover in the northern hemisphere. The Antarctic sea ice will gain mass due to increased precipitation, while the Greenland ice cover will lose mass due to increased runoff.

A rise in surface temperature and in SST will lead to a rise in sea level (SLR) due to thermal expansion of the oceans and loss of mass from glaciers and ice caps. Consequently, the global sea level is projected to rise by 0.09-0.88 m between 1990 and 2100 with substantial regional variations, the projections for the periods 1990-2025 and 1990-2050 being in the range 0.03-0.14 m and 0.05-0.32 m, respectively.

In this article, the impacts of climate change on the coastal and marine environments are analyzed in relation to increase in CO_2, and rise in surface temperature, SST and sea level.

7.2 Rise in Temperature (Surface Temperature and SST)

Formation of deep waters, like North Atlantic Deep Water and Antarctic Bottom Water, takes place in high latitude zones of the oceans, during winter. These waters, as they form at the surface during winter, contain high concentrations of oxygen and nutrients. A rise in sea surface temperature associated with a higher freshwater runoff derived from increased precipitation and melting of ice in high latitudes will result in weakening of the processes of formation of these waters, and their subsequent sinking and thermohaline circulation. Simultaneously, the ability of the oceans for uptake and storage of oxygen and other gases, including the green house gases, will be reduced to a large extent. This will affect not only the cycling of the GHGs but also the circulation and ventilation of the deep waters of the northern Indian Ocean, which depend entirely on these high latitude waters for their supply of oxygen. This will lead to stagnation of the deep waters and contribute to the intensification of anoxia and production of the associated green house gases (CO_2, CH_4 and N_2O). Similarly, the intermediate waters of the northern Indian Ocean depend largely on advection of intermediate waters from the marginal seas, the sub-tropical sub-surface water, and the Indonesian Throughflow, for their oxygen supply. Consequently, a weakening in the formation and circulation of these waters along with their reduced ability to hold the gas, will result in further intensification of the existing sub-oxic environment.

The global warming is expected to result in stronger pressure gradient between the land and the oceans. Consequently, the intensity of the monsoon winds is expected to increase leading to intensification of coastal upwelling and fuelling higher biological production in these areas. This, in turn, will lead to further intensification of the oxygen minimum layer and increased emission rates of green house gases. Additionally, the increased precipitation during monsoon will result in lowering the salinity of coastal surface waters due to higher runoff. This will lead to intensification of the low salinity lens at the surface, thus causing strong stratification having a double effect: (a) a strong stratification will inhibit vertical mixing in the water column, thus preventing the injection of nutrient-rich subsurface waters into the euphotic layer thereby affecting the biological production, and (b) it will cause further intensification of anoxia in coastal waters off the west coast of India, thus leading to higher emission of GHGs. It will also lead to shifts in the geographic distribution of marine biota leading to changes in biodiversity – organisms favouring brackish water conditions may move offshore while the organisms preferring freshwater may thrive.

Increased precipitation associated with the projected increase in the wind intensity of tropical cyclones will lead to increase in erosion, in erosion-prone areas, like the delta regions of Indus and the Bay of Bengal Delta system, causing increased turbidity in coastal surface waters of the northern Indian Ocean. This will cause reduction in biological production through inhibition of light penetration, especially in the Bay of Bengal, thus affecting its ability to sequester atmospheric carbon. However, the ballasting process of the mineral particles (Ittekot et al., 1991), as discussed in Section 1 of this article, could neutralize the effect caused by increased turbidity. Such changes in the properties of the surface waters may lead to alteration in biodiversity at ecosystem and landscape scale, which, in turn will affect the regional and global climate through changes in uptake and release of green house gases and changes in albedo and evapotranspiration.

It is projected that an increase in SST will be accompanied by an increase in the frequency and the intensity of the tropical cyclones and extreme precipitation events. About 5.5% of the average 80 tropical cyclones forming annually in the world oceans occur in the Bay of Bengal, while about 1% occurs in the Arabian Sea (Ali, 1999). This will lead to higher storm surges and more frequent floods as the return periods for the extreme precipitation events will decrease, especially affecting low lying cyclone prone areas of Bangladesh, West Bengal and along the east coast of India.

The landmasses of Central Asia are projected to undergo warming, which will exceed global mean warming by more than 40% (IPCC, 2002). This will weaken the process of winter convective sinking of surface waters in the northern Arabian Sea – a process that helps to inject subsurface nutrients into the euphotic layer – thus affecting its biological productivity.

Increased SST in association with higher nutrient content in surface waters can lead to proliferation of harmful algal blooms. Naqvi et al. (1998) have observed such blooms (*Noctiluca dinoflagellalte*) in the coastal waters off the west coast of India during the summer monsoon. Such blooms can cause economic loss through shellfish closure, impact on tourism, reduction of estuarine fisheries and coastal primary productivity, deterioration of fisheries and mortality of fish and shellfish. Similarly, a rise in SST with simultaneous increase in biological production will lead to enhanced rates of metabolic/microbial processes in the surface mixed layer. This will result in increase in pCO_2 in the mixed layer with the concomitant increase in emissions to the atmosphere under turbulent conditions. While the warming of surface water is expected to cause increased emissions of N_2O and CH_4 gases from north Indian Ocean, due to increased turbulence.

Corals are very sensitive to changes in SST. A 1 °C rise in SST over the seasonal average causes bleaching. It is projected that there will be widespread bleaching of corals over the next 100 years (IPCC, 2002). It is also projected that the amplitudes of El Nino events will show small increase over this period. In such events, a 3 °C rise in SST will result in severe coral mortality, if the high temperature persists over several months.

Increased turbidity due to higher precipitation and soil erosion, associated with lowering of salinity will affect the corals and coral reefs in Srilanka, Gulf of Kutchchh (India), and in the Narikal Jinjira Island in Bangladesh

Changes in temperature may lead to changes in the pattern of flowering, fruiting, root respiration, and growth and turn over rates in mangroves. The soft bodied animals and bivalves from the mangrove habitats are highly sensitive to changes in temperature. Consequently, mangrove biota, such as snails, crabs, shellfish, juvenile fish, etc., will be adversely affected by increase in temperature. Similarly, global warming will cause reduction in the flow of coastal rivers supplying freshwater to the Sunderban forests during the low-flow season, leading to salinity intrusion. Storm surges, with the projected increased frequency, will inundate high levees and back swamps that normally do not get submerged with saline water, and thereby will be affected by salinity.

7.3 Sea Level Rise

According to the IPCC (2002) projections the mean sea level is expected to rise by 0.09-0.88 m during the 21st century, i.e., an average annual rate of 0.8-8.8 $mm.y^{-1}$. However, Gable and Aubrey (1990) have estimated a sea level rise at the rate of 0.67 $mm.y^{-1}$ for the south Asian region, while the estimates of Das (1991) projects a SLR of 3-5 $mm.y^{-1}$ along the Indian coast.

Under the projected sea level rise, the coastal ecosystems will be subjected to increased levels of inundation, storm flooding, coastal erosion, seawater intrusion into fresh groundwater aquifers and encroachment of tidal waters into estuaries, creeks, rivers, aquaculture farms, agricultural lands, etc. Seawater encroachment of freshwater aquifers will greatly affect the population of coastal villages of the region, who depend entirely on ground water (well water) for their freshwater supply for domestic use. As seen in Section 7.2 above, a rise in SST will be accompanied with increased frequency and intensity of tropical cyclones. Bangladesh, a country having very low and flat topography with about 10% of the land hardly 1 m above the sea level, will be the most affected with the projected sea level rise, causing coastal erosion and coastline recession. Ali (1999) has estimated a loss of land due to recession, in Bangladesh, of the order of 5.8 to 11.2 km^2 for a SLR of 0.3 and 0.75 m, respectively. This is in addition to the loss of land due to SLR (for a SLR of 0.3 m the estimated loss of land being about 8,000 km^2).

However, areas showing greatest sensitivity to accelerated sea level rise include coral atolls, reef islands, low lying deltaic and coastal plains, barrier coasts, sandy beaches, coastal wetlands, estuaries and lagoons, where the major impact of accelerated sea level rise will be loss of land due to submergence, e.g. in the Lakshadweep group of islands where the land seldom rises above 2 m. The reefs protecting these islands may lose their value due to submergence leaving the islands exposed to storm surges and erosion. Similarly, the delta regions, such as the Bengal delta and Indus delta, where human interferences – construction of dams, irrigation channels, etc. – in the river system has resulted in reduction of sediment transport (sediment starvation), the deltas may not be able to grow vertically and keep pace with the sea level rise, thus causing submergence.

According to Ellison and Stoddart (1991) mangroves can keep pace with a SLR of 0.08-0.09 m $(100y)^{-1}$, but remain under stress at rates of 0.09-0.12 m $(100y)^{-1}$. Consequently, a SLR at the projected rate will have a devastating effect on low-lying mangrove ecosystems. In case of Sunderbans, increased rainfall/runoff will provide increased freshwater discharge to all major distributaries of the Ganges supplying freshwater to the forest. But increased sea level would cause backwater effect, thereby delaying the discharge process in the estuary. As a consequence, there would be relatively prolonged inundation during the monsoon months (July-September) and also, increased rate of sedimentation/siltation in the back swamps and creeks inside the forest area (Ahmed and Alam, 1998).

Among the corals, the fringing and barrier reefs that grow vertically at a comparatively faster rate will be able to keep pace with the accelerated SLR, provided there is no frequent occurrence of ENSO events. However, the atoll reefs that grow at a comparatively slower rate, will be completely submerged by the year 2030 (the analysis period), if the present rate of SLR persists.

7.4 Effect of CO_2 Rise

An increase in CO_2 levels in the atmosphere will result in a corresponding increase in the dissolved CO_2 concentration in the surface seawater. This will cause a lowering of pH of the surface seawater, thus, further enhancing its pCO_2 level and making the waters more corrosive to skeletal carbonates, and causing further migration of the aragonite saturation depth in the Arabian Sea (Sarma et al., 2002). Similarly, with the expected doubling of the atmospheric CO_2 level, the flux of dimethyl sulphide is expected to increase in the tropics north of the equator (Bopp et al., 2002).

Further, a reduction of pH of surface water will result in the reduction in its $CaCO_3$ content, thus affecting the calcification process in corals and coral reefs. Under these conditions, the vertical growth of corals will be slowed down, and consequently, the corals may get submerged not being able to keep pace with the sea level rise.

8. INFORMATION GAPS AND RESEARCH NEEDS

A considerable amount of work has been carried out in the recent past to study and understand the processes controlling the biogeochemical cycling of nutrients and carbon in the Northern Indian Ocean, especially the Arabian Sea. As a result large amount of data from this region has been collected by the scientific community; however, there are some areas where important information is lacking. For example, not much information is available on the fluxes of nutrients including the biologically active trace metals like iron, through the various pathways, such as, aeolian transport (both dry and wet precipitation), river discharge, land runoff, point discharges from industrial and municipal sources and diffuse groundwater discharges. Such information is important in order to understand the role played by the coastal and marine ecosystems of the region, in the biogeochemical cycling of carbon and nutrients and their control on the global climate through emissions of radiatively important gases.

Limited observations have indicated enrichment of CO_2 in Indian estuaries; however, no major effort has been made to study the carbon fluxes from the major estuarine systems of the sub-continent. Reliable data on such fluxes is badly needed.

More studies are needed to explain the contrasting differences in the biogeochemical processes between the Arabian Sea and the Bay of Bengal – while the Arabian Sea experiences intense denitrification and acts as a perennial source for atmospheric CO_2, the Bay of Bengal, on the other hand, acts as a seasonal sink for the atmospheric CO_2 and does not experience denitrification – inspite of the fact that both are situated at similar latitudes, land-locked in the north and subject to similar physical forcings with high biological production, and both having extensive oxygen minimum zone. While the Arabian Sea has received much international attention, primarily due to high biological production and low oxygen containing subsurface waters, it is only recently that the climatically important Bay of Bengal and the equatorial Indian Ocean started receiving attention particularly for biogeochemical and gas exchange studies.

Episodic events can emit amounts of gases that are many folds higher than those computed using data measured on board ships under relatively calm conditions. Such measurements could, therefore, result in underestimates. Attempts need to be intensified to map the abundance of these gases in surface waters of the key biogeochemical and turbulent regimes during extreme climatic events using automated systems to make realistic estimates of gas emissions during catastrophic events.

Sunderbans is a unique ecosystem in South Asia. The effect of potential reduction of freshwater flows during the dry season and the consequent salinity intrusion, on one hand, and the prolonged inundation during monsoon on the other hand, need to be studied.

The mangroves being complex in nature, a multidimensional long-term approach is required for monitoring the impact of climate variation on this ecosystem. A reliable estimate of their area coverage should be made, using larger scale remotely sensed data. International programmes should be initiated to monitor the mangrove habitats in selected locations, to understand the change in forest structure – density, biomass, annual growth, litter production, associated biota, etc. – and their ability to convert atmospheric CO_2 into biomass, and impacts of increased temperature and CO_2 levels on carbon fixation by the mangroves species.

Effects of ENSO and tropical cyclones on ocean productivity have to be addressed in the future. Similarly, the role of the Indonesian Throughflow in the biogeochemical processes in the Northern Indian Ocean need to be studied.

Effects of construction of dams, irrigation systems and water diversion on coastal ecosystems, such as deltas, mangroves, coral reefs, wetlands, seagrasses and low-lying coastal islands need to be studied, regarding their stability and health.

REFERENCES

Abbas, A.T., 1984. The Ganges Water Dispute. University Press Limited, Dhaka.

Ahmed, A.U., 1998. Ecological Security in a Warmer World: The Case of the Sundarbans. BUP Monograph. Bangladesh Unnayan Parishad, Dhaka, Bangladesh (Mimeo).

Ahmed, A.U. and Alam, M., 1998. Development of Climate Change Scenarios with General Circulation Models. *In:* Vulnerability and Adaptation to Climate Change for Bangladesh. S. Huq, Z. Karim, M. Asaduzzaman and F. Mahtab (eds.). Kluwer Academic Publishers, Dordrecht, The Netherlands. pp. 13-20.

Ahmed, A.U., Alam, M. and Rahman, A.A., 1998 Adaptation to Climate change in Bangladesh: Future Outlook: *In*: Vulnerability and Adaptation to Climate Change for Bangladesh. S. Huq. Z. Karim, M Asaduzzaman and F. Mahtab (eds.). Kluwer Academic Publishers, Dordrecht, The Netherlands. pp. 125-143.

Alam, M., 1972. Tectonic classification of the Bengal Basin. *Geol. Soc. Am. Bull.*, **83,** 519-522.

Alam, M., 1989. Geology and depositional history of Cenozoic sediments of the Bengal Basin of Bangladesh. *Palaeogeography, Palaeoclimatology, Palaeoecology*, **69,** 125-139.

Alam, M.M., 1995. Tide-dominated sedimentation in the upper tertiary succession of the sitapahar anticline, Bangladesh. *In:* Tidal Signatures in Modern and Ancient Sediments, B.W. Fleming and A. Bortholma (eds.). International Association of Sedimentology (Special Publication) 24, pp. 329-341.

Alam, M., 1996. Subsidence of the Ganges-Brahmaputra delta of Bangladesh and associated drainage, sedimentation, and salinity problems. *In:* Sea-level Rise and Coastal Subsidence. J.D. Milliman and B.U. Haq (eds.). Kluwer Academic Publishers, Dordrecht, pp. 169-192.

Alam, M., Alam, M.M., Curray, J.R., Chowdhury, M.L.R. and Gani, M.R., 2003. An overview of the sedimentary geology of the Bengal Basin in relation to the regional tectonic framework and basin-fill history. *Sedimentary Geology*, **155,** 179-208.

Ali, Anwar, 1999. Climate change impacts and adaptation assessment in Bangladesh. *Climate Research*, **12,** 109-116.

Altabet, M.A., Higginson, M.J. and Murray, D.W., 2002. The effect of millennial-scale changes in Arabian Sea denitrification on atmosphere. *Nature*, **415** (6868), 159-162.

Babu, M.T., Prasanna Kumar, S. and Rao, D.P., 1991. A subsurface cyclonic eddy in the Bay of Bengal. *Journal of Marine Research*, **49,** 403-410.

Bange, H.W., Ramesh, R., Rapsomaniks, S. and Andreae, M.O., 1998. Methane in surface waters of the Arabian Sea. *Geophysical Research Letters*, **25,** 3547-3550.

Banse, K., 1987. Seasonality of phytoplankton chlorophyll in the central and northern Arabian Sea. *Deep Sea Research*, **34(5/6)**, 713-723.

Bauer, S., Hitchcock, G.L. and Olson, D.B., 1991. Influence of monsoonally-forced Ekman dynamics upon surface layer depth and plankton biomass distribution in the Arabian Sea. *Deep Sea Research*, **38(5),** 531-553.

Berner, U., Poggenburg, J., Faber, E., Quandfasel, D. and Fische, A., 2003. Methane in the ocean waters of Bay of Bengal: Its sources and exchange with the atmosphere. *Deep Sea Research **II***, **50,** 925-950.

Bhattathiri, P.M.A., Pant, A., Sawant, S., Gauns, M., Matonkar, S.G.P. and Mohanraju, R., 1996. Phytoplankton production and chlorophyll distribution in the eastern and central Arabian Sea in 1994-1995. *Current Science*, **71(11),** 857-862.

Blasco, F., 1975. The Mangroves of India. Institute Francais de Pondichery, India.

Bopp, L., Aumont, O., Belviso, S. and Monfray, P., 2002. Potential impact of climate change on marine dimethyl sulfide emissions. *Tellus,* **55B,** 11-22.

Bouillon, S., Frankignoulle, M., Dehairs, F., Velimirov, B., Eiler, A., Abril, G., Etcheber H. and Vieira Borges, H., 2003. Inorganic and organic carbon biogeochemistry in the Gautami Godavari estuary (Andhra Pradesh, India) during pre-monsoon: The local impact of extensive mangrove forests. *Global Biogeochemical Cycles*, **17,** 1114, doi:10.1029/2002GB002026.

Brock, J.C., McClain, C.R., Luther, M.E. and Hay, W.A., 1991. The phytoplankton bloom in the northwestern Arabian Sea during the southwest monsoon of 1979. *Journal of Geophysical Research*, **96(C-11),** 20,623-20,642.

Brock, J.C. and McClain, C.R., 1992. Interannual variability in phytoplankton blooms observed in the northwestern Arabian Sea during southwest monsoon. *Journal of Geophysical Research*, **97(C-1),** 733-750.

Brock, J.C., Satyendranath, S. and Platt, T., 1994. A model study of seasonal mixed-layer primary production in the Arabian Sea. *Proceedings of the Indian Academy of Sciences* (Earth and Planetary Sciences), **103(2),** 163-176.

Burkill, P.H., Mantoura, R.F.C. and Owens, N.J.P., 1993. Biogeochemical cycling in the northwestern Indian Ocean: A brief overview. *Deep Sea Research* ***II*****, 40(3),** 643-649.

Calvert, S.E., 1987. Oceanographic controls on the accumulation of organic matter in marine sediments. *In:* Marine petroleum source rocks. J. Brooks and A.J. Fleet (eds.). Blackwell Scientific publications, London, U.K., pp. 137-151.

Calvert, S.E., Kallin, R.E., Toolin, L.J., Donahue, D.J., Southon, J.R. and Vogel, J.S., 1991. Low organic carbon accumulation rates in Black Sea sediments. *Nature*, **6230,** 692-695.

Chaffey, D.R., Miller, F.R. and Sandom, J.H., 1985. A Forest Inventory of the Sundarbans, Bangladesh. Main report. Overseas Development Administration, England. 196 pp.

Chaudhuri, A.B. and Choudhury, A., 1994. Mangroves of the Sundarbans, Vol. I India. IUCN-The World Conservation Union, Bangkok, Thailand.

Curray, J.R., 1991. Geological history of the Bengal geosyncline. *J. Assoc. Explor. Geophys.*, **XII,** 209-219.

Curie, R.I., Fisher, A.E. and Hargreaves, P.M., 1973. Arabian Sea upwelling. *In:* The Biology of the Indian Ocean. B. Zeitzschel (ed.). Springer-Verlag Berlin, Heidelberg, pp. 37-52.

Das, P.K. and Radhkrishna, M., 1991. Analysis of Indian tide gauge records. *Proc. Indian Academy of Sciences*, **100,** 177-194.

Demaison, G.J. and Moore, G.T., 1980. Anoxic environments and oil source bed genesis. *Organic Geochemistry*, **2,** 9-31.

Desikacher, S.V., 1974. A review of the tectonic and geological history of eastern India in terms of plate tectonic theory. *Journal of the Geological Society of India*, **15,** 134-149.

de Sousa, S.N., Sawkar, K. and Durga Prasada Rao, P.V.S.S., 1996a. Environmental changes associated with monsoon induced upwelling along the central west coast of India. *Indian Journal of Marine Sciences*, **25,** 115-119.

de Sousa, S.N., Dileep Kumar, M., Sardessai, S., Sarma, V.V.S.S. and Shirodkar, P.V., 1996b. Seasonal variability in oxygen and nutrients in the central and eastern Arabian Sea. *Current Science*, **71(11),** 847-851.

EGIS, 2000. Environmental Baseline of the Gorai River Restoration Project. Bangladesh Water Development Board, Dhaka.

EGIS, 2005. Impact of Sea Level Rise on Land Use Suitability and Adaptation Options. Inception Report submitted to the Ministry of Environment and Forest, under Sustainable Environmental Management Programme (SEMP), May 2005.

Elliott, A.J. and Savidge, G., 1990. Some features of upwelling off Oman. *Journal of Marine Research*, **48,** 319-333.

Ellison, J.C., 1989. Pollen analysis of mangrove sediments as a sea level indicator: Assessment from Tongatapu, Tonga. *Palaeogeography, Palaeoecology, Palaeoclimatology*, **74,** 327-341.

Ellison, J.C. and Stoddart, D.R., 1991. Mangrove ecosystem collapse during predicted sea level rise: Holocene analogues and implications. *Journal of Coastal Research*, **7,** 151-165.

Eysink, W.D., 1983. Basic considerations on the morphology and land accretion potentials in the estuary of the lower Meghna river. Land Reclamation Project, Dhaka, Bangladesh, Technical Report No. 15.

Fieux, M., Andrie, C., Charriaud, E., Ilahude, A.G., Metzl, N., Molcard, R. and Swallow, J.C., 1996. Hydrological and chlorofluoromethane measurements of the Indonesian throughflow entering the Indian Ocean. *Journal of Geophysical Research*, **101(C5),** 12433-12454.

FSI, 1999. State of Forest Report. Forest Survey of India (FSI), Ministry of Environment and Forests, Government of India, Dehra Dun.

Gable, F.J. and Aubrey, D.G., 1990. Potential coastal impacts of contemporary changing climate on South Asian Seas States. *Environmental Management*, **14(1),** 33-46.

George, M.D., Kumar, M.D., Naqvi, S.W.A., Banerjee, S., Narvekar, P.V., de Sousa, S.N. and Jayakumar, D.A., 1994. A study of carbon dioxide system in the northern Indian Ocean during premonsoon. *Marine Chemistry*, **47,** 243-254.

Goldberg, J. and Wilkinson, C., 2004. Global Threats to Coral Reefs: Coral bleaching, global climate change, disease, predator plagues, and invasive species. *In:* Status of Coral Reefs of the World: 2004. C. Wilkinson (ed.). Vol. 1. Australian Institute of Marine Science, 67-92 pp.

Gomes, H.R., Goes, J.I. and Saino, T., 2000. Influence of physical processes and fresh water discharge on the seasonality of phytoplankton regime in the Bay of Bengal. *Continental Shelf Research*, **20,** 313-330.

Gregg, W.W. and Conkright, M.E., 2002. Decadal changes in global ocean chlorophyll. *Geophysical Research Letters*, **29,** 10.1029/2002GL014689.

Gregg, W.W., Conkright, M.E., Ginnoux, P., O'Reilly, J.E. and Casey, N.W., 2003. Ocean primary production and climate: Global decadal changes. *Geophysical Research Letters*, **30,** 10.1029/2003GL016889.

Hoon, V., 1997. Coral reefs of India: Review of their extent, condition, research and management status. Proceedings of Regional Workshop on the Conservative and Sustainable Management of Coral Reefs. M.S. Swaminathan Research Foundation, Chennai, India. Bay of Bengal Programme (FAO), Chennai, India. December 15-17, B1-B35.

Hussain, Z. and Acharya, G. (eds.), 1994. Mangroves of the Sundarbans: Volume Two. IUCN-The World Conservation Union, Bangkok.

IPCC, 2001. Climate changes 2001. The Scientific Basis. Contribution of working Group I to Third Assessment Report of the inter-governmental panel on climate change (IPCC). Cambridge University Press. pp. 881.

IPCC, 2002. IPCC Technical Paper V. Climate change and Biodiversity. April, 2002.

Ittekkot, V., Nair, R.R., Honjo, S., Ramaswamy, V., Bartsch, M., Manginini, S. and Desai, B.N., 1991. Enhanced Particle fluxes in Bay of Bengal induced by injection of fresh water. *Nature*, **351,** 385-387.

Jagtap, T.G., Ansari, Z.A., Coutinho, F.B., Charulata, S. and Gaidhane, D.M., 2003. Impact of climate change on mangrove forests along the southern west coast: A case study from Kasargod, Kerala, India. Proceedings of Workshop on Vulnerability and Adaptations due to Climate Change on Indian Agriculture, Forestry and Natural Ecosystem. N.H. Ravindranath, S. Sharma, A. Garg, S. Bhattacharya and I. Murthy (eds.). NATCOM Publ. New Delhi, 136-149.

Jagtap, T.G. and Komarpant, D.S., 2003. Evaluation of mangrove ecosystem of India for assessing its vulnerability to projected climate change. *In:* Assessment of climate change in India and mitigation policies. S.K. Dash and P. Rao (eds). World Wild Life (WWF), India, pp. 39-51.

Jagtap, T.G., Murthy, P. and Komarpant, D., 2002. Mangrove ecosystems of India: Major biotic constituents, conservation and management. *In:* Wetland Conservation and Management. B.B. Hosetti (ed.). Pointer Publishers, Jaipur, India, pp. 34-64.

Jagtap, T.G. and Nagle, V., 2007. Response and adaptability of mangrove habitats from Indian subcontinent to changing climate. *Ambio*, **36(4),** 328-334

Jayakumar, D.A., Naqvi, S.W.A., Narvekar, P.V. and George, M.D., 2001. Methane in coastal and offshore waters of the Arabian Sea. *Marine Chemistry*, **74(1),** 1-13.

Jokiel, P.L., 1980. Solar ultraviolet radiation and coral reef epifauna. *Science*, **207,** 1069-1071.

Karim, A., 1988. Environmental factors and the distribution of mangroves in Sundarbans with special reference to Heritiera fomes. Buch.-Ham. Ph.D. thesis (unpubl.), University of Calcutta.

Karim, A., 1994. Vegetation. *In:* Mangroves of the Sundarbans. Volume Two: Bangladesh. Hussain, Z. and G. Acharya (eds.). G. IUCN -The World Conservation Union.

Karim, A., 1995. Ecological Implications of the Changing Natural Flow of water on the Ganges River Ecosystem. *In:* Women for Water Sharing. H.J. Moudud (ed.). Academic Publishers, Dhaka.

Karisiddaiah, S.M. and Veerayya, M., 1994. Methane-bearing shallow gas-charged sediments in the eastern Arabian Sea: A probable source for greenhouse gas. Continental Shelf Research, **14,** 1361-1370.

Karisiddaiah, S.M. and Veerayya, M., 1996. Potential distribution of subsurface methane in the sediments of the eastern Arabian Sea and its possible implications. *Journal of Geophysical Research*, **101(D-20)**, 25887-25895.

Khan, A.H., 1993. Farakka barrage: Its impacts on Bangladesh – An Overview (mimeo). Approtech Limited, Dhaka.

Kortzinger, A., Duinker, J.C. and Mintrop, L., 1997. Strong CO_2 emissions from the Arabian Sea during the southwest monsoon. *Geophysical Research Letters*, **24,** 1763-1766.

Kumar, M.D., Naqvi, S.W.A., George, M.D. and Jayakumar, D.A., 1996. A sink for atmospheric carbon dioxide in the Northeast Indian Ocean. *Journal of Geophysical Research* (C Oceans), **101,** 18121-18125.

Liss, P.S. and Merlivat, L., 1986. Air-sea gas exchange rates: Introduction and synthesis. *In:* The role of air-sea exchange in geochemical cycling. P. Buat-Menard (ed.). Massachusetts, D. Reidel, Norwell.

Madhupratap, M., Kumar, S.P., Bhattathiri, P.M.A., Kumar, M.D., Raghukumar, S., Nair, K.K.C. and Ramaiah, N., 1996. Mechanisms of the biological response to winter cooling in the northeastern Arabian Sea. *Nature*, **384,** 349-352.

Menon, M.D. and George, K.C., 1977. On the abundance of zooplankton along the southwest coast of India during the years 1971-75. Proceedings of the Warm Water Zooplankton (National Institute of Oceanography, Goa), 205-213.

Mirza, M.M.Q., 2004. The Ganges Water Diversion: Environmental Effects and Implications. Water Science and Technology Library, Vol. 49, Kluwer Academic Publishers, 364 p.

Mitra, A.P., Kumar, M.D., Rupa Kumar, K.. Abrol, Y.P., Kalra, N., Velayudham, M. and Naqvi, S.W.A., 2002. Global Change and biogeochemical cycles: The South Asia region. *In:* Global-Regional linkages in the earth system. P. Tyson, R. Fuchs, C. Fu, L. Lebel, A. P. Mitra, E. Odada, J. Perry, W. Steffen and H. Virji (eds.). Berlin, Springer Verlag, pp. 75-107.

MOEF, 1995. Forestry Master Plan. Ministry of Environment and Forest (MOEF), Government of the People's Republic of Bangladesh, under Asian Development Bank TA No. 1355-BAN.

Mollah, A.R., 1997. Status of Coral and Associated Resources in Bangladesh. Proceedings of Regional Workshop on the Conservative and Sustainable Management of Coral Reefs. M.S. Swaminathan Research Foundation, Chennai, India, Bay of Bengal Programme (FAO), Chennai, India. December 15-17, 1997. B.37-B51.

Morgan, J.P. and McIntire, W.G., 1959. Quarternary Geology of the Bengal Basin, East Pakistan and India. *Bulletin for the Geological Society of America*, **70,** 319-342.

Morrison, J.M., Codispoti, L.A., Gaurin, S., Jones, B., Manghnani, V. and Sheng, Z. 1998. Seasonal variation of hydrographic and nutrient fields during the US JGOFS Arabian Sea Process Study. *Deep Sea Research II*, **45(10-11),** 2053-2101.

Morrison, J.M., Codispoti, L.A., Smith, S.L., Wishner, K., Flagg, C., Gardner, W.D., Gaurin, S., Naqvi, S.W.A., Manghnani, V., Prosperie, L. and Gundersen, J.S., 1999. Oxygen minimum zone in the Arabian Sea during 1995. *Deep Sea Research II*, **46(8-9),** 1903-1931.

Mukhopadhyay, S.K., Biswas, H., De, T.K., Sen, S. and Jana, T.K., 2002. Seasonal effects on the air-water carbon dioxide exchange in the Hoogly estuary, NE coast of Bay of Bengal, India. *Journal of Environmental Monitoring*, **4,** 549-552.

Muraleedharan, P.M. and Prasanna Kumar, S., 1996. Arabian Sea upwelling – A comparison between open ocean and coastal regions. *Current Science*, **71(11),** 842-846.

Murty, V.S.N., Sarma, Y.V.B., Rao, D.P., Murty, C.S., 1992. Water characteristics, mixing and circulation in the Bay of Bengal during SW monsoon. *Journal of Marine Research*, **50,** 207-228.

Naqvi, S.W.A., 1987. Some aspects of oxygen deficient conditions and denitrification in the Arabian Sea. *Journal of Marine Research*, **45,** 1049-1072.

Naqvi, S.W.A., 1991. Geographical extent of denitrification in the Arabian Sea in relation to some physical processes. *Oceanologica Acta*, **14(3),** 281-290.

Naqvi, S.W.A. and Noronha, R.J., 1991. Nitrous oxide in the Arabian Sea. *Deep Sea Research* (part A) **38,** 871-890.

Naqvi, S.W.A. and Shailaja, M.S., 1993. Activity of the respiratory electron transport system and respiration rates within the oxygen minimum layer of the Arabian Sea. *Deep Sea Research II*, **40(3),** 687-695.

Naqvi, S.W.A., Jayakumar, D.A., Nair, M., Kumar, M.D. and George, M.D., 1994. Nitrous oxide in the western Bay of Bengal. *Marine Chemistry*, **47,** 269-278.

Naqvi, S.W.A., George, M.D., Narvekar, P.V., Jayakumar, D.A., Shailaja, M.S., Sardessai, S., Sarma, V.V.S.S., Shenoy, D.M., Naik, H., Maheswaran, P.A., Krishnakumari, K., Rajesh, G., Sudhir, A.K. and Binu, M.S., 1998. Severe fish mortality associated with 'red tide' observed in the sea off Cochin. *Current Science*, **75(6),** 543-544.

Naqvi, S.W.A., Jayakumar, D.A., Narvekar, P.V., Naik, H., Sharma, V.S.S., D'Sousa, W., Joseph, S. and George, M.D., 2000. Increased marine production of N_2O due to intensifying anoxia on the Indian continental shelf. *Nature*, **408** (6810), 346-349.

Olson, D.B., Hitchcock, G.L., Fine, R.A. and Warren, B.A., 1993. Maintenance of the low-oxygen layer in the central Arabian Sea. *Deep Sea Research II*, **40(3),** 673-685.

Owens, N.J.P., Law, C.S., Mantoura, R.F.C., Burkill, P.H. and Lewellyn, C.A., 1991. Methane flux to the atmosphere from the Arabian Sea. *Nature*, **354**(6351), 293-295.

Paropkari, A.L., Prakash Babu. C. and Mascarenhas, A., 1992. A critical evaluation of depositional parameters controlling the variability of organic carbon in Arabian Sea sediments. *Marine Geology*, **107,** 203-226.

Patra, P.K., Lal, S., Venkataramani, S., Gauns, M. and Sarma, V.V.S.S., 1998. Seasonal variability in distribution and fluxes of methane in the Arabian Sea. *Journal of Geophysical Research* (C), **103,**1167-1176.

Patra, P.K., Maksyutov, S. and Nakazawa, T., 2004. Severe weather conditions in the Arabian Sea and their impact on atmospheric N_2O budget. *Indian Journal of Marine Science*, **33,** 84-94.

Pedersen, T.F. and Calvert, S.E., 1990. Anoxia Vs Productivity: What controls the formation of organic carbon rich sediments and sedimentary rocks? *American Association of Petroleum Geology Bulletin*, **74,** 454-466.

Pet-Soede, L., Wafar, M., Venkatraman, K., Rajan, P.T. and Dan Wilhelmsson, 2000. Status of coral reefs of India following the bleaching event of 1998. *In:* Coral Reef Degradation in the Indian Ocean (CORDIO) Status Report 2000. D. Souter, D. Obura and O. Unden (ed.). CORDIO SAREC Marine Programme, Dept. of Zoology, Univ. of Stockholm, Sweden, pp. 69-73.

Prasanna Kumar, S. and Prasad, T.G., 1996. Winter cooling in the northern Arabian Sea. *Current Science*, **71(11),** 834-841.

Prasanna Kumar, S., Madhupratap, M., Dileep Kumar, M., Muraleedharan, P.M., de Sousa, S.N., Gauns, M. and Sarma, V.V.S.S., 2001. High biological productivity in the central Arabian Sea during the summer monsoon driven by Ekman pumping and lateral advection. *Current Science*, **81(12),** 1633-1638.

Prasanna Kumar, S., Muraleedharan, P.M., Prasad, T.G., Gauns, M., Ramaiah, N., de Sousa, S.N., Sardessai, S. and Madhupratap, M., 2002. Why is Bay of Bengal less productive during summer monsoon compared to the Arabian Sea? *Geophysical Research Letters*, **29,** 24,225.

Prasanna Kumar, S., Jayu Narvekar, Ajoy Kumar, Shaji, C., Anand, P., Sabu, P., Rijomon, G., Josia, J., Jayaraj, K.A., Radhika, A. and Nair, K.K.C., 2004a. Intrusion of the Bay of Bengal water into the Arabian Sea during winter monsoon and associated chemical and biological response. *Geophysical Research Letters*, **31**(L15304), 1-4.

Prasanna Kumar, S., Nuncio, M., Jayu Narvekar, Ajoy Kumar, Sardessai, S., de Sousa, S.N., Mangesh Gauns, Ramaiah, N. and Madhupratap, M., 2004b. Are eddies nature's trigger to enhance biological productivity in the Bay of Bengal? *Geophysical Research Letters*, **31**(L07309), 1-5.

Rajasuria, A., 1997. Coral Reefs of Sri Lanka: Current status and resource management. Proceedings of Regional Workshop on the Conservative and Sustainable Management of Coral Reefs. M.S. Swaminathan Research Foundation, Chennai, India, and Bay of Bengal Programme (FAO), Chennai, India. December 15-17, 1997. B.53-B67.

Rajasuriya, A. and Karunarathna, C., 2000. Status of coral reefs of India following the bleaching event of 1998. *In:* Coral Reef Degradation in the Indian Ocean (CORDIO) Status Report 2000. D. Souter; D. Obura and O. Unden (eds.). CORDIO SAREC Marine Programme, Dept. of Zoology, Univ. of Stockholm, Sweden, pp. 69-73.

Ramachandran, S., Anitha, S., Balamurugan, V., Dharanirajan, K., Ezhil Vendhan, K., Marie Irene Preeti Divien, Senthil Vel, A., Sujjahad Hussaim, I. and Udayaraj, A., 2005. Ecological impact of tsunami on Nicobar Islands (Camorta, Katchal, Nancowry and Trinkat). *Current Science*, **89(1),** 195-200.

Ramaswami, V. and Nair, R.R., 1994. Fluxes of material in the Arabian Sea and Bay of Bengal – Sediment trap studies. *Proceedings of Indian Academy of Science* (Earth and Planetary Science), **103,** 189-210.

RVCC, 2003. Vulnerability Profile: Reducing Vulnerability to Climate Change (RVCC), Care-Bangladesh, Khulna, Bangladesh.

Ryther, J.H. and Menzel, D.W., 1965. On the production, composition and distribution of organic matter in the western Arabian Sea. *Deep Sea Research*, **12,** 199-209.

Ryther, J.H., Hall, J.R., Pease, A.K., Bakum, A. and Jones, M.M., 1966. Primary organic production in relation to the chemistry and hydrography of the western Indian Ocean. *Limnology and Oceanography*, **11,** 371-380.

Sarker, M.H. and Koudstaal, R., 1999. Fate of the Gorai River – A Distributary of the Ganges River, Proceedings of the Regional Seminar on Conflict Management of International River Basins, Bangladesh University of Engineering and Technology (BUET), Dhaka.

Sarma, V.V.S.S., 1998. Variability in forms and fluxes of carbon dioxide in the Arabian Sea. Ph. D. Thesis (Goa University, Goa), pp. 205.

Sarma, V.V.S.S., 2003. Monthly variability in surface pCO_2 and net air-sea carbon dioxide flux in the Arabian Sea. *Journal of Geophysical Research (C Ocean)*, **108,** C 8, 10.1029/2001JC001062.

Sarma, V.V.S.S., Kumar, M.D. and George, M.D., 1998. The central and eastern Arabian Sea as perennial source of atmospheric carbon dioxide. *Tellus*, series B Chemical & Physical Meteorology, **50B, 2,** 179-184.

Sarma, V.V.S.S., Dileep Kumar, M., Gauns, M. and Madhupratap, M., 2000. Seasonal controls on surface pCO_2 in the central and eastern Arabian Sea. Proceedings of Indian Academy of Science (Earth and Planetary Science), **109,** 471-479.

Sarma, V.V.S.S. and Narvekar, P.V., 2001. A study on inorganic carbon components in the Andaman Sea during the post monsoon season. *Oceanologica Acta*, **24,** 125-134.

Sarma, V.V.S.S., Kumar, M.D. and Manerikar, M., 2001. Emission of carbon dioxide from a tropical estuarine system, Goa, India. *Geophysical Research Letters*, **28,** 1239-1242.

Sarma, V.V.S.S., Ono, T. and Saino, T., 2002. Increase of total alkalinity due to shoaling of aragonite saturation horizon in the Pacific and Indian Oceans: Influence of anthropogenic carbon inputs. *Geophysical Research Letters*, **29,** 10.1029/2002GL015135.

Sastry, J.S., Rao, D.P., Murty, V.S.N., Sarma, Y.V.B., Suryanarayana, A. and Babu, M.T., 1985. Water mass structure in the Bay of Bengal. *Mahasagar* (Bulletin of the National Institute of Oceanography), **18,** 153-162.

Savidge, G., Lennon, J. and Mathews, A.J., 1990. A shore based survey of upwelling along the coast of Dhofar region, southern Oman. *Continental Shelf Research* **10(3),** 259-275.

Sen Gupta, R. and Naqvi, S.W.A., 1984. Chemical Oceanography of the Indian Ocean, north of the equator. *Deep Sea Research*, **31,** 671-706.

Shenoy, D.M., 2002. Biogeochemical cycling of dimethyl sulfide in the northern Indian Ocean. Ph.D. Thesis (Goa University, Goa), pp. 102.

Shenoy, D.M., Joseph, S., Kumar, M.D. and George, M.D., 2002. Control and inter-annual variability of dimethyl sulfide in the Indian Ocean. *Journal of Geophysical Research*, **107,** 10.1029/2001JD000371.

Shetye, S.R., Shenoy, S.S.C., Gouveia, A.D., Michael, G.S., Sunder, D. and Nampoothiri, G, 1991. Wind driven coastal upwelling along the western boundary of the Bay of Bengal during the southwest monsoon. *Continental Shelf Research*, **11,** 1397-1408.

Siddiqi, N.A., 1989. Seedling regeneration in relation to time and degree of salinity in the Sundarbans Mangrove Forest. *Bano Biggyan Patrika*, **18:** 9-17.

Siddiqi, N.A., 1994. Natural Regeneration. *In:* Mangroves of the Sundarbans. Volume Two: Bangladesh. Z. Hussain and G. Acharya (eds.). IUCN - The World Conservation Union, Bangkok.

Siegenthaler, U. and Sarmiento, J.L., 1993. Atmospheric carbon dioxide and the ocean. *Nature*, **365,** 119-125.

Smith, S.L. and Codispoti, L.A., 1980. Southwest monsoon of 1979: Chemical and biological response of Somali coastal waters. *Science*, **209,** 597-600.

Smith, S.L., Codispoti, L.A., Morrison, J.M. and Barber, R.T., 1998. An integrated, interdisciplinary investigation of the response of the northwest Indian Ocean to monsoonal forcing. *Deep Sea Research II*, **45(10-11),** 1905-1915.

Subramanian, V., 1993. Sediment load of the Indian rivers. *Current Science*, **64,** 928-930.

Swallow, J.C., 1984. Some aspects of the physical oceanography of the Indian Ocean. *Deep Sea Research*, **31(6-8),** 639-650.

Upstill-Goddard, R.C., Barnes, J. and Owens, N.J.P., 1999. Nitrous oxide and methane during the 1994 SW monsoon in the Arabian Sea/Northwestern Indian Ocean. *Journal of Geophysical Research* (C. Oceans), **104,** 30067-30084.

Wafar, M.V.M., 1990. Global warming and coral reefs. *In:* Sea level variations and its impact on coastal environment. G. Victor Rajamanickam (ed.). Tamil University Publications No. 131 (1990). Tamil University Offset Press, Thanjavur.

Walter, H., 1971. Ecology of Tropical and Sub-Tropical Vegetation. Van Nostrand, New York, 539 pp.

Warren, B.A., 1981. Transindian hydrographic section at latitude 18 °S: Property distributions and circulation in the south Indian Ocean. *Deep Sea Research*, **28,** 759-788.

Warren, B.A., 1994. Context of the suboxic layer in the Arabian Sea. *Proceedings of the Indian Academy of Sciences* (Earth and Planetary Sciences), **103(2),** 301-314.

Wyrtki, K., 1973. Physical Oceanography of the Indian Ocean. *In:* The biology of the Indian Ocean. B. Zeitzschel (ed.). Springer-Verlag. Berlin. Heidelberg, pp. 18-36.

Zahn, R., 2003. Global change – Monsoon linkages. *Nature*, **421(6921),** 324-325.

11 Key Vulnerabilities of Human Society in South Asia to Climate Change and Adaptation Issues and Strategies

Sumana Bhattacharya

1. INTRODUCTION

Most of the area in the South Asian region, namely, Afghanistan, Bangladesh, Bhutan, Nepal, Pakistan, Maldives, Sri Lanka and India, lie in temperate and tropical regions that is influenced predominantly by the monsoons. The physiography of the region is diverse ranging from the Himalayas in the North, the long coast line, a vast desert and tropical forests with rich biodiversity reserve. In the recent years, huge losses have been incurred due to climate related hazards (see Box One) and recovering from such shocks has not been easy as socio-economics of the region is characterized by large population and high levels of poverty and unequal development within the countries mainly due to the non-attainment of its development goals such as eradication of poverty, universal education and a sustainable environment.

Till date, 60% of the population in South Asia is engaged in climate dependent livelihoods such as agriculture of which 75% are poor. About 100 million people are dependent on forest products, availability of which is also climate dependent. Last but not the least, women having least access to wealth constitute a large chunk of the population that are affected maximum by natural disasters. A case in point is the fact that about 90% of the mortality registered due to cyclones in Bangladesh is that of women. These set of population is perennially vulnerable to current climate variability.

Rapid urbanization happening in South Asia is creating strain on its infrastructure and resources (see Fig. 1). However, only few cities are preferred locations due to their economic might. Maximum urbanization has happened in Pakistan (35% of its population lives in urban centres). Additionally, by 2050's South Asia's population will exceed 2.2 billion from the current level of 1.5 billion. Most of this population is likely to reside in cities, for example

Box One: Vulnerability of South Asia

The figure below depicts the damages and population affected due to various climate related extreme events since the 1900s. It is clear that droughts and floods are the two most critical hazards that need to be tackled for people to be affected less. In 2007, monsoon floods and storms in South Asia displaced more than 14 million people in India and seven million in Bangladesh. Over 1,000 people lost their lives across Bangladesh, India, southern Nepal and Pakistan. Further rapid and over-urbanization of certain cities that are economic hubs in these countries add to the vulnerability of the region.

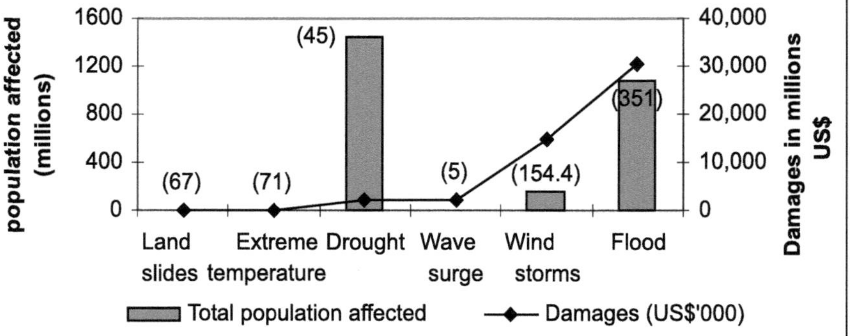

Source: EM-DAT: The OFDA/CRED International Disaster Database. www.em-dat.net - Université catholique de Louvain – Brussels, Belgium.

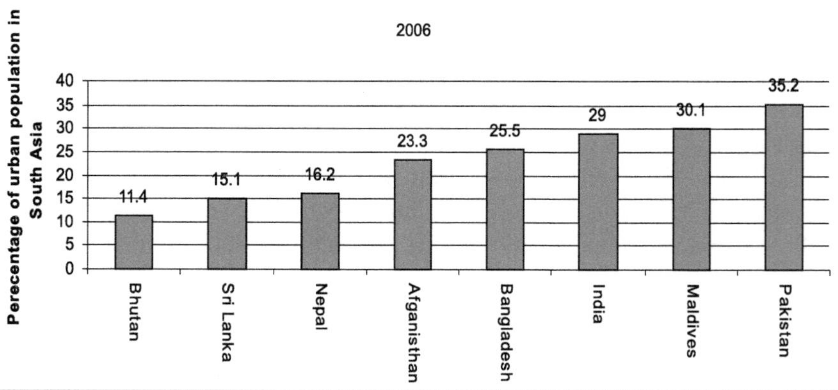

Fig. 1: Percentage of urbanization in South Asia
Source: UNPD, 2007.

55% of population in India will be in its cities. There are particular challenges in making cities climate resilient. Building urban resilience requires improving infrastructure, governance and finance. Without a substantial investment in basic amenities and infrastructure in these large cities, climate change will exacerbate existing vulnerabilities.

With increase in warming it is projected that there will be an increase in the variability, frequency and intensity of extreme events in ways that have not been experienced before. Some of the impacts could be in the form of new challenges (such as sea level rise, glacier melt), others could emerge as old threats made more severe by climate change (flooding or drought). As a result systems such as the water resources, food security, biodiversity, coastal ecosystems, human health and settlements in South Asia are at risk. The increase in population and the depleting resources due to climate change will jeopardize the per capita availability of resources and even small climate shocks may cause irreversible losses and tip a large number of people into destitution and managing cities with huge mass of population will be a challenge.

This chapter reviews the various dimensions of vulnerabilities likely to be associated with climate change in South Asia and the factors driving these vulnerabilities. It reviews the observed and projected climate change in the region and prioritizes the impacts and the vulnerabilities in the context of the economic sectors and habitats such as the cities. The chapter suggests steps necessary to strengthen the adaptation process in the region especially in the context of integration of adaptation into development plans and programmes at national and city scales.

2. OBSERVED AND PROJECTED CHANGES IN CLIMATE

Observed changes in temperature: The entire region is warming and the warming is more on land than on sea. The mean surface temperature has registered an increase in all the countries as observed in the last 50-100 years. In Sri Lanka a gradual increase of about 0.30°C per 100 years has been observed. The warming trend over India has been reported to be about 0.4°C per 100 years.

Observed changes in precipitation: The diverse physiography of the region ranging from coastal zones, hills, mountains, plateaus, forests and deserts cause striking spatial variations in rainfall. In India, long-term time series of summer monsoon rainfall have no discernible trends, but decadal departures are found above and below the long time average alternatively for three consecutive decades. Recent decades have exhibited an increase in extreme rainfall events over northwest and central Indian region during the summer monsoon and the number of rainy days have declined during the monsoon along the east coast (Kolli et al., 2006; Goswami et al., 2006).

Observed sea level rise: Observations along the Indian sea coast indicate a sea level rise of 1.0 millimetre per year (Unni et al., 2006). However, the rise is not equal along the coast as some regions are experiencing subsidence due to tectonic movements also. Sea level changes have direct inundation impacts, and indirect effects such as changes in salinity levels, enhanced storm surge effects, changing sedimentation patterns, and changes in ocean currents.

Projections of temperature: Globally the mean surface temperature is likely to increase by 1 to 2°C (IPCC, 2007a). In South Asia this warming will vary regionally, with already warm areas such as Sri Lanka and the Maldives seeing the lowest rise (about 1°C), while the higher altitude areas of Afghanistan, Bhutan, and Nepal experiencing a rise of 1.5°C to 2.5°C in the moderate. In the central region of South Asia covering entire India, Bangladesh and parts of Pakistan, the temperatures are projected to rise by 2-4°C.

Projections of precipitation: Precipitation projections suggest quantitatively higher and more variable and intense rainfall in South Asia, except in the relatively drier areas of Afghanistan, western India, and Pakistan, which could see even less rainfall. IPCC projections indicate that the number of days for which extreme events last (especially floods and droughts) would increase in duration and severity. This effect will be especially pronounced in South Asia with its reliance on the monsoons – more so than in many other parts of the world.

Projected changes in sea level: Globally it is expected to rise by 0.4 metre by the end of the century. Bangladesh is particularly vulnerable with estimates of sea-level rise varying from 0.30 to 1.5 metres by 2050s under alternative scenarios (Broadus, 1993). The most vulnerable country in South Asia is the Maldives consisting of low lying islands scattered in the Indian Ocean and 80 percent of the country lies below one metre sea level. Even a 0.4 m rise in sea level rise could pose a threat to its existence due to submergence. In addition the projected increase in storm surges and cyclones make the mega-cities lying along the South Asian coastline such as Dhaka, Kolkata, Chennai and Mumbai vulnerable.

Table 1 summarises the globally observed changes in climate, changes observed in the South Asian region and the impacts experienced.

3. PROJECTED IMPACTS OF CLIMATE CHANGE

Glacier melt and retreat: With rising temperatures the ice mass of the Himalayan-Hindu Kush is retreating more rapidly than the global average in some locations. Also leading to glacier lake outbursts due to increase in the volume of lakes caused by melting of glaciers. The October 1994 flash flood on the Pho Chhu river following a glacial lake outburst in the Lunana area was one such example in Bhutan. Possible significant impacts of glacial lake outbursts in the context of Bhutan include perturbation in the quantity of river water used for hydropower generation; destruction of settlements, infrastructure, and agricultural lands; and loss of biodiversity, and even human lives downstream. The Gangotri glacier located in India has been receding since 1780 and in recent years the pace of retreat has accelerated. The

Table 1: Changes in the climate and observed impacts

Indicator	*Globally observed changes*	*Climate change in South Asian Region*	*Impacts*
Surface temperature	Increased by 0.4 to 0.8°C over the 20th century; land areas warmed more than the oceans	The surface mean temperature has increased by 0.4°C over the last 100 years. Warming by 0.3°C is observed in all the regions and cooling by –0.3°C is observed in the northwestern and southwestern regions.	• Decrease in yields in warm areas • Increased insect and pests outbreak • Recession of glaciers • Increase in GLOFs • Heat stress on health • Heat islands in cities • Increased energy demands for cooling
Hot days/ heat index	Increased	More warmer and frequent hot days and nights observed	• Reduced yields • Increased danger of wild fire • Increased water scarcity • Degradation in water quality (algal bloom) • Increased energy demand for cooling in cities
Cold/frost days	Decreased for nearly all land areas during 20th century	Fewer cold days and nights are being observed.	• Reduced yields • Increased energy demand for heating
Continental precipitation	Increased by 5-10% over the 20th century in the northern hemisphere and decreased in some regions, such as north	Monsoon rainfall has been observed to be a trend less over the last 100 years. However, a decrease in total monsoon rainfall by –6% to -8% in north eastern peninsula of India	• Increase in heavy precipitation events in southwest region has impacted the city of

(*Contd.*)

	and west Africa and parts of Mediterranean	and in the north eastern parts of the region and an increase by +10% to +12% in the western coast and central peninsular region with respect to the normal has been observed in this period.	Mumbai leading to large scale damage to property, and life
Heavy precipitation events	Increased in mid and high northern latitudes	Have increased	• Damage to crops • Soil erosion • Water logging in soils • Adverse impacts on quality of surface and ground water • Damage to housing stock in coastal areas especially in cities with economically high value buildings
Frequency and severity of droughts	Increased summer drying and associated incidence of drought in a few areas. In some regions, such as parts of Asia and Africa, the frequency and intensity of droughts have been observed to have increased	Frequency and intensity of droughts has increased	• Land degradation • Lower yields • Crop damage and failure • Livestock mortality • Increased risk of wild fire • More widespread water stress • Affects food security • Malnutrition
Global mean sea level	Global mean sea level has risen by 1.8 +/- 0.3 mm/yr between 1960 and 2000	Increased at an annual average rate of 1-2 mm during the 20th century	• Salinization of soil • Salinization of irrigation water/ground water/ estuaries/fresh water systems • Damage to coastal cities

Source: IPCC, 2007a

receding trends of glacier masses threaten water supplies, livelihoods and the economy of the region. Agriculture and the region's economic structure will need to undergo significant adjustment to cope with these changes.

River runoff: Milly et al. (2005) have estimated that the runoff is likely to increase significantly in the subcontinent (Table 2), except in Afghanistan, with implications for agriculture. The Indus and Ganges/Brahmaputra basins are expected to experience increased runoff driven by precipitation changes and glacial melt. After the glacial melt, however, there could be significant declines in flows. As per the assessment of IPCC (2001), by 2050, the annual runoff in the Brahmaputra is projected to decline by 14 per cent and the Indus by 27 per cent. However, one of the major basins in India, Mahanadi basin is likely to experience higher run off in 2050s with respect to present, as a result of which it might be one of the least water stressed river basin (Gosain et al., 2006). The western Indian rivers, such as Luni and Tapi, are to experience severe water shortage due to trickling of run off in the climate change scenarios. Afghanistan is expected to be impacted the most by a reduction in flows with considerable implications for storage, irrigation, and the development and reliability of hydropower systems. Such outcomes will be further complicated by changes in water use in the basins including diversions, groundwater-surface water interactions, and increased demands for irrigation, hydropower, industrial, and municipal water supplies by the increasing population.

Impacts on agriculture: In the agriculture sector the major impacts of climate change will be on rain fed or un-irrigated crops, which account for nearly 60% of cropland area and on which more than 75 percent of the region's poor are dependent. Long-term changes in temperatures and precipitation have direct implications on evaporative demands and agricultural yields. In arid locations where crops already suffer heat stress, a small increase in temperatures could lead to a dramatic decline in yields. The same temperature increase in, say the cooler Himalayas, could generate an increase in yields. Accordingly, agronomic models project a wide range of changes in crop productivity and loss in revenue that vary by crop type, location and climate scenario (Lal et al., 1998, 1999; Aggarwal and Mall, 2002; Kalra et al., 2004; World Bank, 2006; Cline, 2007).

Table 2: Mean changes in run off in 21st century in South Asian countries

Country	*Mean run off change (%)*
Afghanistan	–20 to –10%
Bangladesh	20 to 30%
Bhutan	10 to 20%
India	30 to 40%
Maldives	20-30%
Nepal	10-20%
Pakistan	>40%

Source: Milly, Dunne, and Vecchia 2005.

A recent study by Cline (2007), using Recardian approach, indicates that climate change can trigger a one per cent loss in agricultural revenues in Nepal to a dramatic 49 per cent decline in average revenue in India by 2080s. The same study using crop models indicate that the crop productivity will decrease between 25% and 36% with respect to current levels of crop productivity in these countries (see Fig. 2). This study does not take into account the changes expected in hydrological characteristics in terms of more intense droughts and floods; changes in surface water availability; or threshold effects in the response of crop growth to temperature changes. Nor do they take into account that, for agriculture that is being practiced in low-lying coastal areas (e.g. Bangladesh, and the Mahanadi delta in India), there is also potential for damage arising from sea level rise and increased salt water intrusion in groundwater aquifers. The combined effects could be devastating for tropical agriculture.

Impacts on forests and forest products: Forests account for about 20–30 per cent of the total land area of India, Nepal, and Sri Lanka and about 68 per cent in Bhutan. These are important to energy, housing, and the livelihoods of many people in rural South Asia. Savannas and dry forests are grazing areas for the region's large population of livestock, which is essential to food security and agricultural drought.

The IPCC projects that carbon fertilization will lead to net primary productivity gains in the medium term (IPCC, 2007b), with the gains

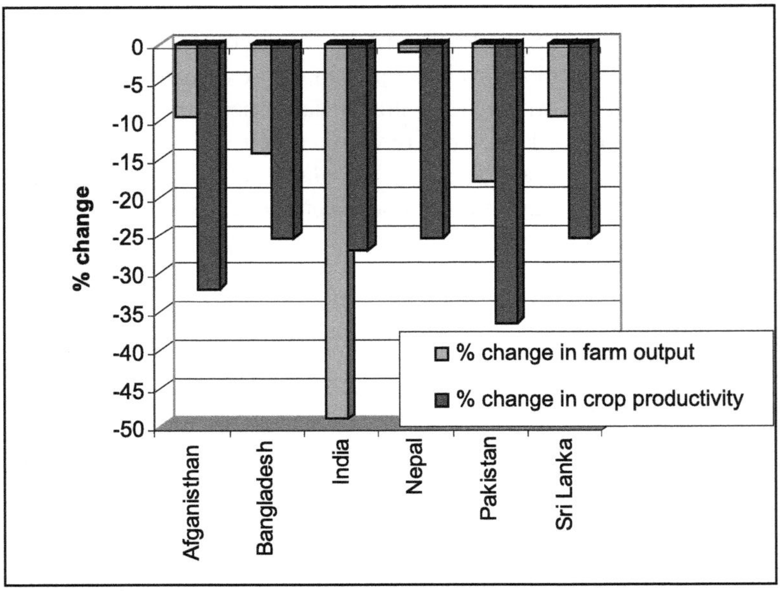

Fig. 2: Estimates of climate change-related impacts on agricultural production by 2080 for selected countries in the South Asia region. *Source:* Cline, 2007.

experienced in some forest types outweighing the losses in others. Vegetation types will shift to higher elevations as a result of global warming, and some vegetation types may disappear in the process, together with dependent species and ecosystems with strict climate niches (CBD, 2003; IPCC, 2007b).

Climate change is expected to lead to a northward shift of vegetation and reduction and loss of alpine tundra cover in the dry temperate and temperate mountains of the region such as Afghanistan and Pakistan. The changes in precipitation in the dry temperate mountains of Pakistan are expected to expand conifer coverage at the expense of alpine vegetation even before the mid-century (Khan et al., 2003). Further, the vegetation is expected to shift towards wetter types in the northeast in Bhutan and the northeastern parts of India and to drier types in the northwest of India by 2085. This will transform the currently dominant land cover into tropical dry forest. Net primary productivity gains are expected for many vegetation types, but reduction in population in several species, and extinction of some species, will inevitably occur. The highly endemic areas of the Western Ghats and central Himalayas are projected to experience forest dieback and loss of biodiversity in the long run (Ravindranath et al., 2006).

Impacts on human health: Rise in temperature and changes in intensity and duration of precipitation will have far reaching effects of human health issues in India leading to the rise in heat stress related diseases and malnutrition due to increase in frequency of droughts, while recurrent floods may cause diarrhea and other diseases.

In the South Asian region, most parts are endemic to malaria with central India being more prone. Malaria has not yet penetrated elevations above 1800 metres and some coastal areas. A study by Bhattacharya et al. (2006) indicates that malaria may penetrate such high altitudes in the future and in 2080s, 10% more states in India may offer climatic opportunities for malaria vector breeding throughout the year with respect to current years. These opportunities are projected to increase by 3 to 5 months in Jammu and Kashmir, and western Rajasthan, while they may reduce by 2 to 3 months in the southern states as temperatures increase.

Heat waves are projected to become more common and of longer duration. This is likely to cause an increase in heatstroke and in the incidence of cardiovascular, cerebrovascular, and respiratory diseases (Hales et al., 2003). In South Asia, heatwaves are associated with high mortality rates in rural areas and, also, among the poor and outdoor laborers (Chaudhury et al., 2000). A large number of heatwave-related deaths have been reported, mainly among vulnerable populations: the poor, the elderly, and laborers such as rickshaw pullers and agricultural workers (Lal, 2003). A heatwave that took place in Andhra Pradesh, India, in 2003, with temperatures rising to over 48°C, caused more than 3000 deaths (Government of Andhra Pradesh, 2004).

The World Bank, in one of its recent reports on vulnerabilities of South Asia (WB, 2009) has synthesized the risks of the region to climate change in terms of High, Low and Medium. According to this report, Bangladesh, Maldives and Srilanka are at very high risk to sea level rise. Glacier retreat is a threat to India, Afghanistan and Bhutan. Rise in temperature is a risk to all sectors of the economy in all the countries. Floods due to climate change are a risk to India especially in its Gangetic plains and western region. Droughts again pose a high level of risk to food security, health and other issues in India, Pakistan and Srilanka (see Table 3).

Table 3: Summary of climate risks by country in South Asia

	Afghanistan	*Bangladesh*	*Bhutan*	*India*	*Nepal*	*Pakistan*	*Maldives*	*Sri Lanka*
Sea level rise	-	H	-	M	-	M	H	H
Glacier retreat	H	M	H	H	H	H	-	-
Rise in. temperature	-	H	H	H	M	H	H	H
Floods	-	H	H	H	H	H	H	
Droughts	L	Some areas	-	H	L	H	-	H

L: likley, H: high; M: moderate
Source: WB, 2009

4. CLIMATE CHANGE AND IMPLICATIONS ON CITIES

The cities in South Asia account for over 25 per cent of regional GDP which reflects upon their ability to attract and retain investment and hence population from other parts of the countries for employment. Currently, the urban population in South Asia ranges from a minimum of 11% in Bhutan to maximum of 35% in Pakistan. By 2020, Mumbai will be the second largest city in the world, closely followed by Delhi, and Dhaka. With Karachi and Kolkata – five of the world's 11 megacities will then be in South Asia.

Therefore, a large chunk of population residing in cities are increasingly becoming vulnerable to a range of climate-related impacts, such as flooding, storm surges, landslides, drought, salt water intrusion, and cyclones, the effects of which are exacerbated by poor-quality and ill-maintained infrastructure, low-quality building stock, and the low resilience of much of the population. In Mumbai, more than half the population are crowded into about 2000 densely populated slums that are at risk from flooding and where settlements lack basic protective infrastructure. Dhaka is already the fastest growing mega-city in the world, drawing an estimated 300,000 to 400,000 mostly poor migrants each year. The poor, with their limited access to development in the cities, are especially vulnerable and bear the greatest burden of such impacts.

Changes in precipitation and temperature. Climate change is predicted to increase the variability of precipitation and raise temperatures across South Asia. This will have a range of outcomes with potential negative impacts on urban areas, including heat waves. The air temperature in cities can be up to 12°C warmer than that in surrounding suburban and rural areas due to absorption of heat by dark-paved surfaces and buildings; lack of vegetation and trees; heat emitted from buildings, vehicles, and air conditioners; and reduced airflow around buildings (CEIDH, 2001). With five of the world's mega cities and many other urban areas situated in South Asia, rising city temperatures have the potential to impact the health of city dwellers and, particularly, of those residing in slums, who usually lack access to air conditioning and other adaptive strategies. The rising temperatures are likely to be accompanied by threat to water supplies and associated health risks such as cholera.

Sea level rise, increase in intensity and frequency of storm surges/ cyclones: Coastal cities with high population and intense economic activities are especially endangered by rising sea levels and more intense weather phenomena, including heavy precipitation events and storm surges. A study conducted by Nicholls et al. (2008) indicates that there is a 74% chance of having one or more of the 136 coastal cities in the world to be affected by a 100-year return event every year (1.5 m sea level rise), and a 99.9% chance of having at least one city being affected by such an event over a 5-year period. As a consequence, even assuming that protection levels will be high in the future, the large exposure in terms of population and assets is likely to translate into regular city-scale disasters at global scale. The study identifies the South Asian cities of Dhaka, Kolkata, Chennai and Mumbai, to be most vulnerable in the region.

A coastal mega city like Mumbai—that has low lying reclaimed land areas, an economy that generates 5% of India's GDP, an ever increasing population which has catapulted from 13 to 17 million between 1991 and 2008, a 268 km of coastline, high value built environment consisting of residential, commercial, industrial and public assets and a rich ecosystem such as wetlands, rivers, coastal ecosystems and groundwater—is especially at threat from climate change and its manifestations in terms of sea level rise, storm surges and cyclones. A study carried out by Ranger et al. (2009) suggests that by the 2080s, in a SRES A2 scenario, the likelihood of the 2005 event could more than double, making it approximately a 1-in-100 yr return period event. It is estimated that total losses (direct plus indirect) associated with a 1-in-100 year event could rise by 200% (i.e. triple) compared with current situation (to \$690 – \$1890 million USD). While these findings are preliminary, the study demonstrates the strong need for effective climate change adaptation in Mumbai. The study further suggests that by improving the drainage system in Mumbai, losses associated with a 1-in-100 year flood event today could be reduced by as much as 70%, and by extending insurance cover to 100% of the population, the indirect effects of flooding could be almost halved.

5. TOWARDS DEVELOPMENT OF NATIONAL STRATEGIES FOR ADAPTATION

5.1 Ranking of Adaptation and Identifying Adaptation Options

Development planners often require a ranking of impacts, to make decisions with regard to how much they should invest in planning or mainstreaming particular response measures. Towards this goal, a ranking of climate change impacts and vulnerabilities for particular sectors for all the South Asian countries has been done here using the method followed by Aggarwala et al. (2003). Further, likely adaptation responses have been identified for each sector that a country can attempt to develop (see Table 4).

Vulnerability is a subjective concept that includes three dimensions: exposure, sensitivity, and adaptive capacity of the affected system (IPCC, 2001b). The sensitivity and adaptive capacity of the affected system in particular depend on a range of socio-economic characteristics of the system. Several measures of social well-being such as income and income inequality, nutritional status, access to lifelines such as insurance and social security, and so on can affect baseline vulnerability to a range of climatic risks. This section instead subjectively ranks biophysical vulnerability based on the following dimensions:

Certainty of impact: This factor uses our knowledge of climate change to assess the likelihood of impacts. Temperatures and sea levels are highly likely to rise and some impacts can be projected based on this. Changes in regional precipitation are less certain. The Intergovernmental Panel on Climate Change (IPCC, 2007a) concluded that higher maximum and minimum temperatures are very likely, more intense precipitation is very likely over most areas, and that more intense droughts, increased cyclone wind speeds and precipitation are likely over some areas.

Timing: When are impacts in a particular sector likely to become severe or critical? Based on available information, we considered whether impacts are likely to become so in the first or second half of this century.

Severity of impact: How large could climate change impacts be? Essentially this factor considers the sensitivity of a sector to climate change.

Importance of the sector: Is the sector particularly critical in terms of its size of economy, cultural or other importance, or its potential to affect other sectors? This factor considers exposure of the sector to climate change, that

Table 4: Country by country priority ranking of vulnerabilities and adaptation options

Sector	*Vulnerability to CC*	*Priority ranking of identified vulnerabilities*	*Adaptation options*
Water	• Reduced run off in some of the river basins	High – Afghanistan, parts of Pakistan and north western region of India	• Enhanced regional cooperation on international rivers and river basins • Improved water resources management • Building climate sensitive infrastructure to avert sea water intrusion and floods in river basins • Research on assessment of impacts of CC on water availability from Himalayan glaciers and the region's large river basins vis a vis water demand in the future; Increased research on new water efficient crops
	• Glacier melting in the Himalayas including lake outburst	High – Nepal, Bhutan, India Medium –Pakistan	
	• Floods	High – India, Bangladesh	
	• Droughts	High – Afghanistan, parts of Pakistan, and 78% of Indian land area	
	• Saline intrusion in coastal aquifers due to sea level rise	High – India and Bangladesh	
Agriculture	• Declining yields of major crops	High – For all South Asian countries	• Promotion of climate resilient cropping patterns and techniques • Agricultural research and extension for promoting climate resilient crop varieties • Improvements in risk management (e.g. climate insurance, contingent credit schemes) • Irrigation development and increased investment in water harvesting infrastructure at required scales that take account of climate risks
	• Agriculture unviable in marginal areas e.g. arid, semi-arid, coastal (saline intrusion affected zones due to sea level rise)	High – Afghanistan, Parts of Pakistan and India, Bangladesh and Sri Lanka	
	• Crop destruction by extreme events	High – Eastern coastal parts of India, coastal parts of Bangladesh, India due to increase in extreme precipitation events	

(*Contd.*)

Health	• Increased incidence of water related diseases	High – India, Pakistan, Bangladesh	• Awareness of the health implications of climate change • Monitoring and surveillance of disease • Improved access to treatment for new disease risk profiles • Improved water supply and sanitation
	• Malaria in newer areas	High – Bhutan, Nepal, and some Himalayan regions of India	
	• Heatstroke	High – India, Pakistan, Bangladesh	
	• Malnutrition	High – All SA countries due to shortage of food arising from damages to crops	
	• Direct health risks, e.g. injury and death caused by extreme events	High – Bangladesh, coastal parts of India	
Forests	• Die back of forests	High – India, Bhutan, Nepal	• Expansion of protected area networks and promotion of ecosystem-based approach in conserving plant species • Mainstreaming of biodiversity and ecosystem management in development projects, climate mitigation, adaptation and risk management • Generation of knowledge and capacity
	• Change in forest types	High – India, Bangladesh	
	• Shift in forest line to higher latitudes	High – India, Afghanistan and Pakistan	
Cities	• Climate related damage upon urban settlements, lives, assets and basic water and sanitation services	High – Mumbai, Kolkata in India; Dhaka in Bangladesh; Karachi in Pakistan	• Developing climate change action plans for cities in consonance with disaster risk management plans
	• Increase in urban vector and waterborne diseases, (associated with urban poverty mainly in slums).	High – Mumbai, Kolkata in India; Dhaka in Bangladesh; Karachi in Pakistan Medium – for present class II cities	

is, how many people, property, or other valuable assets could be affected by climate change. A score of high, medium, or low for each factor is then assigned for each assessed sector for each country.

5.2 Development as a Strategy Towards Adaptation

All development is ultimately about people developing the capabilities that empower them to make choices and to lead lives that they value. Climate change threatens to erode human freedoms and limit choices. In September 2000, at the United Nations Millennium Summit, world leaders agreed to eight specific and measurable development goals—now called the millennium Development Goals (MDGs)—to be achieved by 2015. The first seven goals focus on eradicating extreme poverty and hunger, achieving universal primary education; promoting gender equality and empowering women; reducing child mortality; improving maternal health; combating HIV/AIDS, malaria and other diseases; and ensuring environmental sustainability. The eighth goal calls for the creation of a global partnership for development, with targets for aid, trade, and debt relief. Table 5 enumerates the MDGs and highlights the achievements of the South Asian region as of 2007. Challenges of attaining the goals are at jeopardy due to current global recession as well as the impacts of changing climate (see Table 5).

The Human Development Index is a direct measure of attaining the MDGs as it is a function of per capita GDP, literacy level and life expectancy. These, in turn, are functions of governance, policies, penetration of technology, institutions and risk sharing measures that directly define the adaptive capacities to current climate variability and future climate change. Lower HDI of a nation would mean a lower capacity of its people to combat the impacts of climate change. Figure 3 shows a comparative picture of the human development index of the countries in the South Asian region with respect to low, medium and high development indices defined by the UN in its Human Development Report (HDR, 2008). It is clear that the HDI in South Asia is just above the low average global HDI and has yet far to go in order to reach the current levels of high HDI indices of the developed world. Low HDIs make the region more vulnerable with respect to the developed world to the same stimulus of global warming and climate change. The average HDI of the developed countries now is 0.897. At the current rate at which the Index is growing in South Asia (of CAGR ranging between 1.30 to 1.38), it will take anywhere between 35 and 40 years to reach the high HDI level of the developed world that it currently has (see Fig. 4).

By the time 35-40 years go by, impacts of climate change would have entrenched itself in the region. Therefore attaining atleast a medium level of development at an earlier date as set by the MDGs, would be the key to bracing upto future impacts of climate change.

Table 5: Millennium Development Goals, climate change and South Asia*

Millennium Development Goals and achievements of the South Asian region	*Potential impacts of climate change in South Asia*
Goal 1 Eradicate extreme poverty and hunger • 1990: 49% • 2007: 39%	• Damage to livelihood assets, including homes, water supply, health, and infrastructure, can undermine peoples' ability to earn a living; • Reduction of crop yields affecting food security; • Changes in natural systems and resources, infrastructure and labour productivity may reduce income opportunities and affect economic growth; • Social tensions over resource use can lead to conflict, destabilizing lives and livelihoods and forcing communities to migrate from one country to the other in the region.
Goal 2 Achieve universal primary education • 2000: 79% • 2007: 90%	• Loss of livelihood assets and natural disasters will reduce opportunities for full time education, more children (especially girls) are likely to be taken out of school to help fetch water, earn an income or care for ill family members; • Malnourishment and illness may reduce school attendance and the ability of children to learn when they are in class; • Displacement and migration may reduce access to education.
Goal 3 Promote gender equality and empower women enrolment of girl child in schools • 1999: 84% • 2007: 93%	• Exacerbation in gender inequality as women depend more on the natural environment for their livelihoods, including agricultural production. This may lead to increasingly poor health and less time to engage in decision making and earning additional income; • Women and girls are typically the ones to care for the home and fetch water, fodder, firewood, and often food. During times of climate stress, coping with fewer resources will lead to a greater workload; • Female headed households with few assets are particularly affected by climate related disasters.
Goal 4 Reduce child mortality • 1990: 122 births per thousand • 2007: 77 births per thousand Goal 5 Improve Maternal Health • 1990: 620 live births per 100,000 • 2005: 490 live births per 100,000	• Deaths and illness due to heat-waves, floods, droughts and hurricanes; • Children and pregnant women are particularly susceptible to vector-borne diseases (e.g. malaria and dengue fever) and water-borne diseases (e.g. cholera and dysentery) which may increase and/or spread to new areas – e.g. anaemia resulting from malaria is currently responsible for one quarter of maternal mortality;

(*Contd.*)

Table 5: (*Contd.*)

Millennium Development Goals and achievements of the South Asian region	*Potential impacts of climate change in South Asia*
	• Reduction in the quality and quantity of drinking water exacerbates malnutrition especially among children; • Natural disasters affect food security leading to increased malnutrition and famine.
Goal 6 Combat HIV/AIDS, malaria and other diseases TB: • 1990: 543 per 100,000 persons • 2007: 268 per 100,000 persons	• Water stress and warmer conditions encourage disease; • Households affected by AIDS have lower livelihood assets, and malnutrition accelerates the negative effects of the disease.
Goal 7 Ensure environmental sustainability 0.2 Million ha gained in forest cover between 1990 and 2005 Drinking water through safe sources (piped) is lowest amongst all the regions Open defecation: 779 million people in South Asia, highest amongst all regions even the situation in Sub Saharan Africa seems to be better (72% less than southern Asia)	• Alterations and possible irreversible damage in the quality and productivity of ecosystems and natural resources due to increase in warming; • Decrease in biodiversity and worsening of existing environmental degradation; • Alterations in ecosystem-human interfaces and interactions lead to loss of biodiversity and loss of basic support systems for the livelihood of many people.
Goal 8 Develop a global partnership for development Access to mobile phones and internet: has increased dramatically since 2000 in South Asia. In India the number of mobile users per 100 persons is 36. Access to electricity is the lowest in South Asia (only 706 million in a population of 1.5 billion)	• Climate change is a global issue and a global challenge: responses require global cooperation, International relations amongst neighbours may be strained by climate impacts.

* In 1st column, achievements are mentioned for the South Asian region. In this table the SA region includes Iran, Afghanistan, Pakistan, India, Bangladesh, Nepal, Bhutan and Sri Lanka. *Source:* The Millenium Development Goal Report, 2009.

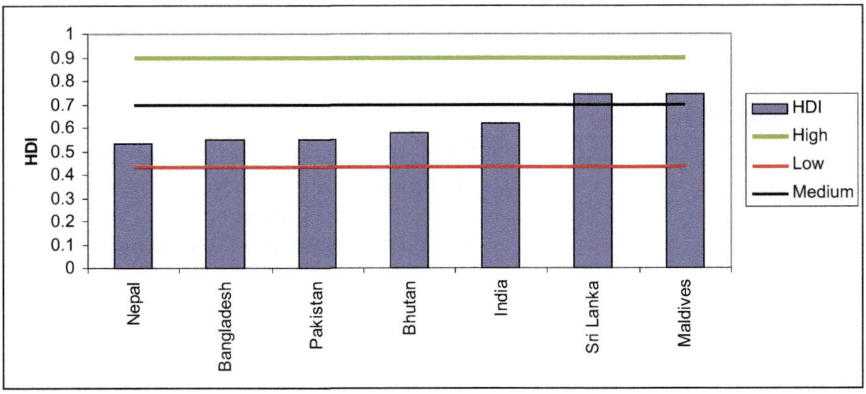

Fig. 3: Human Development Index of the South Asian Region – the index for determining capacity to cope with adversities. The straight lines are the global averages of low, medium and high HDIs.

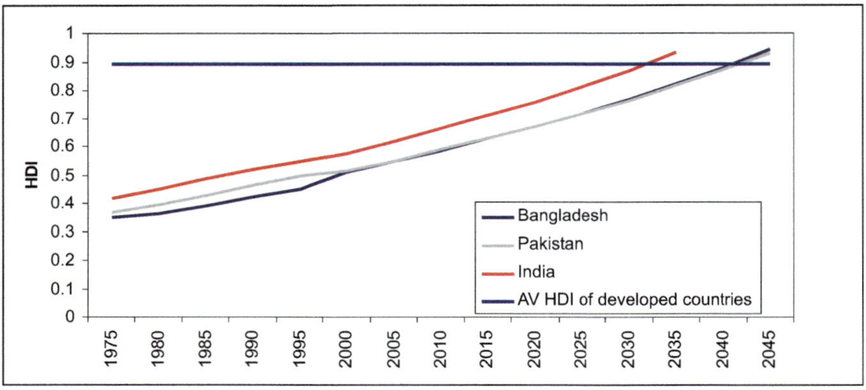

Fig. 4: Current trends of human development Index and future projections. Projections made using the business as usual growth rate of HDI for each country.

6. ADAPTATION AND CITIES

Every mega city in the region has its typical characteristics but is also similar in certain features such as high level of population with respect to other units of governance in the country. They are economic hubs and therefore attract high rates of influx of population to the city every year. Majority of them are coastal cities and are subjected to vagaries of climate variability and likely to be adversely impacted due to climate change. High percentage of urban poor live in low cost housing not adept enough to withstand the extreme climate and its impacts such as extreme heat, poor housing stock susceptible to flooding etc. The cities experience high levels of economic damages even with small shocks of climate. The cities are dependent on natural resources from outside the city and are facing significant water stresses, due to competing demands of various sectors, pollution of surface water sources, and over-exploitation of groundwater.

So far in the South Asian region, only Delhi has recently launched its own Climate Change Agenda Plan (2009). It has set targets for 2012 within the framework of India's National Action Plan on climate change (see Box Two) focusing more on mitigation aspects than adaptation to extreme climate. More consultations need to go in to develop a comprehensive climate action plan for the city taking into account risks to health due to extreme heat conditions, food and water security of the city in the context of enhanced frequency and intensity of droughts etc. On the other hand Mumbai, which experienced in 2005 an event of unprecedented damages of almost 200

Box Two: Delhi's Climate agenda: 2009-2012

For sustaining the eco-system and reducing the carbon footprint, the Delhi Government has identified a set of 65 action points that each department of the administration would have to follow covering a variety of issues like noise pollution, air pollution and water pollution, municipal waste generation and greening.

Though Delhi has taken a lead by switching over to eco-friendly Compressed Natural Gas, a cleaner fuel, transport still remains a critical area. The climate change agenda for Delhi includes action points on imposition of congestion tax to check growth of vehicle population and the possibility of using low sulphur diesel/petrol to reduce harmful emissions. The Department of Environment has been marked to work on the possibility of converting waste oil from commercial establishments into bio-fuel.

On the issue of water pollution, the new agenda calls for promoting decentralized wastewater treatment systems to reduce carriage cost and maintenance of large sewerage system.

The Forest Department has been set a target of bringing 500 sq.km under green cover by 2012 and open a new eco-task force by 2009. Creation of about 250 green buildings, increasing the capacity of solar water heating, building retro fitment for 100 buildings for energy efficiency and encourage the use of renewable energy to five per cent of the total energy produced by 2012 are some of the targets for the Power Department.

The Delhi Jal Board's brief is to increase the efficiency of water use to 20 per cent, check leakages, install water recharge system in 1000 buildings and provide sewer connections, complete 80 per cent work on the Renuka Dam project and restore water bodies.

The climate change agenda for Delhi seeks to strengthen further the efforts like introduction of compact fluorescent lamps, setting up of air ambience fund to provide support for battery-operated vehicles, mandatory use of solar water heaters in all buildings above 500 sq metres, nine new city forests and water to energy projects.

Institutions such as Education Department, the PWD, the Health Department, the Delhi Police and the Industries Department have also been identified as partners in the implementation of the agenda.

billion US$ and 500 fatalities due to flooding from a single rainfall event exceeding 977 mm still does not have a climate action plan that can integrate climate concerns with disaster management.

Managing these combined risks could require significant revision of urban planning practices across city to integrate disaster risk reduction and climate change adaptation measures as well as greenhouse gas mitigation into day-to-day urban development and service delivery activities (Revi, 2007). A risk management framework for South Asian cities, therefore, need to:

- Recognize the short, medium to long term impacts of climate change on cities and the associated vulnerabilities.
- Include innovations in urban planning in the context of climate change to make cities cooler and accommodate urban poor in quality housing with green features.
- Enhance cooperation between the national, state and city level governments for devising climate change policies and define roles for institutions at these levels.
- Look for means and ways to involve city authorities in environmental monitoring and mandatory reporting.
- Adapt and re-frame existing regional and urban policies implemented by national governments to foster climate friendly policies for implementing climate action plans in cities.
- Evaluate budgetary requirements for such actions and devise instruments could be used to finance climate friendly projects.

7. TOWARDS AN INTEGRATED ADAPTATION FRAMEWORK

According to a recent World Bank report on South Asia (WB, 2009), the South Asian region will need to make greater use of the existing range of instruments – knowledge partnerships and capacity building (including climate risk assessments, assistance with global negotiations where required, reports and technical support), as well as priority investments. In a resource constrained environment there will be a need to leverage funds effectively to achieve transformational impacts that could have significant effects in building climate resilient and low-carbon growth economies in the near to medium term.

Therefore to avert the impacts of ongoing and future climate change, it is imperative that national governments across the region set up rapid targets of development in consonance with the concerns of climate change, keeping in mind that the ensuing policies should lead towards:

- Development of national action plans on climate change, keeping the developmental paradigms of the countries as central to that plan. For example, India has come out with a national action plan on climate change in keeping with its developmental aspirations focusing on higher penetration of renewable energy such as solar photovoltaic, energy efficiency in industries and human habitats, and promoting sustainable agriculture, water resources and forests (NAPCC, 2008).

- Alignment of disaster risk reduction management plans with climate change action plan at all administrative levels.
- Development of action plans for cities to combat climate change as cities are increasingly becoming the hubs of heightened levels of economic activities with respect to other units of the country and also housing population, that is increasingly living in squalid and poor conditions.
- Ensuring strong institutions for rapid and effective implementation of policies at national, regional, city and district level encompassing the whole gamut of governance in the country.
- Encouraging research to enable informed policy decisions on climate change for national actions and international negotiations.
- Rapid penetration of new and existing technologies to adapt to the adverse impacts of climate change such as food insecurity that is likely to happen due to crop loss from escalated frequency and intensity of droughts.
- Devise mechanisms to access funds for adaptation such as technology up-gradation, capacity building and awareness through own funds or through bilateral or multilateral sources.
- Float risk sharing measures to hedge the damages that are likely to occur due to climate change through various insurance mechanisms, or through social support groups such as self help groups in most of the rural areas in South Asia, programmes of the government such as the current NREG program (National Rural Employment Guarantee program) of the govt. of India.
- Develop stronger bilateral ties amongst the nations in the region for a smooth access to goods and services across the region for ensuring food security, water security, access to ecosystem goods, energy security, health services etc.

REFERENCES

Aggarwal, P.K. and Mall, R.K., 2002. Climate Change and Rice Yields in Diverse Agro-environments of India. II: Effects of Uncertainties in Scenarios and Crop Models on Impact Assessment. *Climate Change*, **52,** 331-343.

Barnett, T.P., Adam, J.C. and Lettenmaier, D.P., 2005. Potential Impacts of a Warming Climate on Water Availability in Snow-dominated Regions. *Nature*, **438,** 303-309.

Bhattacharya, Sumana, Sharma, C., Dhiman, R.C. and Mitra, A.P., 2006. Climate change and malaria in India. *Current Science*, **90(3),** 369-375.

Chaudhury, S.K., Gore, J.M. and Ray, K.C.S., 2000. Impact of Heat Waves in India. *Current Science*, **79,** 153-155.

Biswas, R., Pal, D. and Mukhopadhyay, S.P., 1999. A Community-Based Study on Health Impact of Flood in a Vulnerable District of West Bengal. *Indian Journal of Public Health*, **43,** 89-90.

Broadus, J., 1993. Possible impacts of, and adjustments to, sea-level rise: the cases of Bangladesh and Egypt. *In:* Climate and Sea Level Change: Observations, Projections and Implications. R.A. Warrick, E.M. Barrow and T.M.L. Wigley. Cambridge, Mass.: Cambridge University Press.

Convention on Biological Diversity (CBD), 2003. Interlinkages between Biological Diversity and Climate Change: Advice on the Integration of Biodiversity Considerations into the Implementation of the United Nations Framework Convention on Climate and Its Kyoto Protocol. CBD Technical Series No. 10.

Cline William, R., 2007. Global Warming and Agriculture: Impact Estimates by country. Centre for Global Development, 2007. (http://www.cgdev.org/content/publications/detail/14090)

Gosain, A.K., Rao, Sandhya and Basu Ray, Debajit, 2006. Climate change impact assessment on hydrology of Indian river basins. *Current Science*, **90(3),** 346-353

Goswami, B.N., Venugopal, V., Sengupta, D., Madhusoodanan, M.S. and Xavier, Prince K., 2006. Increasing Trend of Extreme Rain Events Over India in a Warming Environment. *Science* (www.sciencemag.org), December 2006, Vol. 314.

Government of Andhra Pradesh, 2004. Report of the State Level Committee on Heat Wave Conditions in Andhra Pradesh State. Government of Andhra Pradesh, Revenue (Disaster Management) Department. Hyderabad, India.

Mehra, Rakesh, 2009. Climate Change Agenda for Delhi: 2009-2012. Government of National Capital Region Delhi.

Hales, S., Edwards, S. and Kovats, R.S., 2003. Impacts on Health of Climate Extremes. *In: Climate Change and Human Health: Risks and Responses* (eds.) A.J. McMichael, D.H. Campbell-Lendrum, C.F. Corvalan, K.L. Ebi, A. Githeko, J.D. Scheraga, and A. Woodward., World Health Organization Geneva:.

IPCC, 2007a. Climate Change 2001: The Scientific Basis. Contribution of Working Group II to the Fourth Assessment Report of the Intergovernmental Panel on Climate Change. Cambridge University Press.

IPCC, 2007b. Climate Change 2001: Impacts, Adaptation and Vulnerability. Contribution of Working Group II to the Fourth Assessment Report of the Intergovernmental Panel on Climate Change. Cambridge University Press.

Kalra, N., Chakraborty, D., Ramesh, P.R., Jolly, M. and Sharma, P.K., 2007. Impacts of Climate Change in India: Agricultural Impacts. *In:* Final Report, Joint Indo-UK Programme of Ministry of Environment and Forests, India, and Department for Environment, Food and Rural Affairs (DEFRA), United Kingdom. Indian Agricultural Research Institute (Unit of Simulation and Informatics), New Delhi.

Lal, M., Singh, K.K., Rathore, L.S., Srinivasan, G. and Saseendran, S.A., 1998. Vulnerability of Rice and Wheat Yields in Northwest India to Future Changes in Climate. *Agricultural and Forest Meteorology*, **89,** 101-114.

Lal, M., 2003. Global Climate Change: India's Monsoon and its Variability. *Journal of Environmental Studies and Policy*, **6,** 1-34.

Lal, M., Singh, K.K., Srinivasan, G., Rathore, L.S., Naidu, D. and Tripathi, C.N., 1999. Growth and Yield Responses of Soybean in Madhya Pradesh, India, to Climate Variability and Change. *Agricultural and Forest Meteorology*, **93,** 53-57.

Milly, P.C.D., Dunne, K.A. and Vecchia, V.A., 2005. Global Pattern of Trends in Streamflow and Water Availability in a Changing Climate. *Nature*, **438,** 347-350.

NAPCC, 2008. National Action Plan on Climate Change. Prime Minister's Office, Government of India, New Delhi.

NATCOM, 2004. India's Initial National Communication to the UNFCCC. Ministry of Environment and Forests, Government of India, New Delhi.

Nicholls, R.J. and Lowe, J.A., 2006. Climate stabilization and impacts of sea-level rise. *In:* H.J. Schellnhuber, W. Cramer, N. Nakicenovic, T. Wigley and G. Yohe (Eds), *Avoiding Dangerous Climate Change*, Cambridge University Press, Cambridge.

Ravindranath, N.H., Joshi, N.V., Sukumar, R. and Saxena, A., 2006. Impact of climate change on forests in India. *Current Science*, **90(3),** 354-361.

Rupa Kumar, K., Sahai, A.K., Krishna Kumar, K., Patwardhan, S.K., Mishra, P.K., Revadekar, J.V., Kamala, K. and Pant, G.B., 2006. High-resolution climate change scenarios for India for the 21st century. *Current Science*, **90(3)**.

Unnikrishnan, A.S., Rupa Kumar, K., Fernandes, Sharon E., Michael, G.S. and Patwardhan, S.K., 2006. Sea level changes along the Indian coast: Observations and projections, *Current Science*, **90(3),** 362-368.

Nicola, Ranger, Hallegatte, Stéphane, Bhattacharya, Sumana, Murthy, Bachu, Satya Priya, Dhore, K., Rafique, Farhat, Mathur, P., Naville, Nicolas, Henriet, Fanny, Herweijer, Celine, Pohit, Sanjib and Jan Corfee-Morlot, 2009. A Preliminary Assessment of the Potential Impact of Climate Change on Flood Risk in Mumbai. Submitted for publication in the Special Issue on Cities and Climate Change of *The Journal on Climatic Change.*

Revi, A., 2007. Climate Change Risk: An Adaptation and Mitigation Agenda for Indian Cities. Global Urban Summit, Bellagio, July 2007.

World Bank, 2009. South Asia Region – Towards a climate change strategy. http: //siteresources.worldbank.org/SOUTHASIAEXT/Resources/Publications/448813-

United Nations Population Division, 2007. World Urbanization Prospects: The 2007 Revision Population Database. United Nations Population Division, Department of Economic and Social Affairs (http://esa.un.org/unup/p2k0data.asp).

Index

GPSR Compliance

The European Union's (EU) General Product Safety Regulation (GPSR) is a set of rules that requires consumer products to be safe and our obligations to ensure this.

If you have any concerns about our products, you can contact us on ProductSafety@springernature.com

In case Publisher is established outside the EU, the EU authorized representative is:

Springer Nature Customer Service Center GmbH
Europaplatz 3
69115 Heidelberg, Germany

Batch number: 10755222

Printed by Printforce, the Netherlands